Teacher Edition

Eureka Math Grade 8 Module 4

Special thanks go to the Gordan A. Cain Center and to the Department of Mathematics at Louisiana State University for their support in the development of *Eureka Math*.

For a free *Eureka Math* Teacher Resource Pack, Parent Tip Sheets, and more please visit www.Eureka.tools

Published by the non-profit Great Minds

Copyright © 2015 Great Minds. All rights reserved. No part of this work may be reproduced or used in any form or by any means — graphic, electronic, or mechanical, including photocopying or information storage and retrieval systems — without written permission from the copyright holder. "Great Minds" and "Eureka Math" are registered trademarks of Great Minds.

Printed in the U.S.A.
This book may be purchased from the publisher at eureka-math.org
10 9 8 7 6 5 4 3

ISBN 978-1-63255-622-6

Eureka Math: A Story of Ratios Contributors

Michael Allwood, Curriculum Writer
Tiah Alphonso, Program Manager—Curriculum Production
Catriona Anderson, Program Manager—Implementation Support
Beau Bailey, Curriculum Writer
Scott Baldridge, Lead Mathematician and Lead Curriculum Writer
Bonnie Bergstresser, Math Auditor
Gail Burrill, Curriculum Writer
Beth Chance, Statistician
Joanne Choi, Curriculum Writer
Jill Diniz, Program Director
Lori Fanning, Curriculum Writer
Ellen Fort, Math Auditor
Kathy Fritz, Curriculum Writer
Glenn Gebhard, Curriculum Writer
Krysta Gibbs, Curriculum Writer
Winnie Gilbert, Lead Writer / Editor, Grade 8
Pam Goodner, Math Auditor
Debby Grawn, Curriculum Writer
Bonnie Hart, Curriculum Writer
Stefanie Hassan, Lead Writer / Editor, Grade 8
Sherri Hernandez, Math Auditor
Bob Hollister, Math Auditor
Patrick Hopfensperger, Curriculum Writer
Sunil Koswatta, Mathematician, Grade 8
Brian Kotz, Curriculum Writer
Henry Kranendonk, Lead Writer / Editor, Statistics
Connie Laughlin, Math Auditor
Jennifer Loftin, Program Manager—Professional Development
Nell McAnelly, Project Director
Ben McCarty, Mathematician
Stacie McClintock, Document Production Manager
Saki Milton, Curriculum Writer
Pia Mohsen, Curriculum Writer
Jerry Moreno, Statistician
Ann Netter, Lead Writer / Editor, Grades 6–7
Sarah Oyler, Document Coordinator
Roxy Peck, Statistician, Lead Writer / Editor, Statistics
Terrie Poehl, Math Auditor
Kristen Riedel, Math Audit Team Lead
Spencer Roby, Math Auditor
Kathleen Scholand, Math Auditor
Erika Silva, Lead Writer / Editor, Grade 6–7
Robyn Sorenson, Math Auditor
Hester Sutton, Advisor / Reviewer Grades 6–7
Shannon Vinson, Lead Writer / Editor, Statistics
Allison Witcraft, Math Auditor

Julie Wortmann, Lead Writer / Editor, Grade 7
David Wright, Mathematician, Lead Writer / Editor, Grades 6–7

Board of Trustees

Lynne Munson, President and Executive Director of Great Minds
Nell McAnelly, Chairman, Co-Director Emeritus of the Gordon A. Cain Center for STEM Literacy at Louisiana State University
William Kelly, Treasurer, Co-Founder and CEO at ReelDx
Jason Griffiths, Secretary, Director of Programs at the National Academy of Advanced Teacher Education
Pascal Forgione, Former Executive Director of the Center on K-12 Assessment and Performance Management at ETS
Lorraine Griffith, Title I Reading Specialist at West Buncombe Elementary School in Asheville, North Carolina
Bill Honig, President of the Consortium on Reading Excellence (CORE)
Richard Kessler, Executive Dean of Mannes College the New School for Music
Chi Kim, Former Superintendent, Ross School District
Karen LeFever, Executive Vice President and Chief Development Officer at ChanceLight Behavioral Health and Education
Maria Neira, Former Vice President, New York State United Teachers

This page intentionally left blank

A STORY OF RATIOS

GRADE 8

Mathematics Curriculum

GRADE 8 • MODULE 4

Table of Contents[1]
Linear Equations

Module Overview 3

Topic A: Writing and Solving Linear Equations (8.EE.C.7) 11

    Lesson 1: Writing Equations Using Symbols 13

    Lesson 2: Linear and Nonlinear Expressions in x 22

    Lesson 3: Linear Equations in x 30

    Lesson 4: Solving a Linear Equation 40

    Lesson 5: Writing and Solving Linear Equations 53

    Lesson 6: Solutions of a Linear Equation 65

    Lesson 7: Classification of Solutions 77

    Lesson 8: Linear Equations in Disguise 85

    Lesson 9: An Application of Linear Equations 99

Topic B: Linear Equations in Two Variables and Their Graphs (8.EE.B.5) 111

    Lesson 10: A Critical Look at Proportional Relationships 112

    Lesson 11: Constant Rate 124

    Lesson 12: Linear Equations in Two Variables 140

    Lesson 13: The Graph of a Linear Equation in Two Variables 155

    Lesson 14: The Graph of a Linear Equation—Horizontal and Vertical Lines 171

Mid-Module Assessment and Rubric 188
Topics A through B (assessment 1 day, return 1 day, remediation or further applications 2 days)

Topic C: Slope and Equations of Lines (8.EE.B.5, 8.EE.B.6) 199

    Lesson 15: The Slope of a Non-Vertical Line 201

    Lesson 16: The Computation of the Slope of a Non-Vertical Line 227

    Lesson 17: The Line Joining Two Distinct Points of the Graph $y = mx + b$ Has Slope m 251

    Lesson 18: There Is Only One Line Passing Through a Given Point with a Given Slope 270

---

[1]Each lesson is ONE day, and ONE day is considered a 45-minute period.

Module 4:   Linear Equations

This work is derived from Eureka Math ™ and licensed by Great Minds. ©2015 Great Minds. eureka-math.org

Lesson 19: The Graph of a Linear Equation in Two Variables Is a Line... 295

Lesson 20: Every Line Is a Graph of a Linear Equation 316

Lesson 21: Some Facts About Graphs of Linear Equations in Two Variables 336

Lesson 22: Constant Rates Revisited 351

Lesson 23: The Defining Equation of a Line.. 367

Topic D: Systems of Linear Equations and Their Solutions (8.EE.B.5, 8.EE.C.8) 378

Lesson 24: Introduction to Simultaneous Equations... 380

Lesson 25: Geometric Interpretation of the Solutions of a Linear System 397

Lesson 26: Characterization of Parallel Lines 412

Lesson 27: Nature of Solutions of a System of Linear Equations 426

Lesson 28: Another Computational Method of Solving a Linear System 442

Lesson 29: Word Problems... 460

Lesson 30: Conversion Between Celsius and Fahrenheit 474

Topic E (Optional): Pythagorean Theorem (8.EE.C.8, 8.G.B.7).. 483

Lesson 31: System of Equations Leading to Pythagorean Triples 484

End-of-Module Assessment and Rubric.. 495
Topics C through D (assessment 1 day, return 1 day, remediation or further applications 3 days)

Grade 8 • Module 4
Linear Equations

OVERVIEW

In Module 4, students extend what they already know about unit rates and proportional relationships (**6.RP.A.2**, **7.RP.A.2**) to linear equations and their graphs. Students understand the connections between proportional relationships, lines, and linear equations in this module (**8.EE.B.5**, **8.EE.B.6**). Also, students learn to apply the skills they acquired in Grades 6 and 7 with respect to symbolic notation and properties of equality (**6.EE.A.2**, **7.EE.A.1**, **7.EE.B.4**) to transcribe and solve equations in one variable and then in two variables.

In Topic A, students begin by transcribing written statements using symbolic notation. Then, students write linear and nonlinear expressions leading to linear equations, which are solved using properties of equality (**8.EE.C.7b**). Students learn that not every linear equation has a solution. In doing so, students learn how to transform given equations into simpler forms until an equivalent equation results in a unique solution, no solution, or infinitely many solutions (**8.EE.C.7a**). Throughout Topic A, students must write and solve linear equations in real-world and mathematical situations.

In Topic B, students work with constant speed, a concept learned in Grade 6 (**6.RP.A.3**), but this time with proportional relationships related to average speed and constant speed. These relationships are expressed as linear equations in two variables. Students find solutions to linear equations in two variables, organize them in a table, and plot the solutions on a coordinate plane (**8.EE.C.8a**). It is in Topic B that students begin to investigate the shape of a graph of a linear equation. Students predict that the graph of a linear equation is a line and select points on and off the line to verify their claim. Also in this topic is the standard form of a linear equation, $ax + by = c$, and when $a \neq 0$ and $b \neq 0$, a non-vertical line is produced. Further, when $a = 0$ or $b = 0$, then a vertical or horizontal line is produced.

In Topic C, students know that the slope of a line describes the rate of change of a line. Students first encounter slope by interpreting the unit rate of a graph (**8.EE.B.5**). In general, students learn that slope can be determined using any two distinct points on a line by relying on their understanding of properties of similar triangles from Module 3 (**8.EE.B.6**). Students verify this fact by checking the slope using several pairs of points and comparing their answers. In this topic, students derive $y = mx$ and $y = mx + b$ for linear equations by examining similar triangles. Students generate graphs of linear equations in two variables first by completing a table of solutions and then by using information about slope and y-intercept. Once students are sure that every linear equation graphs as a line and that every line is the graph of a linear equation, students graph equations using information about x- and y-intercepts. Next, students learn some basic facts about lines and equations, such as why two lines with the same slope and a common point are the same line, how to write equations of lines given slope and a point, and how to write an equation given two points. With the concepts of slope and lines firmly in place, students compare two different proportional relationships represented by graphs, tables, equations, or descriptions. Finally, students learn that multiple forms of an equation can define the same line.

Simultaneous equations and their solutions are the focus of Topic D. Students begin by comparing the constant speed of two individuals to determine which has greater speed (**8.EE.C.8c**). Students graph simultaneous linear equations to find the point of intersection and then verify that the point of intersection is in fact a solution to each equation in the system (**8.EE.C.8a**). To motivate the need to solve systems algebraically, students graph systems of linear equations whose solutions do not have integer coordinates. Students learn to solve systems of linear equations by substitution and elimination (**8.EE.C.8b**). Students understand that a system can have a unique solution, no solution, or infinitely many solutions, as they did with linear equations in one variable. Finally, students apply their knowledge of systems to solve problems in real-world contexts, including converting temperatures from Celsius to Fahrenheit.

Optional Topic E is an application of systems of linear equations (**8.EE.C.8b**). Specifically, this system generates Pythagorean triples. First, students learn that a Pythagorean triple can be obtained by multiplying any known triple by a positive integer (**8.G.B.7**). Then, students are shown the Babylonian method for finding a triple that requires the understanding and use of a system of linear equations.

Focus Standards

Understand the connections between proportional relationships, lines, and linear equations.

8.EE.B.5   Graph proportional relationships, interpreting the unit rate as the slope of the graph. Compare two different proportional relationships represented in different ways. *For example,* compare *a distance-time graph to a distance-time equation to determine which of two moving objects has greater speed.*

8.EE.B.6   Use similar triangles to explain why the slope m is the same between any two distinct points on a non-vertical line in the coordinate plane; derive the equation $y = mx$ for a line through the origin and the equation $y = mx + b$ for a line intercepting the vertical axis at b.

Analyze and solve linear equations and pairs of simultaneous linear equations.

8.EE.C.7   Solve linear equations in one variable.

   a. Give examples of linear equations in one variable with one solution, infinitely many solutions, or no solutions. Show which of these possibilities is the case by successively transforming the given equation into simpler forms, until an equivalent equation of the form $x = a$, $a = a$, or $a = b$ results (where a and b are different numbers).

   b. Solve linear equations with rational number coefficients, including equations whose solutions require expanding expressions using the distributive property and collecting like terms.

8.EE.C.8   Analyze and solve pairs of simultaneous linear equations.

   a. Understand that solutions to a system of two linear equations in two variables correspond to points of intersection of their graphs, because points of intersection satisfy both equations simultaneously.

b. Solve systems of two linear equations in two variables algebraically, and estimate solutions by graphing the equations. Solve simple cases by inspection. *For example, $3x + 2y = 5$ and $3x + 2y = 6$ have no solution because $3x + 2y$ cannot simultaneously be 5 and 6.*

c. Solve real-world and mathematical problems leading to two linear equations in two variables. *For example, given coordinates for two pairs of points, determine whether the line through the first pair of points intersects the line through the second pair.*

Foundational Standards

Understand ratio concepts and use ratio reasoning to solve problems.

6.RP.A.2   Understand the concept of a unit rate a/b associated with a ratio $a:b$ with $b \neq 0$, and use rate language in the context of a ratio relationship. *For example, "This recipe has a ratio of 3 cups of flour to 4 cups of sugar, so there is 3/4 cup of flour for each cup of sugar." "We paid $75 for 15 hamburgers, which is a rate of $5 per hamburger."*[2]

6.RP.A.3   Use ratio and rate reasoning to solve real-world and mathematical problems, e.g., by reasoning about tables of equivalent ratios, tape diagrams, double number line diagrams, or equations.

  a. Make tables of equivalent ratios relating quantities with whole-number measurements, find missing values in the tables, and plot the pairs of values on the coordinate plane. Use tables to compare ratios.

  b. Solve unit rate problems including those involving unit pricing and constant speed. *For example, if it took 7 hours to mow 4 lawns, then at that rate, how many lawns could be mowed in 35 hours? At what rate were lawns being mowed?*

Apply and extend previous understandings of arithmetic to algebraic expressions.

6.EE.A.2   Write, read, and evaluate expressions in which letters stand for numbers.

  a. Write expressions that record operations with numbers and with letters standing for numbers. *For example, express the calculation "Subtract y from 5" as $5 - y$.*

  b. Identify parts of an expression using mathematical terms (sum, term, product, factor, quotient, coefficient); view one or more parts of an expression as a single entity. *For example, describe the expression $2(8 + 7)$ as a product of two factors; view $(8 + 7)$ as both a single entity and a sum of two terms.*

---

[2]Expectations for unit rates in this grade are limited to non-complex fractions.

c. Evaluate expressions at specific values of their variables. Include expressions that arise from formulas used in real-world problems. Perform arithmetic operations, including those involving whole-number exponents, in the conventional order when there are no parentheses to specify a particular order (Order of Operations). *For example, use the formulas $V = s^3$ and $\} = 6s^2$ to find the volume and surface area of a cube with side length $s = 1/2$.*

Analyze proportional relationships and use them to solve real-world and mathematical problems.

7.RP.A.2   Recognize and represent proportional relationships between quantities.

a. Decide whether two quantities are in a proportional relationship, e.g., by testing for equivalent ratios in a table or graphing on a coordinate plane and observing whether the graph is a straight line through the origin.

b. Identify the constant of proportionality (unit rate) in tables, graphs, equations, diagrams, and verbal descriptions of proportional relationships.

c. Represent proportional relationships by equations. *For example, if total cost t is proportional to the number n of items purchased at a constant price p, the relationship between the total cost and the number of items can be expressed as $t = pn$.*

d. Explain what a point (x, y) on the graph of a proportional relationship means in terms of the situation, with special attention to the points $(0, 0)$ and $(1, r)$ where r is the unit rate.

Use properties of operations to generate equivalent expressions.

7.EE.A.1   Apply properties of operations as **strategies** to add, subtract, factor, and expand linear expressions with rational coefficients.

Solve real-life and mathematical problems using numerical and algebraic expressions and equations.

7.EE.B.4   Use variables to **represent** quantities in a real-world or mathematical problem, and construct simple equations and inequalities to solve problems by reasoning about the quantities.

a. Solve word problems leading to equations of the form $px + q = r$ and $p(x + q) = r$, where p, q, and r are specific rational numbers. Solve equations of these forms fluently. Compare an algebraic solution to an arithmetic solution, identifying the sequence of the operations used in each approach. *For example, the perimeter of a rectangle is 54 cm. Its length is 6 cm. What is its width?*

Focus Standards for Mathematical Practice

MP.1   **Make sense of problems and persevere in solving them.** Students analyze given constraints to make conjectures about the form and meaning of a solution to a given situation in one-variable and two-variable linear equations, as well as in simultaneous linear equations. Students are systematically guided to understand the meaning of a linear equation in one variable, the natural occurrence of linear equations in two variables with respect to proportional relationships, and the natural emergence of a system of two linear equations when looking at related, continuous proportional relationships.

MP.2   **Reason abstractly and quantitatively.** Students decontextualize and contextualize throughout the module as they represent situations symbolically and make sense of solutions within a context. Students use facts learned about rational numbers in previous grade levels to solve linear equations and systems of linear equations.

MP.3   **Construct viable arguments and critique the reasoning of others.** Students use assumptions, definitions, and previously established facts throughout the module as they solve linear equations. Students make conjectures about the graph of a linear equation being a line and then proceed to prove this claim. While solving linear equations, they learn that they must first assume that a solution exists and then proceed to solve the equation using properties of equality based on the assumption. Once a solution is found, students justify that it is in fact a solution to the given equation, thereby verifying their initial assumption. This process is repeated for systems of linear equations.

MP.4   **Model with mathematics.** Throughout the module, students represent real-world situations symbolically. Students identify important quantities from a context and represent the relationship in the form of an equation, a table, and a graph. Students analyze the various representations and draw conclusions and/or make predictions. Once a solution or prediction has been made, students reflect on whether the solution makes sense in the context presented. One example of this is when students determine how many buses are needed for a field trip. Students must interpret their fractional solution and make sense of it as it applies to the real world.

MP.7   **Look for and make use of structure.** Students use the structure of an equation to make sense of the information in the equation. For example, students write equations that represent the constant rate of motion for a person walking. In doing so, they interpret an equation such as $y = \frac{3}{5}x$ as the total distance a person walks, y, in x amount of time, at a rate of $\frac{3}{5}$. Students look for patterns or structure in tables and show that a rate is constant.

Terminology

New or Recently Introduced Terms

- **Average Speed** (Let a time interval of t hours be given. Suppose that an object travels a total distance of d miles during this time interval. The *average speed of the object in the given time interval* is $\frac{d}{t}$ miles per hour.

- **Constant Speed** (For any positive real number v, an object travels at a *constant speed of v mph* over a fixed time interval if the average speed is always equal to v mph for any smaller time interval of the given time interval.)

- **Horizontal Line** (In a Cartesian plane, a *horizontal line* is either the x-axis or any other line parallel to the x-axis. For example, the graph of the equation $y = -5$ is a horizontal line.)

- **Linear Equation (description)** (A *linear equation* is an equation in which both expressions are linear expressions.)

- **Point-Slope Equation of a Line** (The *point-slope equation of a non-vertical line* in the Cartesian plane that passes through point (x_1, y_1) and has slope m is
$$y - y_1 = m(x - x_1).$$
It can be shown that every non-vertical line is the graph of its point-slope equation, and that every graph of a point-slope equation is a line.)

- **Slope of a Line in a Cartesian Plane** (The *slope* of a non-vertical line in a Cartesian plane that passes through two different points is the number given by the change in y-coordinates divided by the corresponding change in the x-coordinates. For two points (x_1, y_1) and (x_2, y_2) on the line where $x_1 \neq x_2$, the slope of the line m can be computed by the formula
$$m = \frac{y_2 - y_1}{x_2 - x_1}.$$
The slope of a vertical line is not defined. The definition of slope is well-defined after one uses similar triangles to show that expression $\frac{y_2 - y_1}{x_2 - x_1}$ is always the same number for any two distinct points (x_1, y_1) and (x_2, y_2) on the line.)

- **Slope-Intercept Equation of a Line** (The *slope-intercept equation of a non-vertical line* in the Cartesian plane with slope m and y-intercept b is
$$y = mx + b.$$
It can be shown that every non-vertical line is the graph of its slope-intercept equation, and that every graph of a slope-intercept equation is a line.)

- **Solution to a System of Linear Equations (description)** (A *solution to a system of two linear equations in two variables* is an ordered pair of numbers that is a solution to both equations. For example, the solution to the system of linear equations $\begin{cases} x + y = 6 \\ x - y = 4 \end{cases}$ is the ordered pair $(5, 1)$ because substituting 5 in for x and 1 in for y results in two true equations: $5 + 1 = 6$ and $5 - 1 = 4$.)

- **Standard Form of a Linear Equation** (A linear equation in two variables x and y is in *standard form* if it is of the form
$$ax + by = c$$
for real numbers a, b, and c, where a and b are both not zero. The numbers a, b, and c are called *constants*.)
- **System of Linear Equations** (A *system of linear equations* is a set of two or more linear equations. For example, $\begin{cases} x + y = 15 \\ 3x - 7y = -2 \end{cases}$ is a system of linear equations.)
- **Vertical Line** (In a Cartesian plane, a *vertical line* is either the y-axis or any other line parallel to the y-axis. For example, the graph of the equation $x = 3$ is a vertical line.)
- **X-Intercept** (An *x-intercept* of a graph is the x-coordinate of a point where the graph intersects the x-axis. An *x-intercept point* is the coordinate point where the graph intersects the x-axis.

   The x-intercept of a graph of a linear equation can be found by setting $y = 0$ in the equation. Many times the term "x-intercept point" is shortened to just "x-intercept" if it is clear from the context that the term is referring to a point and not a number.
- **Y-Intercept** (A *y-intercept* of a graph is the y-coordinate of a point where the graph intersects the y-axis. A *y-intercept point* is the coordinate point where the graph intersects the y-axis.

   The y-intercept of a graph of a linear equation can be found by setting $x = 0$ in the equation. Many times the term "y-intercept point" is shortened to just "y-intercept" if it is clear from the context that the term is referring to a point and not a number.)

Familiar Terms and Symbols[3]

- Coefficient
- Equation
- Like terms
- Linear Expression
- Solution
- Term
- Unit rate
- Variable

Suggested Tools and Representations

- Scientific calculator
- Online graphing calculator (e.g., https://www.desmos.com/calculator)
- Graph paper
- Straightedge

---

[3]These are terms and symbols students have seen previously.

Assessment Summary

Assessment Type	Administered	Format	Standards Addressed
Mid-Module Assessment Task	After Topic B	Constructed response with rubric	8.EE.C.7, 8.EE.B.5
End-of-Module Assessment Task	After Topic D	Constructed response with rubric	8.EE.B.5, 8.EE.B.6, 8.EE.C.7, 8.EE.C.8

A STORY OF RATIOS

Mathematics Curriculum

GRADE 8 • MODULE 4

Topic A
Writing and Solving Linear Equations

8.EE.C.7

Focus Standard:	8.EE.C.7	Solve linear equations in one variable.
	a.	Give examples of linear equations in one variable with one solution, infinitely many solutions, or no solutions. Show which of these possibilities is the case by successively transforming the given equation into simpler forms, until an equivalent equation of the form $x = a$, $a = a$, or $a = b$ results (where a and b are different numbers).
	b.	Solve linear equations with rational number coefficients, including equations whose solutions require expanding expressions using the distributive property and collecting like terms.
Instructional Days:	9	
Lesson 1:	Writing Equations Using Symbols (P)[1]	
Lesson 2:	Linear and Nonlinear Expressions in x (P)	
Lesson 3:	Linear Equations in x (P)	
Lesson 4:	Solving a Linear Equation (P)	
Lesson 5:	Writing and Solving Linear Equations (P)	
Lesson 6:	Solutions of a Linear Equation (P)	
Lesson 7:	Classification of Solutions (S)	
Lesson 8:	Linear Equations in Disguise (P)	
Lesson 9:	An Application of Linear Equations (S)	

In Lesson 1, students begin by transcribing written statements into symbolic language. Students learn that before they can write a symbolic statement, they must first define the symbols they intend to use. In Lesson 2, students learn the difference between linear expressions in x and nonlinear expressions in x, a distinction that is necessary to know whether or not an equation can be solved (at this point). Also, Lesson 2 contains a

---
[1]Lesson Structure Key: **P**-Problem Set Lesson, **M**-Modeling Cycle Lesson, **E**-Exploration Lesson, **S**-Socratic Lesson

Topic A: Writing and Solving Linear Equations

quick review of terms related to linear equations, such as *constant, term,* and *coefficient.* In Lesson 3, students learn that a linear equation in x is a statement of equality between two linear expressions in x. Students also learn that an equation that contains a variable really is a question: Is there a value of x that makes the linear equation true? In Lesson 4, students begin using properties of equality to rewrite linear expressions, specifically using the distributive property to "combine like terms." Further, students practice substituting numbers into equations to determine if a true number sentence is produced.

In Lesson 5, students practice the skills of the first few lessons in a geometric context. Students transcribe written statements about angles and triangles into symbolic language and use properties of equality to begin solving equations (**8.EE.C.7b**). More work on solving equations occurs in Lesson 6, where the equations are more complicated and require more steps to solve (**8.EE.C.7b**). In Lesson 6, students learn that not every linear equation has a solution (**8.EE.C.7a**). This leads to Lesson 7, where students learn that linear equations either have a unique solution, no solution, or infinitely many solutions (**8.EE.C.7a**). In Lesson 8, students rewrite equations that are not obviously linear equations and then solve them (**8.EE.C.7b**). Finally, in Lesson 9, students take another look at the Facebook problem from Module 1 in terms of linear equations (**8.EE.C.7a**).

Lesson 1: Writing Equations Using Symbols

Student Outcomes

- Students write mathematical statements using symbols to represent numbers.
- Students know that written statements can be written as more than one correct mathematical sentence.

Lesson Notes

The content of this lesson continues to develop the skills and concepts presented in Grades 6 and 7. Specifically, this lesson builds on both **6.EE.B.7** (Solve real-world and mathematical problems by writing and solving equations of the form $x + p = q$ and $px = q$) and **7.EE.B.4** (Solve word problems leading to equations of the form $px + q = r$ and $p(x + q) = r$).

Classwork

Discussion (4 minutes)

Show students the text of a mathematical statement compared to the equation.

The number 1,157 is the sum of the squares of two consecutive odd integers divided by the difference between the two consecutive odd integers.	$1157 = \dfrac{x^2 + (x+2)^2}{(x+2) - x}$

Ask students to write or share aloud (a) how these two are related, (b) which representation they prefer, and (c) why. Then, continue with the discussion that follows.

- Using letters to represent numbers in mathematical statements was introduced by René Descartes in the 1600s. In that era, people used only words to describe mathematical statements. The use of letters, or symbols, to represent numbers not only brought clarity to mathematical statements, it also expanded the horizons of mathematics.
- The reason we want to learn how to write a mathematical statement using symbols is to save time and labor. Imagine having to write the sentence: "The number 1,157 is the sum of the squares of two consecutive odd integers divided by the difference between the two consecutive odd integers." Then, imagine having to write the subsequent sentences necessary to solve it; compare that to the following:

  Let x represent the first odd integer. Then,

  $$1157 = \dfrac{x^2 + (x+2)^2}{(x+2) - x}.$$

- Notice that x is just a number. That means the square of x is also a number, along with the square of the next odd integer and the difference between the numbers. This is a symbolic statement about numbers.
- Writing in symbols is simpler than writing in words, as long as everyone involved is clear about what the symbols mean. This lesson focuses on accurately transcribing written statements into mathematical symbols. When we write mathematical statements using letters, we say we are using symbolic language.

A STORY OF RATIOS   Lesson 1   8•4

- All of the mathematical statements in this lesson are equations. Recall that an equation is a statement of equality between two expressions. Developing equations from written statements forms an important basis for problem solving and is one of the most vital parts of algebra. Throughout this module, there will be work with written statements and symbolic language. We will work first with simple expressions, then with equations that gradually increase in complexity, and finally with systems of equations (more than one equation at a time).

Example 1 (3 minutes)

Throughout Example 1, have students write their thoughts on personal white boards or a similar tool and show their transcription(s).

- We want to express the following statement using symbolic language: A whole number has the property that when the square of half the number is subtracted from five times the number, we get the number itself.
- Do the first step, and hold up your personal white board.
    - *First, we define the variable. Let x be the whole number.*
- Using x to represent the whole number, write "the square of half the number."
    - $\left(\frac{x}{2}\right)^2$ or $\left(\frac{1}{2}x\right)^2$ or $\left(\frac{x^2}{4}\right)$ or $\left(\frac{1}{4}x^2\right)$

Scaffolding:
Alternative statement:
- A whole number has the property that when half the number is added to 15, we get the number itself.
- $\frac{1}{2}x + 15 = x$

Ask students to write their expressions in more than one way. Then, have students share their expressions of "the square of half the number." Elicit the above responses from students (or provide them). Ask students why they are all correct.

- Write the entire statement: A whole number has the property that when the square of half the number is subtracted from five times the number, we get the number itself.
    - $5x - \left(\frac{x}{2}\right)^2 = x$

Challenge students to write this equation in another form. Engage in a conversation about why they are both correct. For example, when a number is subtracted from itself, the difference is zero. For that reason, the equation above can be written as $5x - \left(\frac{x}{2}\right)^2 - x = 0$.

Example 2 (4 minutes)

Throughout Example 2, have students write their thoughts on personal white boards or a similar tool and show their transcription(s).

- We want to express the following statement using symbolic language: Paulo has a certain amount of money. If he spends $6.00, then he has $\frac{1}{4}$ of the original amount left.
- What is the first thing that must be done before we express this situation using symbols?
- *We need to define our variables; that is, we must decide what symbol to use and state what it is going to represent.*

Scaffolding:
Students may need to be reminded that it is not necessary to put the multiplication symbol between a number and a symbol. It is not wrong to include it, but by convention (a common agreement), it is not necessary.

14   Lesson 1:   Writing Equations Using Symbols

- Suppose we decide to use the symbol x. We will let x be the number of dollars Paulo had originally. How do we show Paulo's spending $6.00 using symbols?
  - *To show that Paulo spent $6.00, we write $x - 6$.*
- How do we express, "he has $\frac{1}{4}$ of the original amount?"
  - *We can express it as $\frac{1}{4}x$.*
- Put the parts together to express the following: "Paulo has a certain amount of money. If he spends $6.00, then he has $\frac{1}{4}$ of the original amount left." Use x to represent the amount of money Paulo had originally.
  - $x - 6 = \frac{1}{4}x$

Challenge students to write this equation in another form. Engage in a conversation about why they are both correct. For example, students may decide to show that the six dollars plus what he has left is equal to the amount of money he now has. In symbols, $x = \frac{1}{4}x + 6$.

Example 3 (8 minutes)

Throughout Example 3, have students write their thoughts on personal white boards or a similar tool and show their transcription(s).

- We want to write the following statement using symbolic language: When a fraction of 57 is taken away from 57, what remains exceeds $\frac{2}{3}$ of 57 by 4.

  Scaffolding:
  Alternative statement:
  - When a number is taken away from 57, what remains is four more than 5 times the number.
  - $57 - x = 5x + 4$

- The first step is to clearly state what we want our symbol to represent. If we choose the letter x, then we would say, "Let x be the fraction of 57" because that is the number that is unknown to us in the written statement. It is acceptable to use any letter to represent the unknown number; but regardless of which letter we use to symbolize the unknown number, we must clearly state what it means. This is called *defining our variables*.

- The hardest part of transcription is figuring out exactly how to write the statement so that it is accurately represented in symbols. Begin with the first part of the statement: "When a fraction of 57 is taken away from 57," how can we capture that information in symbols?
  - *Students should write $57 - x$.*

  Scaffolding:
  If students have a hard time thinking about these transcriptions, give them something easier to think about. One number, say 10, exceeds another number, say 6, by 4. Is it accurate to represent this by:

  $10 - 4 = 6$?
  $10 = 6 + 4$?
  $10 - 6 = 4$?

- How do we write $\frac{2}{3}$ of 57?
  - *If we are trying to find $\frac{2}{3}$ of 57, then we multiply $\frac{2}{3} \cdot 57$.*

- Would it be accurate to write $57 - x = \frac{2}{3} \cdot 57$?
  - *No, because we are told that "what remains exceeds $\frac{2}{3}$ of 57 by 4."*

Lesson 1: Writing Equations Using Symbols

15

- Where does the 4 belong? "What remains exceeds $\frac{2}{3}$ of 57 by 4." Think about what the word *exceeds* means in the context of the problem. Specifically, which is bigger: $57 - x$ or $\frac{2}{3}$ of 57? How do you know?
  - $57 - x$ is bigger because $57 - x$ exceeds $\frac{2}{3}$ of 57 by 4. That means that $57 - x$ is 4 more than $\frac{2}{3}$ of 57.
- We know that $57 - x$ is bigger than $\frac{2}{3} \cdot 57$ by 4. What would make the two numbers equal?
  - We either have to subtract 4 from $57 - x$ or add 4 to $\frac{2}{3} \cdot 57$ to make them equal.
- Now, if x is the fraction of 57, then we could write $(57 - x) - 4 = \frac{2}{3} \cdot 57$, or $57 - x = \frac{2}{3} \cdot 57 + 4$. Which is correct?
  - Both transcriptions are correct because both express the written statement accurately.
- Consider this transcription: $(57 - x) - \frac{2}{3} \cdot 57 = 4$. Is it an accurate transcription of the information in the written statement?
  - Yes, because $57 - x$ is bigger than $\frac{2}{3} \cdot 57$ by 4. That means that the difference between the two numbers is 4. If we subtract the smaller number from the bigger number, we have a difference of 4, and that is what this version of the transcription shows.

Example 4 (4 minutes)

Throughout Example 4, have students write their thoughts on personal white boards or a similar tool and show their transcription(s).

- We want to express the following statement using symbolic language: The sum of three consecutive integers is 372.
- Do the first step, and hold up your white board.
  - Let x be the first integer.
- If we let x represent the first integer, what do we need to do to get the next consecutive integer?
  - If x is the first integer, we add 1 to x to get the next integer.
- In symbols, the next integer would be $x + 1$. What do we need to do now to get the next consecutive integer?
  - We need to add 1 to that integer, or $x + 1 + 1$; this is the same as $x + 2$.
- Now, express the statement: The sum of three consecutive integers is 372.
  - $x + x + 1 + x + 2 = 372$

> **Scaffolding:**
> Explain that *consecutive* means one after the next. For example, 18, 19, and 20 are consecutive integers. Consider giving students a number and asking them what the next consecutive integer would be.

Students may also choose to rewrite the above equation as $3x + 3 = 372$. Transforming equations such as is a focus of the next few lessons when students begin to solve linear equations. Ask students why both equations are correct.

Example 5 (3 minutes)

Throughout Example 5, have students write their thoughts on personal white boards or a similar tool and show their transcription(s).

- We want to express the following statement using symbolic language: The sum of three consecutive odd integers is 93.
- Do the first step, and hold up your white board.
  - *Let x be the first odd integer.*
- If we let x represent the first odd integer, what do we need to do to get the next odd integer?
  - *If x is the first odd integer, we add 2 to x to get the next odd integer.*
- In symbols, the next odd integer would be $x + 2$. What do we need to do now to get the next odd integer?
  - *We need to add 2 to that odd integer, or $x + 2 + 2$; this is the same as $x + 4$.*
- Now, express the statement: The sum of three consecutive odd integers is 93.
  - *$x + x + 2 + x + 4 = 93$*
- Represent the statement "The sum of three consecutive odd integers is 93" in another way. Be prepared to explain why both are correct.
  - *Answers will vary. Accept any correct answers and justifications. For example, students may write $(x + x + x) + 6 = 93$ and state that it is equivalent to the equation $x + x + 2 + x + 4 = 93$. Because the associative and commutative properties of addition were applied, those properties do not change the value of an expression.*

Scaffolding:
Consider giving students an odd number and then asking them what the next consecutive odd number is. Then, ask students what they need to add to go from the first number to the second number.

Exercises (10 minutes)

Students complete Exercises 1–5 independently or in pairs.

Exercises

Write each of the following statements using symbolic language.

1. The sum of four consecutive even integers is -28.

   Let x be the first even integer. Then, $x + x + 2 + x + 4 + x + 6 = -28$.

2. A number is four times larger than the square of half the number.

   Let x be the number. Then, $x = 4\left(\frac{x}{2}\right)^2$.

3. Steven has some money. If he spends $\$9.00$, then he will have $\frac{3}{5}$ of the amount he started with.

   Let x be the amount of money (in dollars) Steven started with. Then, $x - 9 = \frac{3}{5}x$.

4. The sum of a number squared and three less than twice the number is 129.

   Let x be the number. Then, $x^2 + 2x - 3 = 129$.

Lesson 1: Writing Equations Using Symbols

A STORY OF RATIOS   Lesson 1   8•4

> 5. Miriam read a book with an unknown number of pages. The first week, she read five less than $\frac{1}{3}$ of the pages. The second week, she read 171 more pages and finished the book. Write an equation that represents the total number of pages in the book.
>
> *Let x be the total number of pages in the book. Then, $\frac{1}{3}x - 5 + 171 = x$.*

Closing (4 minutes)

Summarize, or ask students to summarize, the main points from the lesson:

- We know how to write mathematical statements using symbolic language.
- Written mathematical statements can be represented as more than one correct symbolic statement.
- We must always begin writing a symbolic statement by defining our symbols (variables).
- Complicated statements should be broken into parts or attempted with simple numbers to make the representation in symbolic notation easier.

Lesson Summary

Begin all word problems by defining your variables. State clearly what you want each symbol to represent.

Written mathematical statements can be represented as more than one correct symbolic statement.

Break complicated problems into smaller parts, or try working them with simpler numbers.

Exit Ticket (5 minutes)

A STORY OF RATIOS   Lesson 1   8•4

Name _____   Date _____

Lesson 1: Writing Equations Using Symbols

Exit Ticket

Write each of the following statements using symbolic language.

1. When you square five times a number, you get three more than the number.

2. Monica had some cookies. She gave seven to her sister. Then, she divided the remainder into two halves, and she still had five cookies left.

A STORY OF RATIOS   Lesson 1   8•4

Exit Ticket Sample Solutions

Write each of the following statements using symbolic language.

1. When you square five times a number, you get three more than the number.

   Let x be the number. Then, $(5x)^2 = x + 3$.

2. Monica had some cookies. She gave seven to her sister. Then, she divided the remainder into two halves, and she still had five cookies left.

   Let x be the original amount of cookies. Then, $\frac{1}{2}(x - 7) = 5$.

Problem Set Sample Solutions

Students practice transcribing written statements into symbolic language.

Write each of the following statements using symbolic language.

1. Bruce bought two books. One book costs $4.00 more than three times the other. Together, the two books cost him $72.

   Let x be the cost of the less expensive book. Then, $x + 4 + 3x = 72$.

2. Janet is three years older than her sister Julie. Janet's brother is eight years younger than their sister Julie. The sum of all of their ages is 55 years.

   Let x be Julie's age. Then, $(x + 3) + (x - 8) + x = 55$.

3. The sum of three consecutive integers is $1,623$.

   Let x be the first integer. Then, $x + (x + 1) + (x + 2) = 1623$.

4. One number is six more than another number. The sum of their squares is 90.

   Let x be the smaller number. Then, $x^2 + (x + 6)^2 = 90$.

5. When you add 18 to $\frac{1}{4}$ of a number, you get the number itself.

   Let x be the number. Then, $\frac{1}{4}x + 18 = x$.

6. When a fraction of 17 is taken away from 17, what remains exceeds one-third of seventeen by six.

   Let x be the fraction of 17. Then, $17 - x = \frac{1}{3} \cdot 17 + 6$.

7. The sum of two consecutive even integers divided by four is 189.5.

   Let x be the first even integer. Then, $\frac{x+(x+2)}{4} = 189.5$.

Lesson 1: Writing Equations Using Symbols

8. Subtract seven more than twice a number from the square of one-third of the number to get zero.

   Let x be the number. Then, $\left(\frac{1}{3}x\right)^2 - (2x + 7) = 0$.

9. The sum of three consecutive integers is 42. Let x be the middle of the three integers. Transcribe the statement accordingly.

   $(x - 1) + x + (x + 1) = 42$

Lesson 1: Writing Equations Using Symbols

A STORY OF RATIOS　　　　　　　　　　　　　　　　　　　　　　　　　　　Lesson 2　8•4

Lesson 2: Linear and Nonlinear Expressions in x

Student Outcomes

- Students know the properties of linear and nonlinear expressions in x.
- Students transcribe and identify expressions as linear or nonlinear.

Classwork

Discussion (4 minutes)

- A symbolic statement in x with an equal sign is called an *equation in x*. The equal sign divides the equation into two parts, the left side and the right side. The two sides are called *expressions*.
- For sake of simplicity, we will only discuss expressions in x, but know that we can write expressions in any symbol.
- The following chart contains both linear and nonlinear expressions in x. Sort them into two groups, and be prepared to explain what is different about the two groups.

$5x + 3$	$-8x + \dfrac{7}{9} - 3$	$9 - x^2$
$4x^2 - 9$	$0.31x + 7 - 4.2x$	$\left(\dfrac{x}{2}\right)^3 + 1$
$11(x + 2)$	$-(6 - x) + 15 - 9x$	$7 + x^{-4} + 3x$

Linear expressions are noted in red in the table below.

$5x + 3$	$-8x + \dfrac{7}{9} - 3$	$9 - x^2$
$4x^2 - 9$	$0.31x + 7 - 4.2x$	$\left(\dfrac{x}{2}\right)^3 + 1$
$11(x + 2)$	$-(6 - x) + 15 - 9x$	$7 + x^{-4} + 3x$

MP.3

- Identify which equations you placed in each group. Explain your reasoning for grouping the equations.
  - *Equations that contained an exponent of x other than 1 were put into one group. The other equations were put into another group. That seemed to be the only difference between the types of equations given.*
- Linear expressions in x are special types of expressions. Linear expressions are expressions that are sums of constants and products of a constant and x raised to a power of 0, which simplifies to a value of 1, or a power of 1. Nonlinear expressions are also sums of constants and products of a constant and a power of x. However, nonlinear expressions will have a power of x that is not equal to 1 or 0.

22　　Lesson 2:　Linear and Nonlinear Expressions in x

- The reason we want to be able to distinguish linear expressions from nonlinear expressions is because we will soon be solving linear equations. Nonlinear equations are a set of equations you learn to solve in Algebra I, though we begin to solve simple nonlinear equations later this year (Module 7). We also want to be able to recognize linear equations in order to predict the shapes of their graphs, which is a concept we learn more about later in this module.

Example 1 (3 minutes)

- A linear expression in x is an expression where each term is either a constant or a product of a constant and x. For example, the expression $(57 - x)$ is a linear expression. However, the expression $2x^2 + 9x + 5$ is not a linear expression. Why is $2x^2 + 9x + 5$ not a linear expression in x?
  - *Students should say that $2x^2 + 9x + 5$ is not a linear expression because the terms of linear expressions must either be a constant or the product of a constant and x. The term $2x^2$ does not fit the definition of a linear expression in x.*

Scaffolding:

- *Terms are any product of an integer power of x and a constant, or just a constant.*
- *Constants are fixed numbers.*
- *When a term is the product of a constant and a power of x, the constant is called a coefficient.*

Example 2 (4 minutes)

- Let's examine the expression $4 + 3x^5$ more deeply. To begin, we want to identify the terms of the expression. How many terms are there, and what are they?
  - *There are two terms, 4 and $3x^5$.*
- How many terms comprise just constants, and what are they?
  - *There is one constant term, 4.*
- How many terms have coefficients, and what are they?
  - *There is one term with a coefficient, 3.*
- Is $4 + 3x^5$ a linear or nonlinear expression in x? Why or why not?
  - *The expression $4 + 3x^5$ is a nonlinear expression in x because it is the sum of a constant and the product of a constant and positive integer power of $x > 1$.*

Example 3 (4 minutes)

- How many terms does the expression $7x + 9 + 6 + 3x$ have? What are they?
  - *As is, this expression has 4 terms: $7x$, 9, 6, and $3x$.*
- This expression can be transformed using some of our basic properties of numbers. For example, if we apply the commutative property of addition, we can rearrange the terms from $7x + 9 + 6 + 3x$ to $7x + 3x + 9 + 6$. Then, we can apply the associative property of addition:

$$(7x + 3x) + (9 + 6).$$

Next, we apply the distributive property:

$$(7 + 3)x + (9 + 6).$$

Finally,

$$10x + 15.$$

Lesson 2: Linear and Nonlinear Expressions in x

A STORY OF RATIOS Lesson 2 8•4

- How many terms with coefficients does the expression $10x + 15$ have? What are they?
  - *The expression has one term with a coefficient, $10x$. For this term, the coefficient is 10.*
- Is $10x + 15$ a linear or nonlinear expression in x? Why or why not?
  - *The expression $10x + 15$ is a linear expression in x because it is the sum of constants and products that contain x to the 1st power.*

Example 4 (2 minutes)

- How many terms does the expression $5 + 9x \cdot 7 + 2x^9$ have? What are they?
  - *The expression has three terms: 5, $9x \cdot 7$, and $2x^9$.*
- How many terms with coefficients does the expression $5 + 9x \cdot 7 + 2x^9$ have? What are they?
  - *The term $9x \cdot 7$ can be simplified to $63x$. Then, the expression has two terms with coefficients: $63x$ and $2x^9$. The coefficients are 63 and 2.*
- Is $5 + 9x \cdot 7 + 2x^9$ a linear or nonlinear expression in x? Why or why not?
  - *The expression $5 + 9x \cdot 7 + 2x^9$ is a nonlinear expression in x because it is the sum of constants and products that contain x raised to a power that is greater than 1.*

Example 5 (2 minutes)

- Is $94 + x + 4x^{-6} - 2$ a linear or nonlinear expression in x? Why or why not?
  - *Students may first say that it is neither a linear nor a nonlinear expression in x because of the -2 (linear expressions were described as sums of constants and products of a constant and x raised to a power of 0). Remind them that subtraction can be rewritten as a sum (i.e., $+ (-2)$). The term $4x^{-6}$ contains the reason the expression is nonlinear; x is raised to a power that is not equal to 1 or 0.*

Example 6 (2 minutes)

- Is the expression $x^1 + 9x - 4$ a linear expression in x?
  - *Yes, $x^1 + 9x - 4$ is a linear expression in x because x^1 is the same as x.*
- What powers of x are acceptable in the definition of a linear expression in x?
  - *Only the powers of 0 or 1 are acceptable because x^0 and x^1 are, by definition, just 1 and x, respectively.*

Exercises (14 minutes)

Students complete Exercises 1–12 independently.

> **Exercises**
>
> Write each of the following statements in Exercises 1–12 as a mathematical expression. State whether or not the expression is linear or nonlinear. If it is nonlinear, then explain why.
>
> 1. The sum of a number and four times the number
>
>   *Let x be a number; then, $x + 4x$ is a linear expression.*

Lesson 2: Linear and Nonlinear Expressions in x

2. The product of five and a number

   Let x be a number; then, $5x$ is a linear expression.

3. Multiply six and the reciprocal of the quotient of a number and seven.

   Let x be a number; then, $6 \cdot \frac{7}{x}$ is a nonlinear expression. The expression is nonlinear because the number $\frac{7}{x} = 7 \cdot \frac{1}{x} = 7 \cdot x^{-1}$. The exponent of the x is the reason it is not a linear expression.

4. Twice a number subtracted from four times a number, added to 15

   Let x be a number; then, $15 + (4x - 2x)$ is a linear expression.

5. The square of the sum of six and a number

   Let x be a number; then, $(x + 6)^2$ is a nonlinear expression. When you multiply $(x + 6)^2$, you get $x^2 + 12x + 36$. The x^2 is the reason it is not a linear expression.

6. The cube of a positive number divided by the square of the same positive number

   Let x be a number; then, $\frac{x^3}{x^2}$ is a nonlinear expression. However, if you simplify the expression to x, then it is linear.

7. The sum of four consecutive numbers

   Let x be the first number; then, $x + (x + 1) + (x + 2) + (x + 3)$ is a linear expression.

8. Four subtracted from the reciprocal of a number

   Let x be a number; then, $\frac{1}{x} - 4$ is a nonlinear expression. The term $\frac{1}{x}$ is the same as x^{-1}, which is why this expression is not linear. It is possible that a student may let x be the reciprocal of a number, $\frac{1}{x}$, which would make the expression linear.

9. Half of the product of a number multiplied by itself three times

   Let x be a number; then, $\frac{1}{2} \cdot x \cdot x \cdot x$ is a nonlinear expression. The term $\frac{1}{2} \cdot x \cdot x \cdot x$ is the same as $\frac{1}{2}x^3$, which is why this expression is not linear.

10. The sum that shows how many pages Maria read if she read 45 pages of a book yesterday and $\frac{2}{3}$ of the remaining pages today

    Let x be the number of remaining pages of the book; then, $45 + \frac{2}{3}x$ is a linear expression.

11. An admission fee of $10 plus an additional $2 per game

    Let x be the number of games; then, $10 + 2x$ is a linear expression.

Lesson 2: Linear and Nonlinear Expressions in x

A STORY OF RATIOS　　　　　　　　　　　　　　　　　　　　　　　　　　　　　　　　Lesson 2　8•4

12. Five more than four times a number and then twice that sum

   Let x be the number; then, $2(4x + 5)$ is a linear expression.

Closing (5 minutes)

Summarize, or ask students to summarize, the main points from the lesson:

- We have definitions for linear and nonlinear expressions.
- We know how to use the definitions to identify expressions as linear or nonlinear.
- We can write expressions that are linear and nonlinear.

Lesson Summary

A *linear expression* is an expression that is equivalent to the sum or difference of one or more expressions where each expression is either a number, a variable, or a product of a number and a variable.

A linear expression in x can be represented by terms whose variable x is raised to either a power of 0 or 1. For example, $4 + 3x$, $7x + x - 15$, and $\frac{1}{2}x + 7 - 2$ are all linear expressions in x. A nonlinear expression in x has terms where x is raised to a power that is not 0 or 1. For example, $2x^2 - 9$, $-6x^{-3} + 8 + x$, and $\frac{1}{x} + 8$ are all nonlinear expressions in x.

Exit Ticket (5 minutes)

Name _____ Date_____

Lesson 2: Linear and Nonlinear Expressions in x

Exit Ticket

Write each of the following statements as a mathematical expression. State whether the expression is a linear or nonlinear expression in x.

1. Seven subtracted from five times a number, and then the difference added to nine times a number

2. Three times a number subtracted from the product of fifteen and the reciprocal of a number

3. Half of the sum of two and a number multiplied by itself three times

A STORY OF RATIOS   Lesson 2   8•4

Exit Ticket Sample Solutions

> Write each of the following statements as a mathematical expression. State whether the expression is a linear or nonlinear expression in x.
>
> 1. Seven subtracted from five times a number, and then the difference added to nine times a number
>
>   Let x be a number; then, $(5x - 7) + 9x$. The expression is a linear expression.
>
> 2. Three times a number subtracted from the product of fifteen and the reciprocal of a number
>
>   Let x be a number; then, $15 \cdot \frac{1}{x} - 3x$. The expression is a nonlinear expression.
>
> 3. Half of the sum of two and a number multiplied by itself three times
>
>   Let x be a number; then, $\frac{1}{2}(2 + x^3)$. The expression is a nonlinear expression.

Problem Set Sample Solutions

Students practice writing expressions and identifying them as linear or nonlinear.

> Write each of the following statements as a mathematical expression. State whether the expression is linear or nonlinear. If it is nonlinear, then explain why.
>
> 1. A number decreased by three squared
>
>   Let x be a number; then, $x - 3^2$ is a linear expression.
>
> 2. The quotient of two and a number, subtracted from seventeen
>
>   Let x be a number; then, $17 - \frac{2}{x}$ is a nonlinear expression. The term $\frac{2}{x}$ is the same as $2 \cdot \frac{1}{x}$ and $\frac{1}{x} = x^{-1}$, which is why it is not linear.
>
> 3. The sum of thirteen and twice a number
>
>   Let x be a number; then, $13 + 2x$ is a linear expression.
>
> 4. 5.2 more than the product of seven and a number
>
>   Let x be a number; then, $5.2 + 7x$ is a linear expression.
>
> 5. The sum that represents the number of tickets sold if 35 tickets were sold Monday, half of the remaining tickets were sold on Tuesday, and 14 tickets were sold on Wednesday
>
>   Let x be the remaining number of tickets; then, $35 + \frac{1}{2}x + 14$ is a linear expression.

Lesson 2: Linear and Nonlinear Expressions in x

6. The product of 19 and a number, subtracted from the reciprocal of the number cubed

   Let x be a number; then, $\frac{1}{x^3} - 19x$ is a nonlinear expression. The term $\frac{1}{x^3}$ is the same as x^{-3}, which is why it is not linear.

7. The product of 15 and a number, and then the product multiplied by itself four times

   Let be a number; then, $(15x)^4$ is a nonlinear expression. The expression can be written as $15^4 \cdot x^4$. The exponent of 4 with a base of x is the reason it is not linear.

8. A number increased by five and then divided by two

   Let x be a number; then, $\frac{x+5}{2}$ is a linear expression.

9. Eight times the result of subtracting three from a number

   Let x be a number; then, $8(x-3)$ is a linear expression.

10. The sum of twice a number and four times a number subtracted from the number squared

    Let x be a number; then, $x^2 - (2x + 4x)$ is a nonlinear expression. The term x^2 is the reason it is not linear.

11. One-third of the result of three times a number that is increased by 12

    Let x be a number; then, $\frac{1}{3}(3x + 12)$ is a linear expression.

12. Five times the sum of one-half and a number

    Let x be a number; then, $5\left(\frac{1}{2} + x\right)$ is a linear expression.

13. Three-fourths of a number multiplied by seven

    Let x be a number; then, $\frac{3}{4}x \cdot 7$ is a linear expression.

14. The sum of a number and negative three, multiplied by the number

    Let x be a number; then, $(x + (-3))x$ is a nonlinear expression because $(x + (-3))x = x^2 - 3x$ after using the distributive property. It is nonlinear because the power of x in the term x^2 is greater than 1.

15. The square of the difference between a number and 10

    Let x be a number; then, $(x - 10)^2$ is a nonlinear expression because $(x - 10)^2 = x^2 - 20x + 100$. The term x^2 is a positive power of $x > 1$; therefore, this is not a linear expression.

Lesson 2: Linear and Nonlinear Expressions in x

Lesson 3: Linear Equations in x

Student Outcomes

- Students know that a linear equation is a statement of equality between two expressions.
- Students know that a linear equation in x is actually a question: Can you find all numbers x, if they exist, that satisfy a given equation? Students know that those numbers x that satisfy a given equation are called *solutions*.

Classwork

Concept Development (7 minutes)

- We want to define "linear equation in x." Here are some examples of linear equations in x. Using what you know about the words *linear* (from Lesson 2) and *equation* (from Lesson 1), develop a mathematical definition of "linear equation in x."

> **Scaffolding:**
> Consider developing a word bank or word wall to be used throughout the module.

Show students the examples below, and provide them time to work individually or in small groups to develop an appropriate definition. Once students share their definitions, continue with the definition and discussion that follows.

$x + 11 = 15$	$5 + 3 = 8$	$-\frac{1}{2}x = 22$
$15 - 4x = x + \frac{4}{5}$	$3 - (x + 2) = -12x$	$\frac{3}{4}x + 6(x - 1) = 9(2 - x)$

- When two linear expressions are equal, they can be written as a linear equation in x.
- Consider the following equations. Which are true, and how do you know?

$$4 + 1 = 5$$
$$6 + 5 = 16$$
$$21 - 6 = 15$$
$$6 - 2 = 1$$

  □ The first and third equations are true because the value on the left side is equal to the number on the right side.

- Is $4 + 15x = 49$ true? How do you know?

Have a discussion that leads to students developing a list of values for x that make it false, along with one value of x that makes it true. Then, conclude the discussion by making the two points below.

- A linear equation in x is a statement about equality, but it is also an invitation to find all of the numbers x, if they exist, that make the equation true. Sometimes the question is asked in this way: What number(s) x satisfy the equation? The question is often stated more as a directive: Solve. When phrased as a directive, it is still considered a question. Is there a number(s) x that make the statement true? If so, what is the number(s) x?

- Equations that contain a variable do not have a definitive truth value; in other words, there are values of the variable that make the equation a true statement and values of the variable that make it a false statement. When we say that we have "solved an equation," what we are really saying is that we have found a number (or numbers) x that make the equation true. That number x is called the *solution* to the equation.

Example 1 (4 minutes)

- Here is a linear equation in x: $4 + 15x = 49$. Is there a number x that makes the linear expression $4 + 15x$ equal to the linear expression 49? Suppose you are told this number x has a value of 2, that is, $x = 2$. We replace any instance of x in the linear equation with the value of 2, as shown:

$$4 + 15 \cdot 2 = 49.$$

Next, we evaluate each side of the equation. The left side is

$$4 + 15 \cdot 2 = 4 + 30$$
$$= 34.$$

The right side of the equation is 49. Clearly, $34 \neq 49$. Therefore, the number 2 is not a solution to this equation.

- Is the number 3 a solution to the equation? That is, is this equation a true statement when $x = 3$?
    - *Yes, because the left side of the equation equals the right side of the equation when $x = 3$.*

    The left side is

    $$4 + 15 \cdot 3 = 4 + 45$$
    $$= 49.$$

    The right side is 49. Since $49 = 49$, then we can say that $x = 3$ is a solution to the equation $4 + 15x = 49$.

- 3 is a solution to the equation because it is a value of x that makes the equation a true statement.

> **Scaffolding:**
> Remind students that when a number and a symbol are next to one another, such as $15x$, it is not necessary to use a symbol to represent the multiplication (it is a convention). For clarity, when two numbers are being multiplied, it is necessary to use a multiplication symbol. For example, it is necessary to tell the difference between the number, 152, and the product, $15 \cdot 2$.

Example 2 (4 minutes)

- Here is a linear equation in x: $8x - 19 = -4 - 7x$.
- Is 5 a solution to the equation? That is, is the equation a true statement when $x = 5$?
    - *No, because the left side of the equation does not equal the right side of the equation when $x = 5$.*

    The left side is

    $$8 \cdot 5 - 19 = 40 - 19$$
    $$= 21.$$

    The right side is

    $$-4 - 7 \cdot 5 = -4 - 35$$
    $$= -39.$$

    Since $21 \neq -39$, then $x \neq 5$. That is, 5 is not a solution to the equation.

Lesson 3: Linear Equations in x

- Is 1 a solution to the equation? That is, is this equation a true statement when $x = 1$?
  - Yes. The left side and right side of the equation are equal to the same number when $x = 1$.

  The left side is
  $$8 \cdot 1 - 19 = 8 - 19$$
  $$= -11.$$

  The right side is
  $$-4 - 7 \cdot 1 = -4 - 7$$
  $$= -11.$$

  Since $-11 = -11$, then $x = 1$. That is, 1 is a solution to the equation.

Example 3 (4 minutes)

- Here is a linear equation in x: $3(x + 9) = 4x - 7 + 7x$.
- We can make our work simpler if we use some properties to transform the expression on the right side of the equation into an expression with fewer terms.

Provide students time to transform the equation into fewer terms, and then proceed with the points below.

- For example, notice that on the right side, there are two terms that contain x. First, we will use the commutative property to rearrange the terms to better see what we are doing.
$$4x + 7x - 7$$
Next, we will use the distributive property to collect the terms that contain x.
$$4x + 7x - 7 = (4 + 7)x - 7$$
$$= 11x - 7$$
Finally, the transformed (but still the same) equation can be written as $3(x + 9) = 11x - 7$.

- Is $\frac{5}{4}$ a solution to the equation? That is, is this equation a true statement when $x = \frac{5}{4}$?
  - No, because the left side of the equation does not equal the right side of the equation when $x = \frac{5}{4}$.

  The left side is
  $$3\left(\frac{5}{4} + 9\right) = 3\left(\frac{41}{4}\right)$$
  $$= \frac{123}{4}.$$

  The right side is
  $$11 \cdot \frac{5}{4} - 7 = \frac{55}{4} - 7$$
  $$= \frac{27}{4}.$$

  Since $\frac{123}{4} \neq \frac{27}{4}$, then $x \neq \frac{5}{4}$. That is, $\frac{5}{4}$ is not a solution to the equation.

Example 4 (4 minutes)

- Here is a linear equation in x: $-2x + 11 - 5x = 5 - 6x$.
- We want to check to see if 6 is a solution to the equation; that is, is this equation a true statement when $x = 6$? Before we do that, what would make our work easier?
  - *We could use the commutative and distributive properties to transform the left side of the equation into an expression with fewer terms.*

  $$-2x + 11 - 5x = -2x - 5x + 11$$
  $$= (-2 - 5)x + 11$$
  $$= -7x + 11$$

- The transformed equation can be written as $-7x + 11 = 5 - 6x$. Is 6 a solution to the equation; that is, is this equation a true statement when $x = 6$?
  - *Yes, because the left side of the equation is equal to the right side of the equation when $x = 6$.*

  The left side is
  $$-7x + 11 = -7 \cdot 6 + 11$$
  $$= -42 + 11$$
  $$= -31.$$

  The right side is
  $$5 - 6x = 5 - 6 \cdot 6$$
  $$= 5 - 36$$
  $$= -31.$$

  Since $-31 = -31$, then $x = 6$. That is, 6 is a solution to the equation.

Exercises (12 minutes)

Students complete Exercises 1–7 independently.

Exercises

1. Is the equation a true statement when $x = -3$? In other words, is -3 a solution to the equation $6x + 5 = 5x + 8 + 2x$? Explain.

    If we replace x with the number -3, then the left side of the equation is
    $$6 \cdot (-3) + 5 = -18 + 5$$
    $$= -13,$$

    and the right side of the equation is
    $$5 \cdot (-3) + 8 + 2 \cdot (-3) = -15 + 8 - 6$$
    $$= -7 - 6$$
    $$= -13.$$

    Since $-13 = -13$, then $x = -3$ is a solution to the equation $6x + 5 = 5x + 8 + 2x$.

    Note: Some students may have transformed the equation.

Lesson 3: Linear Equations in x

2. Does $x = 12$ satisfy the equation $16 - \frac{1}{2}x = \frac{3}{4}x + 1$? Explain.

   If we replace x with the number 12, then the left side of the equation is
   $$16 - \frac{1}{2}x = 16 - \frac{1}{2} \cdot (12)$$
   $$= 16 - 6$$
   $$= 10,$$
   and the right side of the equation is
   $$\frac{3}{4}x + 1 = \frac{3}{4} \cdot (12) + 1$$
   $$= 9 + 1$$
   $$= 10.$$
   Since $10 = 10$, then $x = 12$ is a solution to the equation $16 - \frac{1}{2}x = \frac{3}{4}x + 1$.

3. Chad solved the equation $24x + 4 + 2x = 3(10x - 1)$ and is claiming that $x = 2$ makes the equation true. Is Chad correct? Explain.

   If we replace x with the number 2, then the left side of the equation is
   $$24x + 4 + 2x = 24 \cdot 2 + 4 + 2 \cdot 2$$
   $$= 48 + 4 + 4$$
   $$= 56,$$
   and the right side of the equation is
   $$3(10x - 1) = 3(10 \cdot 2 - 1)$$
   $$= 3(20 - 1)$$
   $$= 3(19)$$
   $$= 57.$$
   Since $56 \neq 57$, then $x = 2$ is not a solution to the equation $24x + 4 + 2x = 3(10x - 1)$, and Chad is not correct.

4. Lisa solved the equation $x + 6 = 8 + 7x$ and claimed that the solution is $x = -\frac{1}{3}$. Is she correct? Explain.

   If we replace x with the number $-\frac{1}{3}$, then the left side of the equation is
   $$x + 6 = -\frac{1}{3} + 6$$
   $$= 5\frac{2}{3},$$
   and the right side of the equation is
   $$8 + 7x = 8 + 7 \cdot \left(-\frac{1}{3}\right)$$
   $$= 8 - \frac{7}{3}$$
   $$= \frac{24}{3} - \frac{7}{3}$$
   $$= \frac{17}{3}.$$
   Since $5\frac{2}{3} = \frac{17}{3}$, then $x = -\frac{1}{3}$ is a solution to the equation $x + 6 = 8 + 7x$, and Lisa is correct.

5. Angel transformed the following equation from $6x + 4 - x = 2(x + 1)$ to $10 = 2(x + 1)$. He then stated that the solution to the equation is $x = 4$. Is he correct? Explain.

   No, Angel is not correct. He did not transform the equation correctly. The expression on the left side of the equation $6x + 4 - x = 2(x + 1)$ would transform to

   $$6x + 4 - x = 6x - x + 4$$
   $$= (6 - 1)x + 4$$
   $$= 5x + 4.$$

   If we replace x with the number 4, then the left side of the equation is

   $$5x + 4 = 5 \cdot 4 + 4$$
   $$= 20 + 4$$
   $$= 24,$$

   and the right side of the equation is

   $$2(x + 1) = 2(4 + 1)$$
   $$= 2(5)$$
   $$= 10.$$

   Since $24 \neq 10$, then $x = 4$ is not a solution to the equation $6x + 4 - x = 2(x + 1)$, and Angel is not correct.

6. Claire was able to verify that $x = 3$ was a solution to her teacher's linear equation, but the equation got erased from the board. What might the equation have been? Identify as many equations as you can with a solution of $x = 3$.

   Answers will vary. Ask students to share their equations and justifications as to how they knew $x = 3$ would make a true number sentence.

7. Does an equation always have a solution? Could you come up with an equation that does not have a solution?

   Answers will vary. Expect students to write equations that are false. Ask students to share their equations and justifications as to how they knew the equation they wrote did not have a solution. The concept of "no solution" is introduced in Lesson 6 and solidified in Lesson 7.

Closing (5 minutes)

Summarize, or ask students to summarize, the main points from the lesson:

- We know that equations are statements about equality. That is, the expression on the left side of the equal sign is equal to the expression on the right side of the equal sign.
- We know that a solution to a linear equation in x will be a number and that when all instances of x are replaced with the number, the left side will equal the right side.

Lesson 3: Linear Equations in x

Lesson Summary

An equation is a statement about equality between two expressions. If the expression on the left side of the equal sign has the same value as the expression on the right side of the equal sign, then you have a true equation.

A solution of a linear equation in x is a number, such that when all instances of x are replaced with the number, the left side will equal the right side. For example, 2 is a solution to $3x + 4 = x + 8$ because when $x = 2$, the left side of the equation is

$$3x + 4 = 3(2) + 4$$
$$= 6 + 4$$
$$= 10,$$

and the right side of the equation is

$$x + 8 = 2 + 8$$
$$= 10.$$

Since $10 = 10$, then $x = 2$ is a solution to the linear equation $3x + 4 = x + 8$.

Exit Ticket (5 minutes)

Name _____ Date _____

Lesson 3: Linear Equations in x

Exit Ticket

1. Is 8 a solution to $\frac{1}{2}x + 9 = 13$? Explain.

2. Write three different equations that have $x = 5$ as a solution.

3. Is -3 a solution to the equation $3x - 5 = 4 + 2x$? Explain.

A STORY OF RATIOS   Lesson 3   8•4

Exit Ticket Sample Solutions

1. **Is 8 a solution to $\frac{1}{2}x + 9 = 13$? Explain.**

   If we replace x with the number 8, then the left side is $\frac{1}{2}(8) + 9 = 4 + 9 = 13$, and the right side is 13. Since $13 = 13$, then $x = 8$ is a solution.

2. **Write three different equations that have $x = 5$ as a solution.**

   Answers will vary. Accept equations where $x = 5$ makes a true number sentence.

3. **Is -3 a solution to the equation $3x - 5 = 4 + 2x$? Explain.**

   If we replace x with the number -3, then the left side is $3(-3) - 5 = -9 - 5 = -14$. The right side is $4 + 2(-3) = 4 - 6 = -2$. Since $-14 \neq -2$, then -3 is not a solution of the equation.

Problem Set Sample Solutions

Students practice determining whether or not a given number is a solution to the linear equation.

1. **Given that $2x + 7 = 27$ and $3x + 1 = 28$, does $2x + 7 = 3x + 1$? Explain.**

   No, because a linear equation is a statement about equality. We are given that $2x + 7 = 27$, but $3x + 1 = 28$. Since each linear expression is equal to a different number, $2x + 7 \neq 3x + 1$.

2. **Is -5 a solution to the equation $6x + 5 = 5x + 8 + 2x$? Explain.**

   If we replace x with the number -5, then the left side of the equation is
   $$6 \cdot (-5) + 5 = -30 + 5$$
   $$= -25,$$
   and the right side of the equation is
   $$5 \cdot (-5) + 8 + 2 \cdot (-5) = -25 + 8 - 10$$
   $$= -17 - 10$$
   $$= -27.$$
   Since $-25 \neq -27$, then -5 is not a solution of the equation $6x + 5 = 5x + 8 + 2x$.

   Note: Some students may have transformed the equation.

3. **Does $x = 1.6$ satisfy the equation $6 - 4x = -\frac{x}{4}$? Explain.**

   If we replace x with the number 1.6, then the left side of the equation is
   $$6 - 4 \cdot 1.6 = 6 - 6.4$$
   $$= -0.4,$$
   and the right side of the equation is
   $$-\frac{1.6}{4} = -0.4.$$
   Since $-0.4 = -0.4$, then $x = 1.6$ is a solution of the equation $6 - 4x = -\frac{x}{4}$.

Lesson 3:   Linear Equations in x

4. Use the linear equation $3(x+1) = 3x+3$ to answer parts (a)–(d).

   a. Does $x = 5$ satisfy the equation above? Explain.

   If we replace x with the number 5, then the left side of the equation is
   $$3(5+1) = 3(6)$$
   $$= 18,$$
   and the right side of the equation is
   $$3x + 3 = 3 \cdot 5 + 3$$
   $$= 15 + 3$$
   $$= 18.$$
   Since $18 = 18$, then $x = 5$ is a solution of the equation $3(x+1) = 3x+3$.

   b. Is $x = -8$ a solution of the equation above? Explain.

   If we replace x with the number -8, then the left side of the equation is
   $$3(-8+1) = 3(-7)$$
   $$= -21,$$
   and the right side of the equation is
   $$3x + 3 = 3 \cdot (-8) + 3$$
   $$= -24 + 3$$
   $$= -21.$$
   Since $-21 = -21$, then $x = -8$ is a solution of the equation $3(x+1) = 3x+3$.

   c. Is $x = \frac{1}{2}$ a solution of the equation above? Explain.

   If we replace x with the number $\frac{1}{2}$, then the left side of the equation is
   $$3\left(\frac{1}{2} + 1\right) = 3\left(\frac{1}{2} + \frac{2}{2}\right)$$
   $$= 3\left(\frac{3}{2}\right)$$
   $$= \frac{9}{2},$$
   and the right side of the equation is
   $$3x + 3 = 3 \cdot \left(\frac{1}{2}\right) + 3$$
   $$= \frac{3}{2} + 3$$
   $$= \frac{3}{2} + \frac{6}{2}$$
   $$= \frac{9}{2}.$$
   Since $\frac{9}{2} = \frac{9}{2}$, then $x = \frac{1}{2}$ is a solution of the equation $3(x+1) = 3x+3$.

   d. What interesting fact about the equation $3(x+1) = 3x+3$ is illuminated by the answers to parts (a), (b), and (c)? Why do you think this is true?

   Note to teacher: Ideally, students will notice that the equation $3(x+1) = 3x+3$ is an identity under the distributive law. The purpose of this problem is to prepare students for the idea that linear equations can have more than one solution, which is a topic of Lesson 7.

Lesson 3: Linear Equations in x

Lesson 4: Solving a Linear Equation

Student Outcomes

- Students extend the use of the properties of equality to solve linear equations having rational coefficients.

Classwork

Concept Development (13 minutes)

- To solve an equation means to find all of the numbers x, if they exist, so that the given equation is true.
- In some cases, some simple guesswork can lead us to a solution. For example, consider the following equation:

$$4x + 1 = 13.$$

What number x would make this equation true? That is, what value of x would make the left side equal to the right side? (Give students a moment to guess a solution.)

  □ When $x = 3$, we get a true statement. The left side of the equal sign is equal to 13, and so is the right side of the equal sign.

- In other cases, guessing the correct answer is not so easy. Consider the following equation:

$$3(4x - 9) + 10 = 15x + 2 + 7x.$$

Can you guess a number for x that would make this equation true? (Give students a minute to guess.)

- Guessing is not always an efficient strategy for solving equations. In the last example, there are several terms in each of the linear expressions comprising the equation. This makes it more difficult to easily guess a solution. For this reason, we want to use what we know about the properties of equality to transform equations into equations with fewer terms.
- The ultimate goal of solving any equation is to get it into the form of x (or whatever symbol is being used in the equation) equal to a constant.

Complete the activity described below to remind students of the properties of equality, and then proceed with the discussion that follows.

Give students the equation $4 + 1 = 7 - 2$, and ask them the following questions:

1. Is this equation true?
2. Perform each of the following operations, and state whether or not the equation is still true:
   a. Add three to both sides of the equal sign.
   b. Add three to the left side of the equal sign, and add two to the right side of the equal sign.
   c. Subtract six from both sides of the equal sign.
   d. Subtract three from one side of the equal sign, and subtract three from the other side of the equal sign.
   e. Multiply both sides of the equal sign by ten.

A STORY OF RATIOS

Lesson 4 8•4

MP.8

 f. Multiply the left side of the equation by ten and the right side of the equation by four.
 g. Divide both sides of the equation by two.
 h. Divide the left side of the equation by two and the right side of the equation by five.

3. What do you notice? Describe any patterns you see.

- There are four properties of equality that allow us to transform an equation into the form we want. If A, B, and C are any rational numbers, then:
  - If $A = B$, then $A + C = B + C$.
  - If $A = B$, then $A - C = B - C$.
  - If $A = B$, then $A \cdot C = B \cdot C$.
  - If $A = B$, then $\dfrac{A}{C} = \dfrac{B}{C}$, where C is not equal to zero.

- All four of the properties require us to start off with $A = B$. That is, we have to assume that a given equation has an expression on the left side that is equal to the expression on the right side. Working under that assumption, each time we use one of the properties of equality, we are transforming the equation into another equation that is also true; that is, the left side equals the right side.

Example 1 (3 minutes)

- Solve the linear equation $2x - 3 = 4x$ for the number x.
- Examine the properties of equality. Choose "something" to add, subtract, multiply, or divide on both sides of the equation.

Validate the use of the properties of equality by having students share their thoughts. Then, discuss the "best" choice for the first step in solving the equation with the points below. Be sure to remind students throughout this and the other examples that the goal is to get x equal to a constant; therefore, the "best" choice is one that gets them to that goal most efficiently.

- First, we must assume that there is a number x that makes the equation true. Working under that assumption, when we use the property, if $A = B$, then $A - C = B - C$, we get an equation that is also true.

$$2x - 3 = 4x$$
$$2x - 2x - 3 = 4x - 2x$$

Now, using the distributive property, we get another set of equations that is also true.

$$(2-2)x - 3 = (4-2)x$$
$$0x - 3 = 2x$$
$$-3 = 2x$$

Using another property, if $A = B$, then $\dfrac{A}{C} = \dfrac{B}{C}$, we get another equation that is also true.

$$\dfrac{-3}{2} = \dfrac{2x}{2}$$

Lesson 4: Solving a Linear Equation

41

After simplifying the fraction $\frac{2}{2}$, we have
$$\frac{-3}{2} = x,$$
which is also true.

- The last step is to check to see if $x = -\frac{3}{2}$ satisfies the equation $2x - 3 = 4x$.

  The left side of the equation is equal to $2 \cdot \left(-\frac{3}{2}\right) - 3 = -3 - 3 = -6$.

  The right side of the equation is equal to $4 \cdot \left(-\frac{3}{2}\right) = 2 \cdot (-3) = -6$.

  Since the left side equals the right side, we know we have found the number x that solves the equation $2x - 3 = 4x$.

Example 2 (4 minutes)

- Solve the linear equation $\frac{3}{5}x - 21 = 15$. Keep in mind that our goal is to transform the equation so that it is in the form of x equal to a constant. If we assume that the equation is true for some number x, which property should we use to help us reach our goal, and how should we use it?

Again, provide students time to decide which property is "best" to use first.

  ▫ We should use the property if $\{ = B$, then $\{ + C = B + C$, where the number C is 21.

Note that if students suggest that they subtract 15 from both sides (i.e., where C is -15), remind them that they want the form of x equal to a constant. Subtracting 15 from both sides of the equal sign puts the x and all of the constants on the same side of the equal sign. There is nothing mathematically incorrect about subtracting 15, but it does not get them any closer to reaching the goal.

- If we use $\{ + C = B + C$, then we have the true statement:
$$\frac{3}{5}x - 21 + 21 = 15 + 21$$

and

$$\frac{3}{5}x = 36.$$

  Which property should we use to reach our goal, and how should we use it?

  ▫ We should use the property if $\{ = B$, then $\{ \cdot C = B \cdot C$, where C is $\frac{5}{3}$.

- If we use $\{ \cdot C = B \cdot C$, then we have the true statement:
$$\frac{3}{5}x \cdot \frac{5}{3} = 36 \cdot \frac{5}{3},$$

and by the commutative property and the cancellation law, we have
$$x = 12 \cdot 5 = 60.$$

- Does $x = 60$ satisfy the equation $\frac{3}{5}x - 21 = 15$?
  - Yes, because the left side of the equation is equal to $\frac{180}{5} - 21 = 36 - 21 = 15$. Since the right side is also 15, then we know that 60 is a solution to $\frac{3}{5}x - 21 = 15$.

Example 3 (5 minutes)

- The properties of equality are not the only properties we can use with equations. What other properties do we know that could make solving an equation more efficient?
  - We know the distributive property, which allows us to expand and simplify expressions.
  - We know the commutative and associative properties, which allow us to rearrange and group terms within expressions.
- Now we will solve the linear equation $\frac{1}{5}x + 13 + x = 1 - 9x + 22$. Is there anything we can do to the linear expression on the left side to transform it into an expression with fewer terms?
  - Yes. We can use the commutative and distributive properties:

$$\frac{1}{5}x + 13 + x = \frac{1}{5}x + x + 13$$
$$= \left(\frac{1}{5} + 1\right)x + 13$$
$$= \frac{6}{5}x + 13.$$

- Is there anything we can do to the linear expression on the right side to transform it into an expression with fewer terms?
  - Yes. We can use the commutative property:

$$1 - 9x + 22 = 1 + 22 - 9x$$
$$= 23 - 9x.$$

- Now we have the equation $\frac{6}{5}x + 13 = 23 - 9x$. What should we do now to solve the equation?

Note to Teacher:

There are many ways to solve this equation. Any of the actions listed below are acceptable. In fact, a student could say, "Add 100 to both sides of the equal sign," and that, too, would be an acceptable action. It may not lead directly to the answer, but it is still an action that would make a mathematically correct statement. Make it clear that it does not matter which option students choose or in which order; what matters is that they use the properties of equality to make true statements that lead to a solution in the form of x equal to a constant.

Students should come up with the following four responses as to what should be done next. A case can be made for each of them being the "best" move. In this case, each move gets them one step closer to the goal of having the solution in the form of x equal to a constant. Select one option, and move forward with solving the equation (the notes that follow align to the first choice, subtracting 13 from both sides of the equal sign).

- We should subtract 13 from both sides of the equal sign.
- We should subtract 23 from both sides of the equal sign.
- We should add $9x$ to both sides of the equal sign.
- We should subtract $\frac{6}{5}x$ from both sides of the equal sign.

Lesson 4: Solving a Linear Equation

43

- Let's choose to subtract 13 from both sides of the equal sign. Though all options were generally equal with respect to being the "best" first step, I chose this one because when I subtract 13 on both sides, the value of the constant on the left side is positive. I prefer to work with positive numbers. Then we have

$$\frac{6}{5}x + 13 - 13 = 23 - 13 - 9x$$

$$\frac{6}{5}x = 10 - 9x$$

- What should we do next? Why?
  - We should add $9x$ to both sides of the equal sign. We want our solution in the form of x equal to a constant, and this move puts all terms with an x on the same side of the equal sign.

- Adding $9x$ to both sides of the equal sign, we have

$$\frac{6}{5}x + 9x = 10 - 9x + 9x$$

$$\frac{51}{5}x = 10.$$

- What do we need to do now?
  - We should multiply $\frac{5}{51}$ on both sides of the equal sign.

- Then we have

$$\frac{51}{5}x \cdot \frac{5}{51} = 10 \cdot \frac{5}{51}.$$

By the commutative property and the fact that $\frac{5}{51} \cdot \frac{51}{5} = 1$, we have

$$x = \frac{50}{51}.$$

Note to Teacher:
There are still options. If students say they should subtract $\frac{6}{5}x$ from both sides of the equal sign, remind them of the goal of obtaining the x equal to a constant.

- All of the work we did is only valid if our assumption that $\frac{1}{5}x + 13 + x = 1 - 9x + 22$ is a true statement. Therefore, check to see if $\frac{50}{51}$ makes the original equation a true statement.

$$\frac{1}{5}x + 13 + x = 1 - 9x + 22$$

$$\frac{1}{5}\left(\frac{50}{51}\right) + 13 + \frac{50}{51} = 1 - 9\left(\frac{50}{51}\right) + 22$$

$$\frac{6}{5}\left(\frac{50}{51}\right) + 13 = 23 - 9\left(\frac{50}{51}\right)$$

$$\frac{300}{255} + 13 = 23 - \frac{450}{51}$$

$$\frac{3615}{255} = \frac{723}{51}$$

$$\frac{723}{51} = \frac{723}{51}$$

Since both sides of our equation equal $\frac{723}{51}$, then we know that $\frac{50}{51}$ is a solution of the equation.

A STORY OF RATIOS • Lesson 4 • 8•4

Exercises (10 minutes)

Students work on Exercises 1–5 independently.

> **Exercises**
>
> For each problem, show your work, and check that your solution is correct.
>
> 1. Solve the linear equation $x + x + 2 + x + 4 + x + 6 = -28$. State the property that justifies your first step and why you chose it.
>
>   The left side of the equation can be transformed from $x + x + 2 + x + 4 + x + 6$ to $4x + 12$ using the commutative and distributive properties. Using these properties decreases the number of terms of the equation. Now we have the equation:
>
> $$4x + 12 = -28$$
> $$4x + 12 - 12 = -28 - 12$$
> $$4x = -40$$
> $$\frac{1}{4} \cdot 4x = -40 \cdot \frac{1}{4}$$
> $$x = -10.$$
>
>   The left side of the equation is equal to $(-10) + (-10) + 2 + (-10) + 4 + (-10) + 6$, which is -28. Since the left side is equal to the right side, then $x = -10$ is the solution to the equation.
>
>   Note: Students could use the division property in the last step to get the answer.
>
> 2. Solve the linear equation $2(3x + 2) = 2x - 1 + x$. State the property that justifies your first step and why you chose it.
>
>   Both sides of equation can be rewritten using the distributive property. I have to use it on the left side to expand the expression. I have to use it on the right side to collect like terms.
>
>   The left side is
> $$2(3x + 2) = 6x + 4.$$
>
>   The right side is
> $$2x - 1 + x = 2x + x - 1$$
> $$= 3x - 1.$$
>
>   The equation is
> $$6x + 4 = 3x - 1$$
> $$6x + 4 - 4 = 3x - 1 - 4$$
> $$6x = 3x - 5$$
> $$6x - 3x = 3x - 3x - 5$$
> $$(6 - 3)x = (3 - 3)x - 5$$
> $$3x = -5$$
> $$\frac{1}{3} \cdot 3x = \frac{1}{3} \cdot (-5)$$
> $$x = -\frac{5}{3}.$$
>
>   The left side of the equation is $2(3x + 2)$. Replacing x with $-\frac{5}{3}$ gives $2(3\left(-\frac{5}{3}\right) + 2) = 2(-5 + 2) = 2(-3) = -6$.
>   The right side of the equation is $2x - 1 + x$. Replacing x with $-\frac{5}{3}$ gives $2\left(-\frac{5}{3}\right) - 1 + \left(-\frac{5}{3}\right) = -\frac{10}{3} - 1 - \frac{5}{3} = -6$.
>   Since both sides are equal to -6, then $x = -\frac{5}{3}$ is a solution to $2(3x + 2) = 2x - 1 + x$.
>
>   Note: Students could use the division property in the last step to get the answer.

Lesson 4: Solving a Linear Equation

3. Solve the linear equation $x - 9 = \frac{3}{5}x$. State the property that justifies your first step and why you chose it.

   I chose to use the subtraction property of equality to get all terms with an x on one side of the equal sign.

   $$x - 9 = \frac{3}{5}x$$
   $$x - x - 9 = \frac{3}{5}x - x$$
   $$(1-1)x - 9 = \left(\frac{3}{5} - 1\right)x$$
   $$-9 = -\frac{2}{5}x$$
   $$-\frac{5}{2} \cdot (-9) = -\frac{5}{2} \cdot -\frac{2}{5}x$$
   $$\frac{45}{2} = x$$

   The left side of the equation is $\frac{45}{2} - \frac{18}{2} = \frac{27}{2}$. The right side is $\frac{3}{5} \cdot \frac{45}{2} = \frac{3}{1} \cdot \frac{9}{2} = \frac{27}{2}$. Since both sides are equal to the same number, then $x = \frac{45}{2}$ is a solution to $x - 9 = \frac{3}{5}x$.

4. Solve the linear equation $29 - 3x = 5x + 5$. State the property that justifies your first step and why you chose it.

   I chose to use the addition property of equality to get all terms with an x on one side of the equal sign.

   $$29 - 3x = 5x + 5$$
   $$29 - 3x + 3x = 5x + 3x + 5$$
   $$29 = 8x + 5$$
   $$29 - 5 = 8x + 5 - 5$$
   $$24 = 8x$$
   $$\frac{1}{8} \cdot 24 = \frac{1}{8} \cdot 8x$$
   $$3 = x$$

   The left side of the equal sign is $29 - 3(3) = 29 - 9 = 20$. The right side is equal to $5(3) + 5 = 15 + 5 = 20$. Since both sides are equal, $x = 3$ is a solution to $29 - 3x = 5x + 5$.

   Note: Students could use the division property in the last step to get the answer.

5. Solve the linear equation $\frac{1}{3}x - 5 + 171 = x$. State the property that justifies your first step and why you chose it.

   I chose to combine the constants -5 and 171. Then, I used the subtraction property of equality to get all terms with an x on one side of the equal sign.

   $$\frac{1}{3}x - 5 + 171 = x$$
   $$\frac{1}{3}x + 166 = x$$
   $$\frac{1}{3}x - \frac{1}{3}x + 166 = x - \frac{1}{3}x$$
   $$166 = \frac{2}{3}x$$
   $$166 \cdot \frac{3}{2} = \frac{3}{2} \cdot \frac{2}{3}x$$
   $$83 \cdot 3 = x$$
   $$249 = x$$

   The left side of the equation is $\frac{1}{3} \cdot 249 - 5 + 171 = 83 - 5 + 171 = 78 + 171 = 249$, which is exactly equal to the right side. Therefore, $x = 249$ is a solution to $\frac{1}{3}x - 5 + 171 = x$.

A STORY OF RATIOS — Lesson 4 — 8•4

Closing (5 minutes)

Summarize, or ask students to summarize, the main points from the lesson:

- We know that properties of equality, when used to transform equations, make equations with fewer terms that are simpler to solve.
- When solving an equation, we want the answer to be in the form of the symbol x equal to a constant.

Lesson Summary

The properties of equality, shown below, are used to transform equations into simpler forms. If A, B, C are rational numbers, then:

- If $A = B$, then $A + C = B + C$.   Addition property of equality
- If $A = B$, then $A - C = B - C$.   Subtraction property of equality
- If $A = B$, then $A \cdot C = B \cdot C$.   Multiplication property of equality
- If $A = B$, then $\dfrac{A}{C} = \dfrac{B}{C}$, where C is not equal to zero.   Division property of equality

To solve an equation, transform the equation until you get to the form of x equal to a constant ($x = 5$, for example).

Exit Ticket (5 minutes)

Lesson 4: Solving a Linear Equation

Name _____ Date _____

Lesson 4: Solving a Linear Equation

Exit Ticket

1. Guess a number for x that would make the equation true. Check your solution.

$$5x - 2 = 8$$

2. Use the properties of equality to solve the equation $7x - 4 + x = 12$. State which property justifies your first step and why you chose it. Check your solution.

3. Use the properties of equality to solve the equation $3x + 2 - x = 11x + 9$. State which property justifies your first step and why you chose it. Check your solution.

Exit Ticket Sample Solutions

1. Guess a number for x that would make the equation true. Check your solution.
$$5x - 2 = 8$$

When $x = 2$, the left side of the equation is 8, which is the same as the right side. Therefore, $x = 2$ is the solution to the equation.

2. Use the properties of equality to solve the equation $7x - 4 + x = 12$. State which property justifies your first step and why you chose it. Check your solution.

I used the commutative and distributive properties on the left side of the equal sign to simplify the expression to fewer terms.

$$7x - 4 + x = 12$$
$$7x + x - 4 = 12$$
$$(7 + 1)x - 4 = 12$$
$$8x - 4 = 12$$
$$8x - 4 + 4 = 12 + 4$$
$$8x = 16$$
$$\frac{8}{8}x = \frac{16}{8}$$
$$x = 2$$

The left side of the equation is $7(2) - 4 + 2 = 14 - 4 + 2 = 12$. The right side is also 12. Since the left side equals the right side, $x = 2$ is the solution to the equation.

3. Use the properties of equality to solve the equation $3x + 2 - x = 11x + 9$. State which property justifies your first step and why you chose it. Check your solution.

I used the commutative and distributive properties on the left side of the equal sign to simplify the expression to fewer terms.

$$3x + 2 - x = 11x + 9$$
$$3x - x + 2 = 11x + 9$$
$$(3 - 1)x + 2 = 11x + 9$$
$$2x + 2 = 11x + 9$$
$$2x - 2x + 2 = 11x - 2x + 9$$
$$(2 - 2)x + 2 = (11 - 2)x + 9$$
$$2 = 9x + 9$$
$$2 - 9 = 9x + 9 - 9$$
$$-7 = 9x$$
$$\frac{-7}{9} = \frac{9}{9}x$$
$$-\frac{7}{9} = x$$

The left side of the equation is $3\left(\frac{-7}{9}\right) + 2 - \frac{-7}{9} = -\frac{21}{9} + \frac{18}{9} + \frac{7}{9} = \frac{4}{9}$. The right side is $11\left(-\frac{7}{9}\right) + 9 = \frac{-77}{9} + \frac{81}{9} = \frac{4}{9}$. Since the left side equals the right side, $x = -\frac{7}{9}$ is the solution to the equation.

Lesson 4: Solving a Linear Equation

Problem Set Sample Solutions

Students solve equations using properties of equality.

For each problem, show your work, and check that your solution is correct.

1. Solve the linear equation $x + 4 + 3x = 72$. State the property that justifies your first step and why you chose it.

   I used the commutative and distributive properties on the left side of the equal sign to simplify the expression to fewer terms.

   $$x + 4 + 3x = 72$$
   $$x + 3x + 4 = 72$$
   $$(1 + 3)x + 4 = 72$$
   $$4x + 4 = 72$$
   $$4x + 4 - 4 = 72 - 4$$
   $$4x = 68$$
   $$\frac{4}{4}x = \frac{68}{4}$$
   $$x = 17$$

   The left side is equal to $17 + 4 + 3(17) = 21 + 51 = 72$, which is equal to the right side. Therefore, $x = 17$ is a solution to the equation $x + 4 + 3x = 72$.

2. Solve the linear equation $x + 3 + x - 8 + x = 55$. State the property that justifies your first step and why you chose it.

   I used the commutative and distributive properties on the left side of the equal sign to simplify the expression to fewer terms.

   $$x + 3 + x - 8 + x = 55$$
   $$x + x + x + 3 - 8 = 55$$
   $$(1 + 1 + 1)x + 3 - 8 = 55$$
   $$3x - 5 = 55$$
   $$3x - 5 + 5 = 55 + 5$$
   $$3x = 60$$
   $$\frac{3}{3}x = \frac{60}{3}$$
   $$x = 20$$

   The left side is equal to $20 + 3 + 20 - 8 + 20 = 43 - 8 + 20 = 35 + 20 = 55$, which is equal to the right side. Therefore, $x = 20$ is a solution to $x + 3 + x - 8 + x = 55$.

3. Solve the linear equation $\frac{1}{2}x + 10 = \frac{1}{4}x + 54$. State the property that justifies your first step and why you chose it.

   I chose to use the subtraction property of equality to get all of the constants on one side of the equal sign.

   $$\frac{1}{2}x + 10 = \frac{1}{4}x + 54$$
   $$\frac{1}{2}x + 10 - 10 = \frac{1}{4}x + 54 - 10$$
   $$\frac{1}{2}x = \frac{1}{4}x + 44$$
   $$\frac{1}{2}x - \frac{1}{4}x = \frac{1}{4}x - \frac{1}{4}x + 44$$
   $$\frac{1}{4}x = 44$$
   $$4 \cdot \frac{1}{4}x = 4 \cdot 44$$
   $$x = 176$$

   The left side of the equation is $\frac{1}{2}(176) + 10 = 88 + 10 = 98$. The right side of the equation is $\frac{1}{4}(176) + 54 = 44 + 54 = 98$. Since both sides equal 98, $x = 176$ is a solution to the equation $\frac{1}{2}x + 10 = \frac{1}{4}x + 54$.

4. Solve the linear equation $\frac{1}{4}x + 18 = x$. State the property that justifies your first step and why you chose it.

   I chose to use the subtraction property of equality to get all terms with an x on one side of the equal sign.

   $$\frac{1}{4}x + 18 = x$$
   $$\frac{1}{4}x - \frac{1}{4}x + 18 = x - \frac{1}{4}x$$
   $$18 = \frac{3}{4}x$$
   $$\frac{4}{3} \cdot 18 = \frac{4}{3} \cdot \frac{3}{4}x$$
   $$24 = x$$

   The left side of the equation is $\frac{1}{4}(24) + 18 = 6 + 18 = 24$, which is what the right side is equal to. Therefore, $x = 24$ is a solution to $\frac{1}{4}x + 18 = x$.

5. Solve the linear equation $17 - x = \frac{1}{3} \cdot 15 + 6$. State the property that justifies your first step and why you chose it.

   The right side of the equation can be simplified to 11. Then, the equation is

   $$17 - x = 11,$$

   and $x = 6$. Both sides of the equation equal 11; therefore, $x = 6$ is a solution to the equation $17 - x = \frac{1}{3} \cdot 15 + 6$. I was able to solve the equation mentally without using the properties of equality.

Lesson 4: Solving a Linear Equation

6. Solve the linear equation $\frac{x+x+2}{4} = 189.5$. State the property that justifies your first step and why you chose it.

I chose to use the multiplication property of equality to get all terms with an x on one side of the equal sign.

$$\frac{x+x+2}{4} = 189.5$$
$$x+x+2 = 4(189.5)$$
$$2x+2 = 758$$
$$2x+2-2 = 758-2$$
$$2x = 756$$
$$\frac{2}{2}x = \frac{756}{2}$$
$$x = 378$$

The left side of the equation is $\frac{378+378+2}{4} = \frac{758}{4} = 189.5$, which is equal to the right side of the equation. Therefore, $x = 378$ is a solution to $\frac{x+x+2}{4} = 189.5$.

7. Alysha solved the linear equation $2x - 3 - 8x = 14 + 2x - 1$. Her work is shown below. When she checked her answer, the left side of the equation did not equal the right side. Find and explain Alysha's error, and then solve the equation correctly.

$$2x - 3 - 8x = 14 + 2x - 1$$
$$-6x - 3 = 13 + 2x$$
$$-6x - 3 + 3 = 13 + 3 + 2x$$
$$-6x = 16 + 2x$$
$$-6x + 2x = 16$$
$$-4x = 16$$
$$\frac{-4}{-4}x = \frac{16}{-4}$$
$$x = -4$$

Alysha made a mistake on the fifth line. She added $2x$ to the left side of the equal sign and subtracted $2x$ on the right side of the equal sign. To use the property correctly, she should have subtracted $2x$ on both sides of the equal sign, making the equation at that point:

$$-6x - 2x = 16 + 2x - 2x$$
$$-8x = 16$$
$$\frac{-8}{-8}x = \frac{16}{-8}$$
$$x = -2.$$

A STORY OF RATIOS　　　　　　　　　　　　　　　　　　　　　　　　　　　　　　Lesson 5　8•4

 Lesson 5: Writing and Solving Linear Equations

Student Outcomes

- Students apply knowledge of geometry to writing and solving linear equations.

Lesson Notes

All of the problems in this lesson relate to what students have learned about geometry in recent modules and previous years. The purpose is twofold: first, to reinforce what students have learned about geometry, and second, to demonstrate a need for writing and solving an equation in a context that is familiar. Throughout the lesson, students solve mathematical problems that relate directly to what students have learned about angle relationships, congruence, and the triangle sum theorem. Encourage students to draw diagrams to represent the situations presented in the word problems.

Classwork

Example 1 (5 minutes)

MP.1

- Solve the following problem:

> **Example 1**
>
> One angle is five degrees less than three times the measure of another angle. Together, the angle measures have a sum of $143°$. What is the measure of each angle?

Provide students with time to make sense of the problem and persevere in solving it. They could begin their work by guessing and checking, drawing a diagram, or other methods as appropriate. Then, move to the algebraic method shown below.

- What do we need to do first to solve this problem?
    - *First, we need to define our variable (symbol). Let x be the measure of the first angle in degrees.*
- If x is the measure of the first angle, how do you represent the measure of the second angle?
    - *The second angle is $3x - 5$.*
- What is the equation that represents this situation?
    - *The equation is $x + 3x - 5 = 143$.*
- The equation that represents this situation is $x + 3x - 5 = 143$. Solve for x, and then determine the measure of each angle.

> *Scaffolding:*
>
> Model for students how to use diagrams to help make sense of the problems throughout this lesson. Encourage students to use diagrams to help them understand the situation.

　　Lesson 5:　Writing and Solving Linear Equations　　　　　　53

A STORY OF RATIOS   Lesson 5   8•4

As students share their solutions for this and subsequent problems, ask them a variety of questions to reinforce the concepts of the last few lessons. For example, ask students to discuss whether or not this is a linear equation and how they know, to justify their steps and explain why they chose their particular first step, to explain what the solution means, or to justify how they know their answer is correct.

$$x + 3x - 5 = 143$$
$$(1 + 3)x - 5 = 143$$
$$4x - 5 = 143$$
$$4x - 5 + 5 = 143 + 5$$
$$4x = 148$$
$$x = 37$$

The measure of the first angle is $37°$. The second angle is $3(37°) - 5° = 111° - 5° = 106°$.

- Compare the method you tried at the beginning of the problem with the algebraic method. What advantage does writing and solving an equation have?
  - *Writing and solving an equation is a more direct method than the one I tried before. It allows me to find the answer more quickly.*
- Could we have defined x to be the measure of the second angle? If so, what, if anything, would change?
  - *If we let x be the measure of the second angle, then the equation would change, but the answers for the measures of the angles should remain the same.*
- If x is the measure of the second angle, how would we write the measure of the first angle?
  - *The first angle would be $\frac{x+5}{3}$.*
- The equation that represents the situation is $x + \frac{x+5}{3} = 143$. How should we solve this equation?
  - *We could add the fractions together and then solve for x.*
  - *We could multiply every term by 3 to change the fraction to a whole number.*
- Using either method, solve the equation. Verify that the measures of the angles are the same as before.
  -

$$x + \frac{x+5}{3} = 143$$
$$\frac{3x}{3} + \frac{x+5}{3} = 143$$
$$\frac{3x + x + 5}{3} = 143$$
$$3x + x + 5 = 143(3)$$
$$(3 + 1)x + 5 = 429$$
$$4x + 5 = 429$$
$$4x + 5 - 5 = 429 - 5$$
$$4x = 424$$
$$x = 106$$

Scaffolding:

It may be necessary to remind students how to add fractions by rewriting the term or terms as equivalent fractions and then adding the numerators. Provide support as needed.

OR

$$x + \frac{x+5}{3} = 143$$

$$3\left(x + \frac{x+5}{3} = 143\right)$$

$$3x + x + 5 = 429$$

$$(3+1)x + 5 = 429$$

$$4x + 5 = 429$$

$$4x + 5 - 5 = 429 - 5$$

$$4x = 424$$

$$x = 106$$

So, the measure of the second angle is 106°.

$$\frac{x+5}{3} = \frac{106+5}{3} = \frac{111}{3} = 37$$

The measure of the first angle is 37°.

- Whether we let x represent the measure of the first angle or the second angle does not change our answers. Whether we solve the equation using the first or second method does not change our answers. What matters is that we accurately write the information in the problem and correctly use the properties of equality. You may solve a problem differently from your classmates or teachers. Again, what matters most is that what you do is accurate and correct.

Example 2 (12 minutes)

- Solve the following problem:

> **Example 2**
>
> Given a right triangle, find the degree measure of the angles if one angle is ten degrees more than four times the degree measure of the other angle and the third angle is the right angle.

Give students time to work. As they work, walk around and identify students who are writing and solving the problem in different ways. The instructional goal of this example is to make clear that there are different ways to solve a linear equation as opposed to one "right way." Select students to share their work with the class. If students do not come up with different ways of solving the equation, talk them through the following student work samples.

Again, as students share their solutions, ask them a variety of questions to reinforce the concepts of the last few lessons. For example, ask students to discuss whether or not this is a linear equation and how they know, to justify their steps and explain why they chose their particular first step, to explain what the solution means, or to justify how they know their answer is correct.

Lesson 5: Writing and Solving Linear Equations

Solution One

Let x be the measure of the first angle. Then, the second angle is $4x + 10$. The sum of the measures for the angles for this right triangle is $x + 4x + 10 + 90 = 180$.

$$x + 4x + 10 + 90 = 180$$
$$(1 + 4)x + 100 = 180$$
$$5x + 100 = 180$$
$$5x + 100 - 100 = 180 - 100$$
$$5x = 80$$
$$x = 16$$

The measure of the first angle is $16°$, the measure of the second angle is $4(16°) + 10° = 64° + 10° = 74°$, and the measure of the third angle is $90°$.

Solution Two

Let x be the measure of the first angle. Then, the second angle is $4x + 10$. Since we have a right triangle, we already know that one angle is $90°$, which means that the sum of the other two angles is 90: $x + 4x + 10 = 90$.

$$x + 4x + 10 = 90$$
$$(1 + 4)x + 10 = 90$$
$$5x + 10 = 90$$
$$5x + 10 - 10 = 90 - 10$$
$$5x = 80$$
$$x = 16$$

The measure of the first angle is $16°$, the measure of the second angle is $4(16°) + 10° = 64° + 10° = 74°$, and the measure of the third angle is $90°$.

Solution Three

Let x be the measure of the second angle. Then, the first angle is $\frac{x-10}{4}$. Since we have a right triangle, we already know that one angle is $90°$, which means that the sum of the other two angles is 90: $x + \frac{x-10}{4} = 90$.

$$x + \frac{x - 10}{4} = 90$$
$$4\left(x + \frac{x - 10}{4} = 90\right)$$
$$4x + x - 10 = 360$$
$$(4 + 1)x - 10 = 360$$
$$5x - 10 = 360$$
$$5x - 10 + 10 = 360 + 10$$
$$5x = 370$$
$$x = 74$$

The measure of the second angle is $74°$, the measure of the first angle is $\frac{74° - 10°}{4} = \frac{64°}{4} = 16°$, and the measure of the third angle is $90°$.

Lesson 5: Writing and Solving Linear Equations

A STORY OF RATIOS   Lesson 5   8•4

Solution Four

Let x be the measure of the second angle. Then, the first angle is $\frac{x-10}{4}$. The sum of the three angles is $x + \frac{x-10}{4} + 90 = 180$.

$$x + \frac{x-10}{4} + 90 = 180$$
$$x + \frac{x-10}{4} + 90 - 90 = 180 - 90$$
$$x + \frac{x-10}{4} = 90$$
$$\frac{4x}{4} + \frac{x-10}{4} = 90$$
$$\frac{4x + x - 10}{4} = 90$$
$$4x + x - 10 = 360$$
$$5x - 10 + 10 = 360 + 10$$
$$5x = 370$$
$$x = 74$$

The measure of the second angle is $74°$, the measure of the first angle is $\frac{74° - 10°}{4} = \frac{64°}{4} = 16°$, and the measure of the third angle is $90°$.

Make sure students see at least four different methods of solving the problem. Conclude this example with the statements below.

- Each method is slightly different either in terms of how the variable is defined or how the properties of equality are used to solve the equation. The way you find the answer may be different from your classmates' or your teacher's.
- As long as you are accurate and do what is mathematically correct, you will find the correct answer.

Example 3 (4 minutes)

- A pair of alternate interior angles are described as follows. One angle measure, in degrees, is fourteen more than half a number. The other angle measure, in degrees, is six less than half that number. Are the angles congruent?
- We begin by assuming that the angles are congruent. If the angles are congruent, what does that mean about their measures?
    - *It means that they are equal in measure.*
- Write an expression that describes each angle.
    - *One angle measure is $\frac{x}{2} + 14$, and the other angle measure is $\frac{x}{2} - 6$.*

Lesson 5:   Writing and Solving Linear Equations   57

- If the angles are congruent, we can write the equation as $\frac{x}{2} + 14 = \frac{x}{2} - 6$. We know that our properties of equality allow us to transform equations while making sure that they remain true.

$$\frac{x}{2} + 14 = \frac{x}{2} - 6$$
$$\frac{x}{2} - \frac{x}{2} + 14 = \frac{x}{2} - \frac{x}{2} - 6$$
$$14 \neq -6$$

Therefore, our assumption was not correct, and the angles are not congruent.

Exercises (16 minutes)

Students complete Exercises 1–6 independently or in pairs.

Exercises

For each of the following problems, write an equation and solve.

1. A pair of congruent angles are described as follows: The degree measure of one angle is three more than twice a number, and the other angle's degree measure is 54.5 less than three times the number. Determine the measure of the angles in degrees.

   Let x be the number. Then, the measure of one angle is $3 + 2x$, and the measure of the other angle is $3x - 54.5$. Because the angles are congruent, their measures are equal. Therefore,

   $$3 + 2x = 3x - 54.5$$
   $$3 + 2x - 2x = 3x - 2x - 54.5$$
   $$3 = x - 54.5$$
   $$3 + 54.5 = x - 54.5 + 54.5$$
   $$57.5 = x$$

   Replacing x with 57.5 in $3 + 2x$ gives $3 + 2(57.5) = 3 + 115 = 118$; therefore the measure of the angles is $118°$.

2. The measure of one angle is described as twelve more than four times a number. Its supplement is twice as large. Find the measure of each angle in degrees.

   Let x be the number. Then, the measure of one angle is $4x + 12$. The other angle is $2(4x + 12) = 8x + 24$. Since the angles are supplementary, their sum must be $180°$.

   $$4x + 12 + 8x + 24 = 180$$
   $$12x + 36 = 180$$
   $$12x + 36 - 36 = 180 - 36$$
   $$12x = 144$$
   $$x = 12$$

   Replacing x with 12 in $4x + 12$ gives $4(12) + 12 = 48 + 12 = 60$. Replacing x with 12 in $2(4x + 12)$ gives $2(4(12) + 12) = 2(48 + 12) = 2(60) = 120$. Therefore, the measures of the angles are $60°$ and $120°$.

3. A triangle has angles described as follows: The measure of the first angle is four more than seven times a number, the measure of the second angle is four less than the first, and the measure of the third angle is twice as large as the first. What is the measure of each angle in degrees?

Let x be the number. The measure of the first angle is $7x + 4$. The measure of the second angle is $7x + 4 - 4 = 7x$. The measure of the third angle is $2(7x + 4) = 14x + 8$. The sum of the angles of a triangle must be $180°$.

$$7x + 4 + 7x + 14x + 8 = 180$$
$$28x + 12 = 180$$
$$28x + 12 - 12 = 180 - 12$$
$$28x = 168$$
$$x = 6$$

Replacing x with 6 in $7x + 4$ gives $7(6) + 4 = 42 + 4 = 46$. Replacing x with 6 in $7x$ gives $7(6) = 42$. Replacing x with 6 in $14x + 8$ gives $14(6) + 8 = 84 + 8 = 92$. Therefore, the measures of the angles are $46°$, $42°$, and $92°$.

4. One angle measures nine more than six times a number. A sequence of rigid motions maps the angle onto another angle that is described as being thirty less than nine times the number. What is the measure of the angle in degrees?

Let x be the number. Then, the measure of one angle is $6x + 9$. The measure of the other angle is $9x - 30$. Since rigid motions preserve the measures of angles, then the measure of these angles is equal.

$$6x + 9 = 9x - 30$$
$$6x + 9 - 9 = 9x - 30 - 9$$
$$6x = 9x - 39$$
$$6x - 9x = 9x - 9x - 39$$
$$-3x = -39$$
$$x = 13$$

Replacing x with 13 in $6x + 9$ gives $6(13) + 9 = 78 + 9 = 87$. Therefore, the angle measure is $87°$.

5. A right triangle is described as having an angle of measure six less than negative two times a number, another angle measure that is three less than negative one-fourth the number, and a right angle. What are the measures of the angles in degrees?

Let x be a number. Then, the measure of one angle is $-2x - 6$. The measure of the other angle is $-\frac{x}{4} - 3$. The sum of the two angles must be $90°$.

$$-2x - 6 + \left(-\frac{x}{4}\right) - 3 = 90$$
$$\left(-\frac{8x}{4}\right) + \left(-\frac{x}{4}\right) - 9 = 90$$
$$\left(-\frac{9x}{4}\right) - 9 + 9 = 90 + 9$$
$$-\frac{9x}{4} = 99$$
$$-9x = 396$$
$$x = -44$$

Replacing x with -44 gives $-2x - 6$ gives $-2(-44) - 6 = 88 - 6 = 82$. Replacing x with -44 in $-\frac{x}{4} - 3$ gives $90 - 82 = 8$. Therefore, the angle measures are $82°$ and $8°$.

Lesson 5: Writing and Solving Linear Equations

> 6. One angle is one less than six times the measure of another. The two angles are complementary angles. Find the measure of each angle in degrees.
>
> Let x be the measure of the first angle. Then, the measure of the second angle is $6x - 1$. The sum of the measures will be 90 because the angles are complementary.
>
> $$x + 6x - 1 = 90$$
> $$7x - 1 = 90$$
> $$7x - 1 + 1 = 90 + 1$$
> $$7x = 91$$
> $$x = 13$$
>
> The first angle is x and therefore measures $13°$. Replacing x with 13 in $6x - 1$ gives $6(13) - 1 = 78 - 1 = 77$. Therefore, the second angle measure is $77°$.

Closing (4 minutes)

Summarize, or ask students to summarize, the main points from the lesson:

- We know that an algebraic method for solving equations is more efficient than guessing and checking.
- We know how to write and solve equations that relate to angles, triangles, and geometry, in general.
- We know that drawing a diagram can sometimes make it easier to understand a problem and that there is more than one way to solve an equation.

Exit Ticket (4 minutes)

Name _____   Date _____

Lesson 5: Writing and Solving Linear Equations

Exit Ticket

For each of the following problems, write an equation and solve.

1. Given a right triangle, find the measures of all the angles, in degrees, if one angle is a right angle and the measure of the second angle is six less than seven times the measure of the third angle.

2. In a triangle, the measure of the first angle is six times a number. The measure of the second angle is nine less than the first angle. The measure of the third angle is three times the number more than the measure of the first angle. Determine the measure of each angle in degrees.

Exit Ticket Sample Solutions

For each of the following problems, write an equation and solve.

1. Given a right triangle, find the measures of all of the angles, in degrees, if one angle is a right angle and the measure of the second angle is six less than seven times the measure of the third angle.

   Let x represent the measure of the third angle. Then, $7x - 6$ can represent the measure of the second angle. The sum of the two angles in the right triangle will be $90°$.

   $$7x - 6 + x = 90$$
   $$8x - 6 = 90$$
   $$8x - 6 + 6 = 90 + 6$$
   $$8x = 96$$
   $$\frac{8}{8}x = \frac{96}{8}$$
   $$x = 12$$

   The third angle is x and therefore measures $12°$. Replacing x with 12 in $7x - 6$ gives $7(12) - 6 = 84 - 6 = 78$. Therefore, the measure of the second angle is $78°$. The measure of the third angle is $90°$.

2. In a triangle, the measure of the first angle is six times a number. The measure of the second angle is nine less than the first angle. The measure of the third angle is three times the number more than the measure of the first angle. Determine the measure of each angle in degrees.

   Let x be the number. Then, the measure of the first angle is $6x$, the measure of the second angle is $6x - 9$, and the measure of the third angle is $3x + 6x$. The sum of the measures of the angles in a triangle is $180°$.

   $$6x + 6x - 9 + 3x + 6x = 180$$
   $$21x - 9 = 180$$
   $$21x - 9 + 9 = 180 + 9$$
   $$21x = 189$$
   $$\frac{21}{21}x = \frac{189}{21}$$
   $$x = 9$$

   Replacing x with 9 in $6x$ gives $6(9) = 54$. Replacing x with 9 in $6x - 9$ gives $6(9) - 9 = 54 - 9 = 45$. Replacing x with 9 in $3x + 6x$ gives $54 + 3(9) = 54 + 27 = 81$. Therefore, the angle measures are $54°$, $45°$, and $81°$.

Note to teacher: There are several ways to solve problems like these. For example, a student may let x be the measure of the first angle and write the measure of the other angles accordingly. Either way, make sure that students are defining their symbols and correctly using the properties of equality to solve.

A STORY OF RATIOS   Lesson 5   8•4

Problem Set Sample Solutions

Students practice writing and solving linear equations.

For each of the following problems, write an equation and solve.

1. The measure of one angle is thirteen less than five times the measure of another angle. The sum of the measures of the two angles is $140°$. Determine the measure of each angle in degrees.

   Let x be the measure of the one angle. Then, the measure of the other angle is $5x - 13$.

   $$x + 5x - 13 = 140$$
   $$6x - 13 = 140$$
   $$6x - 13 + 13 = 140 + 13$$
   $$6x = 153$$
   $$x = 25.5$$

   Since one angle measure is x, it is $25.5°$. Replacing x with 25.5 in $5x - 13$ gives $5(25.5) - 13 = 140 - 25.5 = 114.5$. Therefore, the other angle measures $114.5°$.

2. An angle measures seventeen more than three times a number. Its supplement is three more than seven times the number. What is the measure of each angle in degrees?

   Let x be the number. Then, the measure of one angle is $3x + 17$. The measure of the other angle is $7x + 3$. Since the angles are supplementary, the sum of their measures will be 180.

   $$3x + 17 + 7x + 3 = 180$$
   $$10x + 20 = 180$$
   $$10x + 20 - 20 = 180 - 20$$
   $$10x = 160$$
   $$x = 16$$

   Replacing x with 16 in $3x + 17$ gives $3(16) + 17 = 65$. Replacing x with 16 in $7x + 3$ gives $(16) + 3 = 112 + 3 = 115$. Therefore, the angle measures are $65°$ and $115°$.

3. The angles of a triangle are described as follows: $\angle A$ is the largest angle; its measure is twice the measure of $\angle B$. The measure of $\angle C$ is 2 less than half the measure of $\angle B$. Find the measures of the three angles in degrees.

   Let x be the measure of $\angle B$. Then, the measure of $\angle A$ is $2x$, and the measure of $\angle C$ is $\frac{x}{2} - 2$. The sum of the measures of the angles must be $180°$.

   $$x + 2x + \frac{x}{2} - 2 = 180$$
   $$3x + \frac{x}{2} - 2 + 2 = 180 + 2$$
   $$3x + \frac{x}{2} = 182$$
   $$\frac{6x}{2} + \frac{x}{2} = 182$$
   $$\frac{7x}{2} = 182$$
   $$7x = 364$$
   $$x = 52$$

   Since x is the measure of $\angle B$, then $\angle B$ is $52°$. Replacing x with 52 in $2x$ gives $2(52) = 104$. Therefore, the measure of $\angle A$ is $104°$. Replacing x with 52 in $\frac{x}{2} - 2$ gives $\frac{52}{2} - 2 = 26 - 2 = 24$. Therefore, the measure of $\angle C$ is $24°$.

Lesson 5:   Writing and Solving Linear Equations   63

This work is derived from Eureka Math ™ and licensed by Great Minds. ©2015 Great Minds. eureka-math.org
G8-M4-TE-B3-1.3.0-07.2015

4. A pair of corresponding angles are described as follows: The measure of one angle is five less than seven times a number, and the measure of the other angle is eight more than seven times the number. Are the angles congruent? Why or why not?

   Let x be the number. Then, the measure of one angle is $7x - 5$, and the measure of the other angle is $7x + 8$. Assume they are congruent, which means their measures are equal.

   $$7x - 5 = 7x + 8$$
   $$7x - 7x - 5 = 7x - 7x + 8$$
   $$-5 \neq 8$$

   Since $-5 \neq 8$, the angles are not congruent.

5. The measure of one angle is eleven more than four times a number. Another angle is twice the first angle's measure. The sum of the measures of the angles is $195°$. What is the measure of each angle in degrees?

   Let x be the number. The measure of one angle can be represented with $4x + 11$, and the other angle's measure can be represented as $2(4x + 11) = 8x + 22$.

   $$4x + 11 + 8x + 22 = 195$$
   $$12x + 33 = 195$$
   $$12x + 33 - 33 = 195 - 33$$
   $$12x = 162$$
   $$x = 13.5$$

   Replacing x with 13.5 in $4x + 11$ gives $4(13.5) + 11 = 54 + 11 = 65$. Replacing x with 13.5 in $2(4x + 11)$ gives $2(4(13.5) + 11) = 2(54 + 11) = 2(65) = 130$. Therefore, the measures of the angles are $65°$ and $130°$.

6. Three angles are described as follows: $\angle B$ is half the size of $\angle A$. The measure of $\angle C$ is equal to one less than two times the measure of $\angle B$. The sum of $\angle A$ and $\angle B$ is $114°$. Can the three angles form a triangle? Why or why not?

   Let x represent the measure of $\angle A$. Then, the measure of $\angle B$ is $\frac{x}{2}$, and the measure of $\angle C$ is $2\left(\frac{x}{2}\right) - 1 = x - 1$.

   The sum of the measures of $\angle A$ and $\angle B$ is 114.

   $$x + \frac{x}{2} = 114$$
   $$\frac{3x}{2} = 114$$
   $$3x = 228$$
   $$x = 76$$

   Since x is the measure of $\angle A$, then $\angle A$ is $76°$. Replacing x with 76 in $\frac{x}{2}$ gives $\frac{76}{2} = 38$; therefore, the measure of $\angle B$ is $38°$. Replacing x with 76 in $x - 1$ gives $76 - 1 = 75$, therefore the measure of $\angle C$ is $75°$. The sum of the three angle measures is $76° + 38° + 75° = 189°$. Since the sum of the measures of the interior angles of a triangle must equal $180°$, these angles do not form a triangle. The sum is too large.

A STORY OF RATIOS   Lesson 6   8•4

Lesson 6: Solutions of a Linear Equation

Student Outcomes

- Students transform equations into simpler forms using the distributive property.
- Students learn that not every linear equation has a solution.

Lesson Notes

The distributive property can be used to both expand and simplify expressions. Students have already used the distributive property to "collect like terms." For example, $2x + 6x = (2 + 6)x = 8x$. Students have also used the distributive property to expand expressions. For example, $2(x + 5) = 2x + 10$. In this lesson, students continue to use the distributive property to solve more complicated equations. Also highlighted in this lesson is a common error that students make when using the distributive property, which is multiplying a factor to terms that are not part of the group. For example, in the expression $3(x + 1) - 5$, students should know that they do not distribute the factor 3 to the term -5 because it is not in the group $(x + 1)$.

Classwork

Example 1 (4 minutes)

- What value of x would make the linear equation $4x + 3(4x + 7) = 4(7x + 3) - 3$ true? What is the "best" first step, and why?

Have a discussion with students about what the "best" first step is and why. Make clear that the distributive property allows students to better see and work with the terms of the linear equation. Proceed with the following discussion.

- In order to find out what that solution might be, we must use the distributive property. The left side of the equation has the following expression:

$$4x + 3(4x + 7).$$

Where and how will the distributive property be used?

- We will need to use the distributive property to expand $3(4x + 7)$ and then again to collect like terms.

- Our work for now is just on the left side of the equation; the right side will remain unchanged for the moment.

$$4x + 3(4x + 7) = 4(7x + 3) - 3$$
$$4x + 12x + 21 = 4(7x + 3) - 3$$
$$(4 + 12)x + 21 = 4(7x + 3) - 3$$
$$16x + 21 = 4(7x + 3) - 3$$

- Now we need to rewrite the right side. Again, we will use the distributive property. The left side of the equation will remain unchanged.

$$16x + 21 = 4(7x + 3) - 3$$
$$16x + 21 = 28x + 12 - 3$$

Lesson 6:   Solutions of a Linear Equation   65

A STORY OF RATIOS　　　　　　　　　　　　　　　　　　　　　　　　　　　Lesson 6　8•4

- Notice that we did not apply the distributive property to the term -3. Since it was not part of the group $(7x + 3)$, it is not multiplied by 4.

$$16x + 21 = 28x + 9$$

- Now we have transformed the given equation into the following form: $16x + 21 = 28x + 9$. Solve the equation.
  - Student work:

$$16x + 21 = 28x + 9$$
$$16x - 16x + 21 = 28x - 16x + 9$$
$$21 = 12x + 9$$
$$21 - 9 = 12x + 9 - 9$$
$$12 = 12x$$
$$x = 1$$

- Is $x = 1$ really a solution to the equation $4x + 3(4x + 7) = 4(7x + 3) - 3$? How do you know?
  - Yes, $x = 1$ is a solution because $4 + 3(11) = 37$ and $4(10) - 3 = 37$. Since both expressions are equal to 37, then $x = 1$ is a solution to the equation.

Example 2 (4 minutes)

- What value of x would make the following linear equation true: $20 - (3x - 9) - 2 = -(-11x + 1)$? Since we have a group of terms that is preceded by a "$-$" sign, we will simplify this first. The "$-$" sign means we need to take the opposite of each of the terms within the group (i.e., parentheses).

- We begin with the left side of the equation:

$$20 - (3x - 9) - 2 = -(-11x + 1).$$

We need only to take the opposite of the terms within the grouping symbols. Is the term -2 within the grouping symbol?
  - No

- For that reason, we need only find the opposite of $3x - 9$. What is the opposite of $3x - 9$?
  - The opposite of $3x - 9$ is $-3x + 9$.

- The left side of the equation is rewritten as

$$20 - (3x - 9) - 2 = -(-11x + 1)$$
$$20 - 3x + 9 - 2 = -(-11x + 1)$$
$$20 + 9 - 2 - 3x = -(-11x + 1)$$
$$27 - 3x = -(-11x + 1).$$

- Now we rewrite the right side of the equation: $-(-11x + 1)$.

$$27 - 3x = -(-11x + 1)$$
$$27 - 3x = 11x - 1$$

> **Scaffolding:**
> The equation in this example can be modified to $20 - (3x - 9 + 1) = 10$ to meet the needs of diverse learners. Also, consider having students fold a piece of paper in half, solve on the left side, and justify their steps on the right side.

- The transformed equation is $27 - 3x = 11x - 1$.

$$27 - 3x = 11x - 1$$
$$27 - 3x + 3x = 11x + 3x - 1$$
$$27 = 14x - 1$$
$$27 + 1 = 14x - 1 + 1$$
$$28 = 14x$$
$$2 = x$$

- Check: The left side is $20 - (3x - 9) - 2 = 20 - (3(2) - 9) - 2 = 20 - (-3) - 2 = 20 + 3 - 2 = 21$. The right side is $-(-11x + 1) = -(-11(2) + 1) = -(-22 + 1) = -(-21) = 21$. Since $21 = 21$, $x = 2$ is the solution.

Example 3 (4 minutes)

- What value of x would make the following linear equation true: $\frac{1}{2}(4x + 6) - 2 = -(5x + 9)$? Begin by transforming both sides of the equation into a simpler form.
  - *Student work:*

$$\frac{1}{2}(4x + 6) - 2 = -(5x + 9)$$
$$2x + 3 - 2 = -5x - 9$$

Scaffolding:

The equation in this example can be modified to $2x + 1 = -(5x + 9)$ to meet the needs of diverse learners.

Make sure that students do not distribute the factor $\frac{1}{2}$ to the term -2 and that they have, in general, transformed the equation correctly.

- Now that we have the simpler equation, $2x + 3 - 2 = -5x - 9$, complete the solution.
  - *Student work:*

$$2x + 3 - 2 = -5x - 9$$
$$2x + 1 = -5x - 9$$
$$2x + 1 - 1 = -5x - 9 - 1$$
$$2x + 0 = -5x - 10$$
$$2x = -5x - 10$$
$$2x + 5x = -5x + 5x - 10$$
$$(2 + 5)x = (-5 + 5)x - 10$$
$$7x = 0x - 10$$
$$7x = -10$$
$$\frac{7}{7}x = -\frac{10}{7}$$
$$x = -\frac{10}{7}$$

Lesson 6: Solutions of a Linear Equation

- Check: The left side is

$$\frac{1}{2}(4x+6) - 2 = \frac{1}{2}\left(4\left(-\frac{10}{7}\right)+6\right) - 2 = \frac{1}{2}\left(-\frac{40}{7}+6\right) - 2 = \frac{1}{2}\left(\frac{2}{7}\right) - 2 = \frac{1}{7} - 2 = -\frac{13}{7}.$$

The right side is

$$-(5x+9) = -\left(5\left(-\frac{10}{7}\right)+9\right) = -\left(-\frac{50}{7}+9\right) = -\left(\frac{13}{7}\right) = -\frac{13}{7}.$$

Since $-\frac{13}{7} = -\frac{13}{7}$, $x = -\frac{10}{7}$ is the solution.

Example 4 (11 minutes)

- Consider the following equation: $2(x+1) = 2x - 3$. What value of x makes the equation true?
    - *Student work:*

    $$2(x+1) = 2x - 3$$
    $$2x + 2 = 2x - 3$$
    $$2x - 2x + 2 = 2x - 2x - 3$$
    $$2 = -3$$

- How should we interpret $2 = -3$?

MP.3 Lead a discussion with the conclusion that since $2 \neq -3$, then the equation has no solution. Allow students time to try to find a value of x that would make it true by guessing and checking. After they realize that there is no such number x, make it clear to students that some equations have no solution. Ask the following question.

- Why do you think this happened?
    - *We know that an equation is a statement of equality. The linear expressions were such that they could not be equal to each other, no matter what value was substituted for x.*

- What value of x would make the following linear equation true: $9(4 - 2x) - 3 = 4 - 6(3x - 5)$? Transform the equation by simplifying both sides.
    - *Student work:*

    $$9(4 - 2x) - 3 = 4 - 6(3x - 5)$$
    $$36 - 18x - 3 = 4 - 18x + 30$$

Scaffolding:
The equation in this example can be modified to $9(4 - 2x) - 3 = -18x$ to meet the needs of diverse learners.

Be sure to check that students did not subtract $4 - 6$ on the right side and then distribute -2. This is a common error. Remind students that they must multiply first and then add or subtract, just like they would to simplify expressions using the correct order of operations.

A STORY OF RATIOS　　　　　　　　　　　　　　　　　　　　　　　　　　Lesson 6　8•4

- The transformed equation is $36 - 18x - 3 = 4 - 18x + 30$. Now, complete the solution.
  - *Student work:*
  $$36 - 18x - 3 = 4 - 18x + 30$$
  $$33 - 18x = 34 - 18x$$
  $$33 - 18x + 18x = 34 - 18x + 18x$$
  $$33 = 34$$

  Like the last problem, there is no value of x that can be substituted into the equation to make it true. Therefore, this equation has no solution.

- Write at least one equation that has no solution. It does not need to be complicated, but the result should be similar to the last two problems. The result from the first equation was $2 \neq -3$, and the second was $33 \neq 34$.

Have students share their equations and verify that they have no solution.

Example 5 (Optional – 4 minutes)

- So far, we have used the distributive property to simplify expressions when solving equations. In some cases, we can use the distributive property to make our work even simpler. Consider the following equation:
$$3x + 15 = -6.$$

Notice that each term has a common factor of 3. We will use the distributive property and what we know about the properties of equality to solve this equation quickly.

$$3x + 15 = -6$$
$$3(x + 5) = 3 \cdot (-2)$$

Notice that the expressions on both sides of the equal sign have a factor of 3. We can use the multiplication property of equality, *if $\{ = B$, then $\{ \cdot C = B \cdot C$ as follows:*

$$3(x + 5) = 3 \cdot (-2)$$
$$x + 5 = -2$$
$$x + 5 - 5 = -2 - 5$$
$$x = -7.$$

- This is not something that we can expect to do every time we solve an equation, but it is good to keep an eye out for it.

Lesson 6:　Solutions of a Linear Equation　　　　　　　　　　　　　　　　69

A STORY OF RATIOS — Lesson 6   8•4

Exercises (12 minutes)

Students complete Exercises 1–6 independently.

Exercises

Find the value of x that makes the equation true.

1. $17 - 5(2x - 9) = -(-6x + 10) + 4$

$$17 - 5(2x - 9) = -(-6x + 10) + 4$$
$$17 - 10x + 45 = 6x - 10 + 4$$
$$62 - 10x = 6x - 6$$
$$62 - 10x + 10x = 6x + 10x - 6$$
$$62 = 16x - 6$$
$$62 + 6 = 16x - 6 + 6$$
$$68 = 16x$$
$$\frac{68}{16} = \frac{16}{16}x$$
$$\frac{68}{16} = x$$
$$\frac{17}{4} = x$$

2. $-(x - 7) + \frac{5}{3} = 2(x + 9)$

$$-(x - 7) + \frac{5}{3} = 2(x + 9)$$
$$-x + 7 + \frac{5}{3} = 2x + 18$$
$$-x + \frac{26}{3} = 2x + 18$$
$$-x + x + \frac{26}{3} = 2x + x + 18$$
$$\frac{26}{3} = 3x + 18$$
$$\frac{26}{3} - 18 = 3x + 18 - 18$$
$$-\frac{28}{3} = 3x$$
$$\frac{1}{3} \cdot \frac{-28}{3} = \frac{1}{3} \cdot 3x$$
$$-\frac{28}{9} = x$$

Lesson 6: Solutions of a Linear Equation

3. $\frac{4}{9} + 4(x-1) = \frac{28}{9} - (x-7x) + 1$

$$\frac{4}{9} + 4(x-1) = \frac{28}{9} - (x-7x) + 1$$
$$\frac{4}{9} - \frac{4}{9} + 4(x-1) = \frac{28}{9} - \frac{4}{9} - (x-7x) + 1$$
$$4x - 4 = \frac{24}{9} - x + 7x + 1$$
$$4x - 4 = \frac{33}{9} + 6x$$
$$4x - 4 + 4 = \frac{33}{9} + \frac{36}{9} + 6x$$
$$4x = \frac{69}{9} + 6x$$
$$4x - 6x = \frac{69}{9} + 6x - 6x$$
$$-2x = \frac{23}{3}$$
$$\frac{1}{-2} \cdot -2x = \frac{1}{-2} \cdot \frac{23}{3}$$
$$x = -\frac{23}{6}$$

4. $5(3x+4) - 2x = 7x - 3(-2x+11)$

$$5(3x+4) - 2x = 7x - 3(-2x+11)$$
$$15x + 20 - 2x = 7x + 6x - 33$$
$$13x + 20 = 13x - 33$$
$$13x - 13x + 20 = 13x - 13x - 33$$
$$20 \neq -33$$

This equation has no solution.

5. $7x - (3x+5) - 8 = \frac{1}{2}(8x+20) - 7x + 5$

$$7x - (3x+5) - 8 = \frac{1}{2}(8x+20) - 7x + 5$$
$$7x - 3x - 5 - 8 = 4x + 10 - 7x + 5$$
$$4x - 13 = -3x + 15$$
$$4x - 13 + 13 = -3x + 15 + 13$$
$$4x = -3x + 28$$
$$4x + 3x = -3x + 3x + 28$$
$$7x = 28$$
$$x = 4$$

6. Write at least three equations that have no solution.

Answers will vary. Verify that the equations written have no solution.

A STORY OF RATIOS • Lesson 6 • 8•4

Closing (5 minutes)

Summarize, or ask students to summarize, the following main points from the lesson:

- We know how to transform equations into simpler forms using the distributive property.
- We now know that there are some equations that do not have solutions.

Lesson Summary

The distributive property is used to expand expressions. For example, the expression $2(3x - 10)$ is rewritten as $6x - 20$ after the distributive property is applied.

The distributive property is used to simplify expressions. For example, the expression $7x + 11x$ is rewritten as $(7 + 11)x$ and $18x$ after the distributive property is applied.

The distributive property is applied only to terms within a group:
$$4(3x + 5) - 2 = 12x + 20 - 2.$$
Notice that the term -2 is not part of the group and, therefore, not multiplied by 4.

When an equation is transformed into an untrue sentence, such as $5 \neq 11$, we say the equation has *no solution*.

Exit Ticket (5 minutes)

Name _____ Date _____

Lesson 6: Solutions of a Linear Equation

Exit Ticket

Transform the equation if necessary, and then solve to find the value of x that makes the equation true.

1. $5x - (x + 3) = \frac{1}{3}(9x + 18) - 5$

2. $5(3x + 9) - 2x = 15x - 2(x - 5)$

Exit Ticket Sample Solutions

Transform the equation if necessary, and then solve to find the value of x that makes the equation true.

1. $5x - (x + 3) = \frac{1}{3}(9x + 18) - 5$

$$5x - (x + 3) = \frac{1}{3}(9x + 18) - 5$$
$$5x - x - 3 = 3x + 6 - 5$$
$$4x - 3 = 3x + 1$$
$$4x - 3x - 3 = 3x - 3x + 1$$
$$x - 3 = 1$$
$$x - 3 + 3 = 1 + 3$$
$$x = 4$$

2. $5(3x + 9) - 2x = 15x - 2(x - 5)$

$$5(3x + 9) - 2x = 15x - 2(x - 5)$$
$$15x + 45 - 2x = 15x - 2x + 10$$
$$13x + 45 = 13x + 10$$
$$13x - 13x + 45 = 13x - 13x + 10$$
$$45 \neq 10$$

Since $45 \neq 10$, the equation has no solution.

Problem Set Sample Solutions

Students practice using the distributive property to transform equations and solve.

Transform the equation if necessary, and then solve it to find the value of x that makes the equation true.

1. $x - (9x - 10) + 11 = 12x + 3\left(-2x + \frac{1}{3}\right)$

$$x - (9x - 10) + 11 = 12x + 3\left(-2x + \frac{1}{3}\right)$$
$$x - 9x + 10 + 11 = 12x - 6x + 1$$
$$-8x + 21 = 6x + 1$$
$$-8x + 8x + 21 = 6x + 8x + 1$$
$$21 = 14x + 1$$
$$21 - 1 = 14x + 1 - 1$$
$$20 = 14x$$
$$\frac{20}{14} = \frac{14}{14}x$$
$$\frac{10}{7} = x$$

2. $7x + 8\left(x + \frac{1}{4}\right) = 3(6x - 9) - 8$

$$7x + 8\left(x + \frac{1}{4}\right) = 3(6x - 9) - 8$$
$$7x + 8x + 2 = 18x - 27 - 8$$
$$15x + 2 = 18x - 35$$
$$15x - 15x + 2 = 18x - 15x - 35$$
$$2 = 3x - 35$$
$$2 + 35 = 3x - 35 + 35$$
$$37 = 3x$$
$$\frac{37}{3} = \frac{3}{3}x$$
$$\frac{37}{3} = x$$

3. $-4x - 2(8x + 1) = -(-2x - 10)$

$$-4x - 2(8x + 1) = -(-2x - 10)$$
$$-4x - 16x - 2 = 2x + 10$$
$$-20x - 2 = 2x + 10$$
$$-20x + 20x - 2 = 2x + 20x + 10$$
$$-2 = 22x + 10$$
$$-2 - 10 = 22x + 10 - 10$$
$$-12 = 22x$$
$$-\frac{12}{22} = \frac{22}{22}x$$
$$-\frac{6}{11} = x$$

4. $11(x + 10) = 132$

$$11(x + 10) = 132$$
$$\left(\frac{1}{11}\right)11(x + 10) = \left(\frac{1}{11}\right)132$$
$$x + 10 = 12$$
$$x + 10 - 10 = 12 - 10$$
$$x = 2$$

5. $37x + \frac{1}{2} - \left(x + \frac{1}{4}\right) = 9(4x - 7) + 5$

$$37x + \frac{1}{2} - \left(x + \frac{1}{4}\right) = 9(4x - 7) + 5$$
$$37x + \frac{1}{2} - x - \frac{1}{4} = 36x - 63 + 5$$
$$36x + \frac{1}{4} = 36x - 58$$
$$36x - 36x + \frac{1}{4} = 36x - 36x - 58$$
$$\frac{1}{4} \neq -58$$

This equation has no solution.

6. $3(2x - 14) + x = 15 - (-9x - 5)$

$$3(2x - 14) + x = 15 - (-9x - 5)$$
$$6x - 42 + x = 15 + 9x + 5$$
$$7x - 42 = 20 + 9x$$
$$7x - 7x - 42 = 20 + 9x - 7x$$
$$-42 = 20 + 2x$$
$$-42 - 20 = 20 - 20 + 2x$$
$$-62 = 2x$$
$$-31 = x$$

7. $8(2x + 9) = 56$

$$8(2x + 9) = 56$$
$$\left(\frac{1}{8}\right)8(2x + 9) = \left(\frac{1}{8}\right)56$$
$$2x + 9 = 7$$
$$2x + 9 - 9 = 7 - 9$$
$$2x = -2$$
$$\left(\frac{1}{2}\right)2x = \left(\frac{1}{2}\right) - 2$$
$$x = -1$$

A STORY OF RATIOS   Lesson 7   8•4

Lesson 7: Classification of Solutions

Student Outcomes

- Students know the conditions for which a linear equation has a unique solution, no solution, or infinitely many solutions.

Lesson Notes

Part of the discussion on page 78 in this lesson is optional. The key parts of the discussion are those that point out the characteristics of the constants and coefficients of an equation that allow students to see the structure in equations that lead to a unique solution, no solution, or infinitely many solutions. Go through the discussion with students, or use the activity on page 80 of this lesson. The activity requires students to examine groups of equations and make conclusions about the nature of their solutions based on what they observe, leading to the same result as the discussion.

Classwork

Exercises 1–3 (6 minutes)

Students complete Exercises 1–3 independently in preparation for the discussion that follows.

Exercises

Solve each of the following equations for x.

1. $7x - 3 = 5x + 5$

$$7x - 3 = 5x + 5$$
$$7x - 3 + 3 = 5x + 5 + 3$$
$$7x = 5x + 8$$
$$7x - 5x = 5x - 5x + 8$$
$$2x = 8$$
$$x = 4$$

2. $7x - 3 = 7x + 5$

$$7x - 3 = 7x + 5$$
$$7x - 7x - 3 = 7x - 7x + 5$$
$$-3 \neq 5$$

This equation has no solution.

Lesson 7:   Classification of Solutions   77

3. $7x - 3 = -3 + 7x$

$$7x - 3 = -3 + 7x$$
$$7x - 3 + 3 = -3 + 3 + 7x$$
$$7x = 7x$$

OR

$$7x - 3 = -3 + 7x$$
$$7x - 7x - 3 = -3 + 7x - 7x$$
$$-3 = -3$$

Discussion (15 minutes)

Display the three equations so that students can easily see and compare them throughout the discussion.

| $7x - 3 = 5x + 5$ | $7x - 3 = 7x + 5$ | $7x - 3 = -3 + 7x$ |

- Was there anything new or unexpected about Exercise 1?
  - *No. We solved the equation for x, and $x = 4$.*

Continue with the discussion of Exercise 1, or complete the activity described at the end of the Discussion on page 80. Be sure to revisit Exercises 2 and 3, and classify the solution to those equations.

- What can you say about the coefficients of terms with x in the equation?
  - *They are different.*
- Do you think any number other than 4 would make the equation true?
  - *I don't think so, but I would have to try different numbers to find out.*
- Instead of having to check every single number to see if it makes the equation true, we can look at a general form of an equation to show that there can be only one solution.
- Given a linear equation that has been simplified on both sides, $ax + b = cx + d$, where $a, b, c,$ and d are constants and $a \neq c$, we can use our normal properties of equality to solve for x.

$$ax + b = cx + d$$
$$ax + b - b = cx + d - b$$
$$ax = cx + d - b$$
$$ax - cx = cx - cx + d - b$$
$$(a - c)x = d - b$$
$$\frac{a - c}{a - c}x = \frac{d - b}{a - c}$$
$$x = \frac{d - b}{a - c}$$

- The only value of x that will make the equation true is the division of the difference of the constants by the difference of the coefficients of x. In other words, if the coefficients of x are different on each side of the equal sign, then the equation will have one solution.

- As an aside, if the coefficients of x are different and the value of the constants are the same, the only solution is $x = 0$. Consider the following equation:
$$2x + 12 = x + 12$$
$$2x + 12 - 12 = x + 12 - 12$$
$$2x = x$$
$$2x - x = x - x$$
$$x = 0.$$

- The only value of x that will make $2x + 12 = x + 12$ true is $x = 0$.
- Was there anything new or unexpected about Exercise 2?
  - No. We tried to solve the equation and got an untrue statement, $-3 \neq 5$; therefore, the equation had no solutions.
- What can you say about the coefficients of the terms with x in the second equation?
  - The coefficients were the same.
- What can you say about the constants in the second equation?
  - The constants were different.
- It is time to begin formalizing what we have observed as we have solved equations. To begin with, we have learned that some linear equations have no solutions. That is, when we solve an equation using our properties of equality and end up with a statement like $a = b$, where a and b are different numbers, we say that the equation has no solution.
- To produce and/or recognize an equation with no solutions, consider the equation with both sides simplified. The coefficients of the variable x will be the same on both sides of the equal sign, and the value of the constants on each side will be different. This type of equation has no solution because there is no value of x in an equation such as $x + a = x + b$ that will make the equation true.

Give students a moment to compare the equation $x + a = x + b$ with Exercise 2.

- Was there anything new or unexpected about Exercise 3?
  - Yes. When solving the equation, we got $-3 = -3$, which did not give us a value for x.
  - Yes. When solving the equation, we got to $7x = 7x$, and we weren't sure what to do next or what it meant.

Provide students with time to discuss in small groups how they should interpret the meaning of such statements. Students have yet to work with equations that yield such results, and they need the time to make sense of the statements. If necessary, ask the questions in the next two bullet points to guide their thinking. Give them time to try different values of x in both equations to verify that it is always true. Next, have students share their thoughts and interpretations and resume the discussion below.

- Is $-3 = -3$ a true statement? Is it always true?
  - Yes
- If $7x = 7x$, will we always have an equality, no matter which number we choose for x?
  - Yes. This is an identity because the same expression is on both sides of the equal sign.
- What can you say about the coefficients of the terms with x in the equation in Exercise 3?
  - The coefficients are the same.

Lesson 7: Classification of Solutions

- What can you say about the constants in the equation?
  - *The constants are the same.*
- We now know that there are equations that produce statements that are always true. That is, -3 will always equal -3, 0 will always equal 0, and $7x$ will always equal $7x$, no matter what number x happens to be (because 7 times a number is just a number). When we solve an equation using our properties of equality and end up with a statement like $a = a$ where a is a number, then we say that the equation has infinitely many solutions.
- To produce and/or recognize an equation with infinitely many solutions, the coefficients of the variable x and the value of the constants will be the same in the simplified expressions on both sides of the equal sign. This type of equation has infinitely many solutions because *any value* of x in an equation like $x + a = x + a$ will make the equation true.

Give students a moment to compare the equation $x + a = x + a$ with Exercise 3.

Activity: Display the table below. Provide students time to work independently to solve the equations; then, allow them to work in small groups to generalize what they notice about each of the equations. That is, they should answer the question: What can we see in an equation that will tell us about the solution to the equation?

$3x + 4 = 8x - 9$	$6x + 5 = 8 + 6x$	$10x - 4 = -4 + 10x$
$-4x - 5 = 6 - 11x$	$12 - 15x = -2 - 15x$	$-2x + 5 = -2x + 5$
$9 + \frac{1}{2}x = 5x - 1$	$\frac{5}{4}x - 1 = 1 + \frac{5}{4}x$	$7 + 9x = 9x + 7$

MP.8

The goal of the activity is for students to make the following three observations: (1) In equations where both sides are simplified and the coefficients of x and the constants in the equation are unique, (i.e., $ax + b = cx + d$), there is a unique solution, (i.e., $x = \frac{d-b}{a-c}$), as in the first column. (2) Similarly, students should observe that when an equation is of the form $x + b = x + c$, there is no solution, as in the second column. (3) Lastly, when an equation is of the form $x + a = x + a$, then there are infinitely many solutions, as in the third column.

Exercises 4–10 (14 minutes)

Students complete Exercises 4–10 independently or in pairs.

> Give a brief explanation as to what kind of solution(s) you expect the following linear equations to have. Transform the equations into a simpler form if necessary.
>
> 4. $11x - 2x + 15 = 8 + 7 + 9x$
>
>   *If I use the distributive property on the left side, I notice that the coefficients of the x are the same, specifically 9, and when I simplify the constants on the right side, I notice that they are the same. Therefore, this equation has infinitely many solutions.*
>
> 5. $3(x - 14) + 1 = -4x + 5$
>
>   *If I use the distributive property on the left side, I notice that the coefficients of x are different. Therefore, the equation has one solution.*

6. $-3x + 32 - 7x = -2(5x + 10)$

   If I use the distributive property on the each side of the equation, I notice that the coefficients of x are the same, but the constants are different. Therefore, this equation has no solutions.

7. $\frac{1}{2}(8x + 26) = 13 + 4x$

   If I use the distributive property on the left side, I notice that the coefficients of x are the same, specifically 4, and the constants are also the same, 13. Therefore, this equation has infinitely many solutions.

8. Write two equations that have no solutions.

   Answers will vary. Verify that students have written equations where the coefficients of x on each side of the equal sign are the same and that the constants on each side are unique.

9. Write two equations that have one unique solution each.

   Answers will vary. Accept equations where the coefficients of x on each side of the equal sign are unique.

10. Write two equations that have infinitely many solutions.

    Answers will vary. Accept equations where the coefficients of x and the constants on each side of the equal sign are the same.

Closing (5 minutes)

Summarize, or ask students to summarize, the main points from the lesson:

- We know that equations will either have a unique solution, no solution, or infinitely many solutions.
- We know that equations with no solution will, after being simplified on both sides, have coefficients of x that are the same on both sides of the equal sign and constants that are different.
- We know that equations with infinitely many solutions will, after being simplified on both sides, have coefficients of x and constants that are the same on both sides of the equal sign.

Lesson Summary

There are three classifications of solutions to linear equations: one solution (unique solution), no solution, or infinitely many solutions.

Equations with no solution will, after being simplified, have coefficients of x that are the same on both sides of the equal sign and constants that are different. For example, $x + b = x + c$, where b and c are constants that are not equal. A numeric example is $8x + 5 = 8x - 3$.

Equations with infinitely many solutions will, after being simplified, have coefficients of x and constants that are the same on both sides of the equal sign. For example, $x + a = x + a$, where a is a constant. A numeric example is $6x + 1 = 1 + 6x$.

Exit Ticket (5 minutes)

Lesson 7: Classification of Solutions

Name _____ Date _____

Lesson 7: Classification of Solutions

Exit Ticket

Give a brief explanation as to what kind of solution(s) you expect the following linear equations to have. Transform the equations into a simpler form if necessary.

1. $3(6x + 8) = 24 + 18x$

2. $12(x + 8) = 11x - 5$

3. $5x - 8 = 11 - 7x + 12x$

Exit Ticket Sample Solutions

> Give a brief explanation as to what kind of solution(s) you expect the following linear equations to have. Transform the equations into a simpler form if necessary.
>
> 1. $3(6x + 8) = 24 + 18x$
>
>   *If I use the distributive property on the left side, I notice that the coefficients of x are the same, and the constants are the same. Therefore, this equation has infinitely many solutions.*
>
> 2. $12(x + 8) = 11x - 5$
>
>   *If I use the distributive property on the left side, I notice that the coefficients of x are different, and the constants are different. Therefore, this equation has a unique solution.*
>
> 3. $5x - 8 = 11 - 7x + 12x$
>
>   *If I collect the like terms on the right side, I notice that the coefficients of x are the same, but the constants are different. Therefore, this equation has no solution.*

Problem Set Sample Solutions

Students apply their knowledge of solutions to linear equations by writing equations with unique solutions, no solutions, and infinitely many solutions.

> 1. Give a brief explanation as to what kind of solution(s) you expect for the linear equation $18x + \frac{1}{2} = 6(3x + 25)$. Transform the equation into a simpler form if necessary.
>
>   *If I use the distributive property on the right side of the equation, I notice that the coefficients of x are the same, but the constants are different. Therefore, this equation has no solutions.*
>
> 2. Give a brief explanation as to what kind of solution(s) you expect for the linear equation $8 - 9x = 15x + 7 + 3x$. Transform the equation into a simpler form if necessary.
>
>   *If I collect the like terms on the right side of the equation, then I notice that the coefficients of x are different, and so are the constants. Therefore, this equation will have a unique solution.*
>
> 3. Give a brief explanation as to what kind of solution(s) you expect for the linear equation $5(x + 9) = 5x + 45$. Transform the equation into a simpler form if necessary.
>
>   *This is an identity under the distributive property. Therefore, this equation will have infinitely many solutions.*
>
> 4. Give three examples of equations where the solution will be unique; that is, only one solution is possible.
>
>   *Accept equations where the coefficients of x on each side of the equal sign are unique.*
>
> 5. Solve one of the equations you wrote in Problem 4, and explain why it is the only solution.
>
>   *Verify that students solved one of the equations. They should have an explanation that includes the statement that there is only one possible number that could make the equation true. They may have referenced one of the simpler forms of their transformed equation to make their case.*

Lesson 7: Classification of Solutions

A STORY OF RATIOS Lesson 7 8•4

6. Give three examples of equations where there will be no solution.

   Accept equations where the coefficients of x on each side of the equal sign are the same, and the constants on each side are unique.

7. Attempt to solve one of the equations you wrote in Problem 6, and explain why it has no solution.

   Verify that students have solved one of the equations. They should have an explanation that includes the statement about getting a false equation (e.g., $6 \neq 10$).

8. Give three examples of equations where there will be infinitely many solutions.

   Accept equations where the coefficients of x and constants on each side of the equal sign are the same.

9. Attempt to solve one of the equations you wrote in Problem 8, and explain why it has infinitely many solutions.

   Verify that students have solved one of the equations. They should have an explanation that includes the statement about the linear expressions being exactly the same, an identity; therefore, any rational number x would make the equation true.

Lesson 8: Linear Equations in Disguise

Student Outcomes

- Students rewrite and solve equations that are not obviously linear equations using properties of equality.

Lesson Notes

In this lesson, students learn that some equations that may not look like linear equations are, in fact, linear. This lesson on solving rational equations is included because of the types of equations students see in later topics of this module related to slope. Students recognize these equations as proportions. It is not necessary to refer to these types of equations as equations that contain rational expressions. They can be referred to simply as proportions since students are familiar with this terminology. Expressions of this type are treated carefully in algebra as they involve a discussion about why the denominator of such expressions cannot be equal to zero. That discussion is not included in this lesson.

Classwork

Concept Development (3 minutes)

- Some linear equations may not look like linear equations upon first glance. A simple example that you should recognize is

$$\frac{x}{5} = \frac{6}{12}.$$

- What do we call this kind of problem, and how do we solve it?
  - *This is a proportion. We can solve this by multiplying both sides of the equation by 5. We can also solve it by multiplying each numerator with the other fraction's denominator.*

Students may not think of multiplying each numerator with the other fraction's denominator because multiplying both sides by 5 requires fewer steps and uses the multiplication property of equality that has been used to solve other equations. If necessary, state the theorem, and give a brief explanation.

THEOREM: Given rational numbers A, B, C, and D, so that $B \neq 0$ and $D \neq 0$, the property states the following:

$$\text{If } \frac{A}{B} = \frac{C}{D}, \text{ then } AD = BC.$$

- To find the value of x, we can multiply each numerator by the other fraction's denominator.

$$\frac{x}{5} = \frac{6}{12}$$
$$12x = 6(5)$$

- It should be more obvious now that we have a linear equation. We can now solve it as usual using the properties of equality.

$$12x = 30$$
$$x = \frac{30}{12}$$
$$x = \frac{5}{2}$$

- In this lesson, our work will be similar, but the numerator and/or the denominator of the fractions may contain more than one term. However, the way we solve these kinds of problems remains the same.

Example 1 (5 minutes)

- Given a linear equation in disguise, we will *try* to solve it. To do so, we must first assume that the following equation is true for some number x.

$$\frac{x-1}{2} = \frac{x+\frac{1}{3}}{4}$$

- We want to make this equation look like the linear equations we are used to. For that reason, we will multiply both sides of the equation by 2 and 4, as we normally do with proportions:

$$2\left(x + \frac{1}{3}\right) = 4(x-1).$$

- Is this a linear equation? How do you know?
  - *Yes, this is a linear equation because the expressions on the left and right of the equal sign are linear expressions.*

- Notice that the expressions that contained more than one term were put in parentheses. We do that so we do not make a mistake and forget to use the distributive property.

- Now that we have a linear equation, we will use the distributive property and solve as usual.

$$2\left(x + \frac{1}{3}\right) = 4(x-1)$$
$$2x + \frac{2}{3} = 4x - 4$$
$$2x - 2x + \frac{2}{3} = 4x - 2x - 4$$
$$\frac{2}{3} = 2x - 4$$
$$\frac{2}{3} + 4 = 2x - 4 + 4$$
$$\frac{14}{3} = 2x$$
$$\frac{1}{2} \cdot \frac{14}{3} = \frac{1}{2} \cdot 2x$$
$$\frac{7}{3} = x$$

Lesson 8: Linear Equations in Disguise

A STORY OF RATIOS   Lesson 8   8•4

- How can we verify that $\frac{7}{3}$ is the solution to the equation?
  - *We can replace x with $\frac{7}{3}$ in the original equation.*

$$\frac{x-1}{2} = \frac{x+\frac{1}{3}}{4}$$

$$\frac{\frac{7}{3}-1}{2} = \frac{\frac{7}{3}+\frac{1}{3}}{4}$$

$$\frac{\frac{4}{3}}{2} = \frac{\frac{8}{3}}{4}$$

$$4\left(\frac{4}{3}\right) = 2\left(\frac{8}{3}\right)$$

$$\frac{16}{3} = \frac{16}{3}$$

- Since $\frac{7}{3}$ made the equation true, we know it is a solution to the equation.

Example 2 (4 minutes)

- Can we solve the following equation? Explain.

$$\frac{\frac{1}{5}-x}{7} = \frac{2x+9}{3}$$

  - *We need to multiply each numerator with the other fraction's denominator.*
- So,

$$\frac{\frac{1}{5}-x}{7} = \frac{2x+9}{3}$$

$$7(2x+9) = 3\left(\frac{1}{5}-x\right).$$

- What would be the next step?
  - *Use the distributive property.*

Lesson 8: Linear Equations in Disguise

A STORY OF RATIOS — Lesson 8 — 8•4

- Now we have

$$7(2x + 9) = 3\left(\frac{1}{5} - x\right)$$

$$14x + 63 = \frac{3}{5} - 3x$$

$$14x + 3x + 63 = \frac{3}{5} - 3x + 3x$$

$$17x + 63 = \frac{3}{5}$$

$$17x + 63 - 63 = \frac{3}{5} - 63$$

$$17x = \frac{3}{5} - \frac{315}{5}$$

$$17x = -\frac{312}{5}$$

$$\frac{1}{17}(17x) = \left(-\frac{312}{5}\right)\frac{1}{17}$$

$$x = -\frac{312}{85}.$$

- Is this a linear equation? How do you know?
  - *Yes, this is a linear equation because the left and right side are linear expressions.*

Example 3 (5 minutes)

- Can we solve the following equation? If so, go ahead and solve it. If not, explain why not.

> **Example 3**
>
> **Can this equation be solved?**
>
> $$\frac{6+x}{7x+\frac{2}{3}} = \frac{3}{8}$$

Give students a few minutes to work. Provide support to students as needed.

- *Yes, we can solve the equation because we can multiply each numerator with the other fraction's denominator and then use the distributive property to begin solving it.*

Lesson 8: Linear Equations in Disguise

A STORY OF RATIOS   Lesson 8   8•4

$$\frac{6+x}{7x+\frac{2}{3}} = \frac{3}{8}$$
$$(6+x)8 = \left(7x+\frac{2}{3}\right)3$$
$$48 + 8x = 21x + 2$$
$$48 + 8x - 8x = 21x - 8x + 2$$
$$48 = 13x + 2$$
$$48 - 2 = 13x + 2 - 2$$
$$46 = 13x$$
$$\frac{46}{13} = x$$

Example 4 (5 minutes)

- Can we solve the following equation? If so, go ahead and solve it. If not, explain why not.

Example 4

Can this equation be solved?

$$\frac{7}{3x+9} = \frac{1}{8}$$

Give students a few minutes to work. Provide support to students as needed.

▫ *Yes, we can solve the equation because we can multiply each numerator with the other fraction's denominator and then use the distributive property to begin solving it.*

$$\frac{7}{3x+9} = \frac{1}{8}$$
$$7(8) = (3x+9)1$$
$$56 = 3x + 9$$
$$56 - 9 = 3x + 9 - 9$$
$$47 = 3x$$
$$\frac{47}{3} = x$$

Lesson 8:   Linear Equations in Disguise

Example 5 (5 minutes)

Example 5

In the diagram below, △ ABC ~ △ A′B′C′. Using what we know about similar triangles, we can determine the value of x.

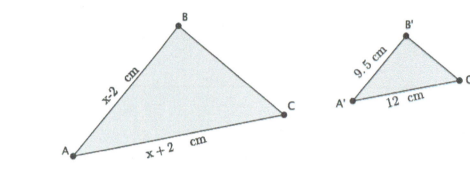

- Begin by writing the ratios that represent the corresponding sides.

$$\frac{x-2}{9.5} = \frac{x+2}{12}$$

It is possible to write several different proportions in this case. If time, discuss this fact with students.

- Now that we have the ratios, solve for x and find the lengths of $\overline{AB}$ and $\overline{AC}$.

$$\frac{x-2}{9.5} = \frac{x+2}{12}$$
$$(x-2)12 = 9.5(x+2)$$
$$12x - 24 = 9.5x + 19$$
$$12x - 24 + 24 = 9.5x + 19 + 24$$
$$12x = 9.5x + 43$$
$$12x - 9.5x = 9.5x - 9.5x + 43$$
$$2.5x = 43$$
$$x = \frac{43}{2.5}$$
$$x = 17.2$$

$|AB| = 15.2$ cm, and $|AC| = 19.2$ cm.

Lesson 8

Exercises (8 minutes)

Students complete Exercises 1–4 independently.

Exercises

Solve the following equations of rational expressions, if possible.

1. $\dfrac{2x+1}{9} = \dfrac{1-x}{6}$

$$\dfrac{2x+1}{9} = \dfrac{1-x}{6}$$
$$9(1-x) = (2x+1)6$$
$$9 - 9x = 12x + 6$$
$$9 - 9x + 9x = 12x + 9x + 6$$
$$9 = 21x + 6$$
$$9 - 6 = 21x + 6 - 6$$
$$3 = 21x$$
$$\dfrac{3}{21} = \dfrac{21}{21}x$$
$$\dfrac{1}{7} = x$$

2. $\dfrac{5+2x}{3x-1} = \dfrac{6}{7}$

$$\dfrac{5+2x}{3x-1} = \dfrac{6}{7}$$
$$(5+2x)7 = (3x-1)6$$
$$35 + 14x = 18x - 6$$
$$35 - 35 + 14x = 18x - 6 - 35$$
$$14x = 18x - 41$$
$$14x - 18x = 18x - 18x - 41$$
$$-4x = -41$$
$$\dfrac{-4}{-4}x = \dfrac{-41}{-4}$$
$$x = \dfrac{41}{4}$$

3. $\dfrac{x+9}{12} = \dfrac{-2x - \frac{1}{2}}{3}$

$$\dfrac{x+9}{12} = \dfrac{-2x - \frac{1}{2}}{3}$$
$$12\left(-2x - \dfrac{1}{2}\right) = (x+9)3$$
$$-24x - 6 = 3x + 27$$
$$-24x + 24x - 6 = 3x + 24x + 27$$
$$-6 = 27x + 27$$
$$-6 - 27 = 27x + 27 - 27$$
$$-33 = 27x$$
$$\dfrac{-33}{27} = \dfrac{27}{27}x$$
$$-\dfrac{11}{9} = x$$

Lesson 8: Linear Equations in Disguise

A STORY OF RATIOS — Lesson 8 — 8•4

4. $\dfrac{8}{3-4x} = \dfrac{5}{2x+\frac{1}{4}}$

$$\frac{8}{3-4x} = \frac{5}{2x+\frac{1}{4}}$$
$$8\left(2x+\frac{1}{4}\right) = (3-4x)5$$
$$16x + 2 = 15 - 20x$$
$$16x + 2 - 2 = 15 - 2 - 20x$$
$$16x = 13 - 20x$$
$$16x + 20x = 13 - 20x + 20x$$
$$36x = 13$$
$$\frac{36}{36}x = \frac{13}{36}$$
$$x = \frac{13}{36}$$

Closing (5 minutes)

Summarize, or ask students to summarize, the main points from the lesson:

- We know that proportions that have more than one term in the numerator and/or denominator can be solved the same way we normally solve a proportion.
- When multiplying a fraction with more than one term in the numerator and/or denominator by a number, we should put the expressions with more than one term in parentheses so that we are less likely to forget to use the distributive property.

Lesson Summary

Some proportions are linear equations in disguise and are solved the same way we normally solve proportions.

When multiplying a fraction with more than one term in the numerator and/or denominator by a number, put the expressions with more than one term in parentheses so that you remember to use the distributive property when transforming the equation. For example:

$$\frac{x+4}{2x-5} = \frac{3}{5}$$
$$5(x+4) = 3(2x-5).$$

The equation $5(x+4) = 3(2x-5)$ is now clearly a linear equation and can be solved using the properties of equality.

Exit Ticket (5 minutes)

Lesson 8: Linear Equations in Disguise

Exit Ticket

Solve the following equations for x.

1. $\dfrac{5x-8}{3} = \dfrac{11x-9}{5}$

2. $\dfrac{x+11}{7} = \dfrac{2x+1}{-8}$

3. $\dfrac{-x-2}{-4} = \dfrac{3x+6}{2}$

Exit Ticket Sample Solutions

Solve the following equations for x.

1. $\dfrac{5x-8}{3} = \dfrac{11x-9}{5}$

$$\dfrac{5x-8}{3} = \dfrac{11x-9}{5}$$
$$5(5x-8) = 3(11x-9)$$
$$25x - 40 = 33x - 27$$
$$25x - 25x - 40 = 33x - 25x - 27$$
$$-40 = 8x - 27$$
$$-40 + 27 = 8x - 27 + 27$$
$$-13 = 8x$$
$$-\dfrac{13}{8} = x$$

2. $\dfrac{x+11}{7} = \dfrac{2x+1}{-8}$

$$\dfrac{x+11}{7} = \dfrac{2x+1}{-8}$$
$$7(2x+1) = -8(x+11)$$
$$14x + 7 = -8x - 88$$
$$14x + 7 - 7 = -8x - 88 - 7$$
$$14x = -8x - 95$$
$$14x + 8x = -8x + 8x - 95$$
$$22x = -95$$
$$x = -\dfrac{95}{22}$$

3. $\dfrac{-x-2}{-4} = \dfrac{3x+6}{2}$

$$\dfrac{-x-2}{-4} = \dfrac{3x+6}{2}$$
$$-4(3x+6) = 2(-x-2)$$
$$-12x - 24 = -2x - 4$$
$$-12x - 24 + 24 = -2x - 4 + 24$$
$$-12x = -2x + 20$$
$$-12x + 2x = -2x + 2x + 20$$
$$-10x = 20$$
$$x = -2$$

Problem Set Sample Solutions

Students practice solving equations with rational expressions, if a solution is possible.

> Solve the following equations of rational expressions, if possible. If an equation cannot be solved, explain why.
>
> 1. $\dfrac{5}{6x-2} = \dfrac{-1}{x+1}$
>
> $$\dfrac{5}{6x-2} = \dfrac{-1}{x+1}$$
> $$5(x+1) = -1(6x-2)$$
> $$5x + 5 = -6x + 2$$
> $$5x + 5 - 5 = -6x + 2 - 5$$
> $$5x = -6x - 3$$
> $$5x + 6x = -6x + 6x - 3$$
> $$11x = -3$$
> $$x = -\dfrac{3}{11}$$
>
> 2. $\dfrac{4-x}{8} = \dfrac{7x-1}{3}$
>
> $$\dfrac{4-x}{8} = \dfrac{7x-1}{3}$$
> $$8(7x - 1) = (4 - x)3$$
> $$56x - 8 = 12 - 3x$$
> $$56x - 8 + 8 = 12 + 8 - 3x$$
> $$56x = 20 - 3x$$
> $$56x + 3x = 20 - 3x + 3x$$
> $$59x = 20$$
> $$\dfrac{59}{59}x = \dfrac{20}{59}$$
> $$x = \dfrac{20}{59}$$
>
> 3. $\dfrac{3x}{x+2} = \dfrac{5}{9}$
>
> $$\dfrac{3x}{x+2} = \dfrac{5}{9}$$
> $$9(3x) = (x+2)5$$
> $$27x = 5x + 10$$
> $$27x - 5x = 5x - 5x + 10$$
> $$22x = 10$$
> $$\dfrac{22}{22}x = \dfrac{10}{22}$$
> $$x = \dfrac{5}{11}$$

Lesson 8: Linear Equations in Disguise

4. $\dfrac{\frac{1}{2}x+6}{3} = \dfrac{x-3}{2}$

$$\dfrac{\frac{1}{2}x+6}{3} = \dfrac{x-3}{2}$$
$$3(x-3) = 2\left(\dfrac{1}{2}x+6\right)$$
$$3x - 9 = x + 12$$
$$3x - 9 + 9 = x + 12 + 9$$
$$3x = x + 21$$
$$3x - x = x - x + 21$$
$$2x = 21$$
$$x = \dfrac{21}{2}$$

5. $\dfrac{7-2x}{6} = \dfrac{x-5}{1}$

$$\dfrac{7-2x}{6} = \dfrac{x-5}{1}$$
$$6(x-5) = (7-2x)1$$
$$6x - 30 = 7 - 2x$$
$$6x - 30 + 30 = 7 + 30 - 2x$$
$$6x = 37 - 2x$$
$$6x + 2x = 37 - 2x + 2x$$
$$8x = 37$$
$$\dfrac{8}{8}x = \dfrac{37}{8}$$
$$x = \dfrac{37}{8}$$

6. $\dfrac{2x+5}{2} = \dfrac{3x-2}{6}$

$$\dfrac{2x+5}{2} = \dfrac{3x-2}{6}$$
$$2(3x-2) = 6(2x+5)$$
$$6x - 4 = 12x + 30$$
$$6x - 4 + 4 = 12x + 30 + 4$$
$$6x = 12x + 34$$
$$6x - 12x = 12x - 12x + 34$$
$$-6x = 34$$
$$x = -\dfrac{34}{6}$$
$$x = -\dfrac{17}{3}$$

Lesson 8: Linear Equations in Disguise

7. $\dfrac{6x+1}{3} = \dfrac{9-x}{7}$

$$\dfrac{6x+1}{3} = \dfrac{9-x}{7}$$
$$(6x+1)7 = 3(9-x)$$
$$42x + 7 = 27 - 3x$$
$$42x + 7 - 7 = 27 - 7 - 3x$$
$$42x = 20 - 3x$$
$$42x + 3x = 20 - 3x + 3x$$
$$45x = 20$$
$$\dfrac{45}{45}x = \dfrac{20}{45}$$
$$x = \dfrac{4}{9}$$

8. $\dfrac{\frac{1}{3}x - 8}{12} = \dfrac{-2 - x}{15}$

$$\dfrac{\frac{1}{3}x - 8}{12} = \dfrac{-2 - x}{15}$$
$$12(-2 - x) = \left(\dfrac{1}{3}x - 8\right)15$$
$$-24 - 12x = 5x - 120$$
$$-24 - 12x + 12x = 5x + 12x - 120$$
$$-24 = 17x - 120$$
$$-24 + 120 = 17x - 120 + 120$$
$$96 = 17x$$
$$\dfrac{96}{17} = \dfrac{17}{17}x$$
$$\dfrac{96}{17} = x$$

9. $\dfrac{3-x}{1-x} = \dfrac{3}{2}$

$$\dfrac{3-x}{1-x} = \dfrac{3}{2}$$
$$(3-x)2 = (1-x)3$$
$$6 - 2x = 3 - 3x$$
$$6 - 2x + 2x = 3 - 3x + 2x$$
$$6 = 3 - x$$
$$6 - 3 = 3 - 3 - x$$
$$3 = -x$$
$$-3 = x$$

Lesson 8: Linear Equations in Disguise

10. In the diagram below, △ABC ~ △A'B'C'. Determine the lengths of $\overline{AC}$ and $\overline{BC}$.

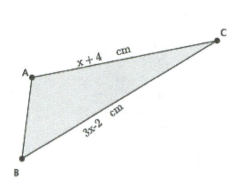
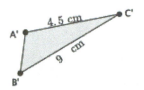

$$\frac{x+4}{4.5} = \frac{3x-2}{9}$$
$$9(x+4) = 4.5(3x-2)$$
$$9x + 36 = 13.5x - 9$$
$$9x + 36 + 9 = 13.5x - 9 + 9$$
$$9x + 45 = 13.5x$$
$$9x - 9x + 45 = 13.5x - 9x$$
$$45 = 4.5x$$
$$10 = x$$

$|AC| = 14$ cm, and $|BC| = 28$ cm.

Lesson 9: An Application of Linear Equations

Student Outcomes

- Students know how to rewrite an exponential expression that represents a series as a linear equation.

Lesson Notes

The purpose of this lesson is to expose students to applications of linear equations. The Discussion revisits the Facebook problem from Module 1 but this time in the context of a linear equation. This is an opportunity to highlight MP.1 (Make sense of problems and persevere in solving them), as the Discussion requires students to work with equations in ways they have not before. If the Discussion is too challenging for students, use Exercises 3–11, which is a series of more accessible applications of linear equations.

Classwork

Discussion (30 minutes)

- In Module 1, you saw the following problem:

    You sent a photo of you and your family on vacation to seven Facebook friends. If each of them sends it to five of their friends, and each of those friends sends it to five of their friends, and those friends send it to five more, how many people (not counting yourself) will see your photo? Assume that no friend received the photo twice.

    In Module 1, you were asked to express your answer in exponential notation. The solution is given here:

    (1) The number of friends you sent a photo to is 7.
    (2) The number of friends 7 people sent the photo to is $7 \cdot 5$.
    (3) The number of friends $7 \cdot 5$ people sent the photo to is $(7 \cdot 5) \cdot 5$.
    (4) The number of friends $(7 \cdot 5) \cdot 5$ people sent the photo to is $((7 \cdot 5) \cdot 5) \cdot 5$.

    Therefore, the total number of people who received the photo is

    $$7 + 7 \cdot 5 + 7 \cdot 5^2 + 7 \cdot 5^3.$$

- Let's refer to "you sending the photo" as the first step. Then, "your friends sending the photo to their friends" is the second step, and so on. In the original problem, there were four steps. Assuming the trend continues, how would you find the sum after 10 steps?
    - *We would continue the pattern until we got to the 10th step.*
- What if I asked you how many people received the photo after 100 steps?
    - *It would take a long time to continue the pattern to the 100th step.*
- We want to be able to answer the question for any number of steps. For that reason, we will work toward expressing our answer as a linear equation.

- For convenience, let's introduce some symbols. Since we are talking about steps, we will refer to the sum after step one as S_1, the sum after step two as S_2, the sum after step three as S_3, and so on. Thus:

$$S_1 = 7 \tag{1}$$
$$S_2 = 7 + 7 \cdot 5 \tag{2}$$
$$S_3 = 7 + 7 \cdot 5 + 7 \cdot 5^2 \tag{3}$$
$$S_4 = 7 + 7 \cdot 5 + 7 \cdot 5^2 + 7 \cdot 5^3 \tag{4}$$

- What patterns do you notice within each of the equations (1)–(4)?
  - They contain some of the same terms. For example, equation (2) is the same as (1), except equation (2) has the term $7 \cdot 5$. Similarly, equation (3) is the same as (2), except equation (3) has the term $7 \cdot 5^2$.

- What you noticed is true. However, we want to generalize in a way that does not require us to know one step before getting to the next step. Let's see what other hidden patterns there are.

- Let's begin with equation (2):

$$S_2 = 7 + 7 \cdot 5$$
$$S_2 - 7 = 7 \cdot 5$$
$$S_2 - 7 + 7 \cdot 5^2 = 7 \cdot 5 + 7 \cdot 5^2$$
$$S_2 - 7 + 7 \cdot 5^2 = 5(7 + 7 \cdot 5).$$

> **Scaffolding:**
> Talk students through the manipulation of the equation. For example, "We begin by subtracting 7 from both sides. Next, we will add the number 7 times 5^2 to both sides of the equation. Then, using the distributive property …."

Notice that the grouping on the right side of the equation is exactly S_2, so we have the following:

$$S_2 - 7 + 7 \cdot 5^2 = 5S_2.$$

This equation is a linear equation in S_2. It is an equation we know how to solve (pretend S_2 is an x if that helps).

- Let's do something similar with equation (3):

$$S_3 = 7 + 7 \cdot 5 + 7 \cdot 5^2$$
$$S_3 - 7 = 7 \cdot 5 + 7 \cdot 5^2$$
$$S_3 - 7 + 7 \cdot 5^3 = 7 \cdot 5 + 7 \cdot 5^2 + 7 \cdot 5^3$$
$$S_3 - 7 + 7 \cdot 5^3 = 5(7 + 7 \cdot 5 + 7 \cdot 5^2).$$

Again, the grouping on the right side of the equation is exactly the equation we began with, S_3, so we have the following:

$$S_3 - 7 + 7 \cdot 5^3 = 5S_3.$$

This is a linear equation in S_3.

- Let's work together to do something similar with equation (4):

$$S_4 = 7 + 7 \cdot 5 + 7 \cdot 5^2 + 7 \cdot 5^3.$$

- What did we do first in each of the equations (2) and (3)?
  - We subtracted 7 from both sides of the equation.

- Now we have the following:

$$S_4 - 7 = 7 \cdot 5 + 7 \cdot 5^2 + 7 \cdot 5^3.$$

Lesson 9: An Application of Linear Equations

- What did we do next?
  - We added $7 \cdot 5$ raised to a power to both sides of the equation. When it was the second step, the power of 5 was 2. When it was the third step, the power of 5 was 3. Now that it is the fourth step, the power of 5 should be 4.
- Now we have
$$S_4 - 7 + 7 \cdot 5^4 = 7 \cdot 5 + 7 \cdot 5^2 + 7 \cdot 5^3 + 7 \cdot 5^4.$$
- What did we do after that?
  - We used the distributive property to rewrite the right side of the equation.
- Now we have
$$S_4 - 7 + 7 \cdot 5^4 = 5(7 + 7 \cdot 5 + 7 \cdot 5^2 + 7 \cdot 5^3).$$
- What do we do now?
  - We substitute the grouping on the right side of the equation with S_4.
- Finally, we have a linear equation in S_4:
$$S_4 - 7 + 7 \cdot 5^4 = 5S_4.$$
- Let's look at the linear equations all together.
$$S_2 - 7 + 7 \cdot 5^2 = 5S_2$$
$$S_3 - 7 + 7 \cdot 5^3 = 5S_3$$
$$S_4 - 7 + 7 \cdot 5^4 = 5S_4$$

What do you think the equation would be for S_{10}?
  - According to the pattern, it would be $S_{10} - 7 + 7 \cdot 5^{10} = 5S_{10}$.

This equation is solved differently from the equations that have been solved in order to show students the normal form of an equation for the summation of a geometric series. In general, if $a \neq 1$, then
$$1 + a + a^2 + \cdots + a^{n-1} + a^n = \frac{1 - a^{n+1}}{1 - a}.$$

- Now let's solve the equation. (Note, that we do not simplify $(1 - 5)$ for the reason explained above.)
$$S_{10} - 7 + 7 \cdot 5^{10} = 5S_{10}$$
$$S_{10} - 5S_{10} - 7 + 7 \cdot 5^{10} = 5S_{10} - 5S_{10}$$
$$S_{10}(1 - 5) - 7 + 7 \cdot 5^{10} = 0$$
$$S_{10}(1 - 5) - 7 + 7 + 7 \cdot 5^{10} - 7 \cdot 5^{10} = 7 - 7 \cdot 5^{10}$$
$$S_{10}(1 - 5) = 7(1 - 5^{10})$$
$$S_{10} = \frac{7(1 - 5^{10})}{(1 - 5)}$$
$$S_{10} = 17\,089\,842$$

After 10 steps, 17,089,842 will see the photo!

Lesson 9: An Application of Linear Equations

Exercises 1–2 (5 minutes)

Students complete Exercises 1–2 independently.

Exercises

1. Write the equation for the 15th step.

$$S_{15} - 7 + 7 \cdot 5^{15} = 5S_{15}$$

2. How many people would see the photo after 15 steps? Use a calculator if needed.

$$S_{15} - 7 + 7 \cdot 5^{15} = 5S_{15}$$
$$S_{15} - 5S_{15} - 7 + +7 \cdot 5^{15} = 5S_{15} - 5S_{15}$$
$$S_{15}(1-5) - 7 + 7 \cdot 5^{15} = 0$$
$$S_{15}(1-5) - 7 + 7 + 7 \cdot 5^{15} - 7 \cdot 5^{15} = 7 - 7 \cdot 5^{15}$$
$$S_{15}(1-5) = 7(1-5^{15})$$
$$S_{15} = \frac{7(1-5^{15})}{(1-5)}$$
$$S_{15} = 53\,405\,761\,717$$

Exercises 3–11 as an Alternative to Discussion (30 minutes)

Students should be able to complete the following problems independently as they are an application of skills learned to this point, namely, transcription and solving linear equations in one variable. Have students work on the problems one at a time and share their work with the whole class, or assign the entire set and allow students to work at their own pace. Provide correct solutions at the end of the lesson.

3. Marvin paid an entrance fee of $5 plus an additional $1.25 per game at a local arcade. Altogether, he spent $26.25. Write and solve an equation to determine how many games Marvin played.

   Let x represent the number of games he played.

   $$5 + 1.25x = 26.25$$
   $$1.25x = 21.25$$
   $$x = \frac{21.25}{1.25}$$
   $$x = 17$$

   Marvin played 17 games.

4. The sum of four consecutive integers is -26. What are the integers?

   Let x be the first integer.

   $$x + (x+1) + (x+2) + (x+3) = -26$$
   $$4x + 6 = -26$$
   $$4x = -32$$
   $$x = -8$$

   The integers are $-8, -7, -6,$ and -5.

Lesson 9: An Application of Linear Equations

5. A book has x pages. How many pages are in the book if Maria read 45 pages of a book on Monday, $\frac{1}{2}$ the book on Tuesday, and the remaining 72 pages on Wednesday?

   Let x be the number of pages in the book.

   $$x = 45 + \frac{1}{2}x + 72$$
   $$x = 117 + \frac{1}{2}x$$
   $$\frac{1}{2}x = 117$$
   $$x = 234$$

   The book has 234 pages.

6. A number increased by 5 and divided by 2 is equal to 75. What is the number?

   Let x be the number.

   $$\frac{x+5}{2} = 75$$
   $$x + 5 = 150$$
   $$x = 145$$

   The number is 145.

7. The sum of thirteen and twice a number is seven less than six times a number. What is the number?

   Let x be the number.

   $$13 + 2x = 6x - 7$$
   $$20 + 2x = 6x$$
   $$20 = 4x$$
   $$5 = x$$

   The number is 5.

8. The width of a rectangle is 7 less than twice the length. If the perimeter of the rectangle is 43.6 inches, what is the area of the rectangle?

   Let x represent the length of the rectangle.

   $$2x + 2(2x - 7) = 43.6$$
   $$2x + 4x - 14 = 43.6$$
   $$6x - 14 = 43.6$$
   $$6x = 57.6$$
   $$x = \frac{57.6}{6}$$
   $$x = 9.6$$

   The length of the rectangle is 9.6 inches, and the width is 12.2 inches, so the area is 117.12 in^2.

9. Two hundred and fifty tickets for the school dance were sold. On Monday, 35 tickets were sold. An equal number of tickets were sold each day for the next five days. How many tickets were sold on one of those days?

   Let x be the number of tickets sold on one of those days.

   $$250 = 35 + 5x$$
   $$215 = 5x$$
   $$43 = x$$

   43 tickets were sold on each of the five days.

10. Shonna skateboarded for some number of minutes on Monday. On Tuesday, she skateboarded for twice as many minutes as she did on Monday, and on Wednesday, she skateboarded for half the sum of minutes from Monday and Tuesday. Altogether, she skateboarded for a total of three hours. How many minutes did she skateboard each day?

    Let x be the number of minutes she skateboarded on Monday.

    $$x + 2x + \frac{2x + x}{2} = 180$$
    $$\frac{2x}{2} + \frac{4x}{2} + \frac{2x + x}{2} = 180$$
    $$\frac{9x}{2} = 180$$
    $$9x = 360$$
    $$x = 40$$

    Shonna skateboarded 40 minutes on Monday, 80 minutes on Tuesday, and 60 minutes on Wednesday.

11. In the diagram below, $\triangle ABC \sim \triangle A'B'C'$. Determine the length of $\overline{AC}$ and $\overline{BC}$.

    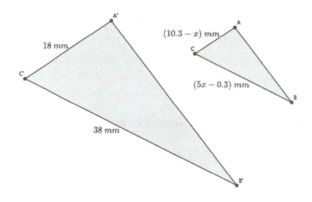

    $$\frac{18}{10.3 - x} = \frac{38}{5x - 0.3}$$
    $$18(5x - 0.3) = 38(10.3 - x)$$
    $$90x - 5.4 = 391.4 - 38x$$
    $$128x - 5.4 = 391.4$$
    $$128x = 396.8$$
    $$x = \frac{396.8}{128}$$
    $$x = 3.1$$

    The length of $\overline{AC}$ is 7.2 mm, and the length of $\overline{BC}$ is 15.2 mm.

Closing (5 minutes)

Summarize, or ask students to summarize, the main points from the lesson:

- We can rewrite equations to develop a pattern and make predictions.
- We know that for problems like these, we can generalize equations so that we do not have to do each step to get our answer.
- We learned how equations can be used to solve problems.

Exit Ticket (5 minutes)

Name _____ Date _____

Lesson 9: An Application of Linear Equations

Exit Ticket

1. Rewrite the equation that would represent the sum in the fifth step of the Facebook problem:

$$S_5 = 7 + 7 \cdot 5 + 7 \cdot 5^2 + 7 \cdot 5^3 + 7 \cdot 5^4.$$

2. The sum of four consecutive integers is 74. Write an equation, and solve to find the numbers.

Exit Ticket Sample Solutions

1. Rewrite the equation that would represent the sum in the fifth step of the Facebook problem:
$S_5 = 7 + 7 \cdot 5 + 7 \cdot 5^2 + 7 \cdot 5^3 + 7 \cdot 5^4.$

$$S_5 = 7 + 7 \cdot 5 + 7 \cdot 5^2 + 7 \cdot 5^3 + 7 \cdot 5^4$$
$$S_5 - 7 = 7 \cdot 5 + 7 \cdot 5^2 + 7 \cdot 5^3 + 7 \cdot 5^4$$
$$S_5 - 7 + 7 \cdot 5^5 = 7 \cdot 5 + 7 \cdot 5^2 + 7 \cdot 5^3 + 7 \cdot 5^4 + 7 \cdot 5^5$$
$$S_5 - 7 + 7 \cdot 5^5 = 5(7 + 7 \cdot 5 + 7 \cdot 5^2 + 7 \cdot 5^3 + 7 \cdot 5^4)$$
$$S_5 - 7 + 7 \cdot 5^5 = 5(S_5)$$
$$S_5 - 5S_5 - 7 + 7 \cdot 5^5 = 0$$
$$S_5 - 5S_5 = 7 - (7 \cdot 5^5)$$
$$(1 - 5)S_5 = 7 - (7 \cdot 5^5)$$
$$(1 - 5)S_5 = 7(1 - 5^5)$$
$$S_5 = \frac{7(1 - 5^5)}{1 - 5}$$

2. The sum of four consecutive integers is 74. Write an equation, and solve to find the numbers.

   Let x be the first number.

$$x + (x + 1) + (x + 2) + (x + 3) = 74$$
$$4x + 6 = 74$$
$$4x = 68$$
$$x = 17$$

   The numbers are 17, 18, 19, and 20.

Problem Set Sample Solutions

Assign the problems that relate to the elements of the lesson that were used with students.

1. You forward an e-card that you found online to three of your friends. They liked it so much that they forwarded it on to four of their friends, who then forwarded it on to four of their friends, and so on. The number of people who saw the e-card is shown below. Let S_1 represent the number of people who saw the e-card after one step, let S_2 represent the number of people who saw the e-card after two steps, and so on.

$$S_1 = 3$$
$$S_2 = 3 + 3 \cdot 4$$
$$S_3 = 3 + 3 \cdot 4 + 3 \cdot 4^2$$
$$S_4 = 3 + 3 \cdot 4 + 3 \cdot 4^2 + 3 \cdot 4^3$$

Lesson 9: An Application of Linear Equations

a. Find the pattern in the equations.

$$S_2 = 3 + 3 \cdot 4$$
$$S_2 - 3 = 3 \cdot 4$$
$$S_2 - 3 + 3 \cdot 4^2 = 3 \cdot 4 + 3 \cdot 4^2$$
$$S_2 - 3 + 3 \cdot 4^2 = 4(3 + 3 \cdot 4)$$
$$S_2 - 3 + 3 \cdot 4^2 = 4S_2$$

$$S_3 = 3 + 3 \cdot 4 + 3 \cdot 4^2$$
$$S_3 - 3 = 3 \cdot 4 + 3 \cdot 4^2$$
$$S_3 - 3 + 3 \cdot 4^3 = 3 \cdot 4 + 3 \cdot 4^2 + 3 \cdot 4^3$$
$$S_3 - 3 + 3 \cdot 4^3 = 4(3 + 3 \cdot 4 + 3 \cdot 4^2)$$
$$S_3 - 3 + 3 \cdot 4^3 = 4S_3$$

$$S_4 = 3 + 3 \cdot 4 + 3 \cdot 4^2 + 3 \cdot 4^3$$
$$S_4 - 3 = 3 \cdot 4 + 3 \cdot 4^2 + 3 \cdot 4^3$$
$$S_4 - 3 + 3 \cdot 4^4 = 3 \cdot 4 + 3 \cdot 4^2 + 3 \cdot 4^3 + 3 \cdot 4^4$$
$$S_4 - 3 + 3 \cdot 4^4 = 4(3 + 3 \cdot 4 + 3 \cdot 4^2 + 3 \cdot 4^3)$$
$$S_4 - 3 + 3 \cdot 4^4 = 4S_4$$

b. Assuming the trend continues, how many people will have seen the e-card after 10 steps?

$$S_{10} - 3 + 3 \cdot 4^{10} = 4S_{10}$$
$$S_{10} - 4S_{10} - 3 + 3 \cdot 4^{10} = 0$$
$$S_{10}(1 - 4) = 3 - 3 \cdot 4^{10}$$
$$S_{10}(1 - 4) = 3(1 - 4^{10})$$
$$S_{10} = \frac{3(1 - 4^{10})}{(1 - 4)}$$
$$S_{10} = 1\,048\,575$$

After 10 steps, $1,048,575$ people will have seen the e-card.

c. How many people will have seen the e-card after n steps?

$$S_n = \frac{3(1 - 4^n)}{(1 - 4)}$$

For each of the following questions, write an equation, and solve to find each answer.

2. Lisa has a certain amount of money. She spent $39 and has $\frac{3}{4}$ of the original amount left. How much money did she have originally?

Let x be the amount of money Lisa had originally.

$$x - 39 = \frac{3}{4}x$$
$$-39 = -\frac{1}{4}x$$
$$156 = x$$

Lisa had $156 originally.

3. The length of a rectangle is 4 more than 3 times the width. If the perimeter of the rectangle is 18.4 cm, what is the area of the rectangle?

   Let x represent the width of the rectangle.

   $$2(4 + 3x) + 2x = 18.4$$
   $$8 + 6x + 2x = 18.4$$
   $$8 + 8x = 18.4$$
   $$8x = 10.4$$
   $$x = \frac{10.4}{8}$$
   $$x = 1.3$$

   The width of the rectangle is 1.3 cm, and the length is 7.9 cm, so the area is 10.27 cm^2.

4. Eight times the result of subtracting 3 from a number is equal to the number increased by 25. What is the number?

   Let x be the number.

   $$8(x - 3) = x + 25$$
   $$8x - 24 = x + 25$$
   $$7x - 24 = 25$$
   $$7x = 49$$
   $$x = 7$$

   The number is 7.

5. Three consecutive odd integers have a sum of 3. What are the numbers?

   Let x be the first odd number.

   $$x + (x + 2) + (x + 4) = 3$$
   $$3x + 6 = 3$$
   $$3x = -3$$
   $$x = -1$$

   The three consecutive odd integers are -1, 1, and 3.

6. Each month, Liz pays $35 to her phone company just to use the phone. Each text she sends costs her an additional $0.05. In March, her phone bill was $72.60. In April, her phone bill was $65.85. How many texts did she send each month?

   Let x be the number of texts she sent in March.

   $$35 + 0.05x = 72.60$$
   $$0.05x = 37.6$$
   $$x = \frac{37.6}{0.05}$$
   $$x = 752$$

   She sent 752 texts in March.

   Let y be the number of texts she sent in April.

   $$35 + 0.05y = 65.85$$
   $$0.05y = 30.85$$
   $$y = \frac{30.85}{0.05}$$
   $$y = 617$$

   She sent 617 texts in April.

7. Claudia is reading a book that has 360 pages. She read some of the book last week. She plans to read 46 pages today. When she does, she will be $\frac{4}{5}$ of the way through the book. How many pages did she read last week?

   Let x be the number of pages she read last week.

   $$x + 46 = \frac{4}{5}(360)$$
   $$x + 46 = 288$$
   $$x = 242$$

   Claudia read 242 pages last week.

8. In the diagram below, $\triangle ABC \sim \triangle A'B'C'$. Determine the measure of $\angle A$.

   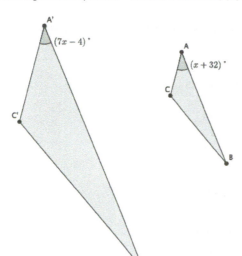

   $$7x - 4 = x + 32$$
   $$6x - 4 = 32$$
   $$6x = 36$$
   $$x = 6$$

   The measure of $\angle A$ is $38°$.

9. In the diagram below, $\triangle ABC \sim \triangle A'B'C'$. Determine the measure of $\angle A$.

   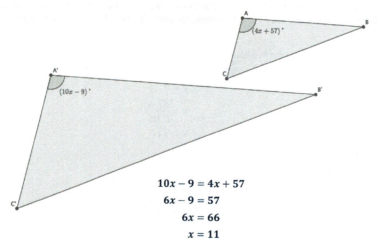

   $$10x - 9 = 4x + 57$$
   $$6x - 9 = 57$$
   $$6x = 66$$
   $$x = 11$$

   The measure of $\angle A$ is $101°$.

A STORY OF RATIOS

Mathematics Curriculum

GRADE 8 • MODULE 4

Topic B
Linear Equations in Two Variables and Their Graphs

8.EE.B.5

Focus Standard:	8.EE.B.5	Graph proportional relationships, interpreting the unit rate as the slope of the graph. Compare two different proportional relationships represented in different ways. *For example, compare a distance-time graph to a distance-time equation to determine which of two moving objects has greater speed.*
Instructional Days:	5	
Lesson 10:	A Critical Look at Proportional Relationships (S)[1]	
Lesson 11:	Constant Rate (P)	
Lesson 12:	Linear Equations in Two Variables (E)	
Lesson 13:	The Graph of a Linear Equation in Two Variables (S)	
Lesson 14:	The Graph of a Linear Equation—Horizontal and Vertical Lines (S)	

Topic B begins with students working with proportional relationships related to average speed and constant speed. In Lesson 10, students use information that is organized in the form of a table to write linear equations. In Lesson 11, students learn how to apply the concept of constant rate to a variety of contexts requiring two variables (**8.EE.B.5**). Lesson 12 introduces students to the standard form of an equation in two variables. At this point, students use a table to help them find and organize solutions to a linear equation in two variables. Students then use the information from the table to begin graphing solutions on a coordinate plane. In Lesson 13, students begin to question whether or not the graph of a linear equation is a line, as opposed to something that is curved. Lesson 14 presents students with equations in standard form, $ax + by = c$, where $a = 0$ or $b = 0$, which produces lines that are either vertical or horizontal.

---

[1]Lesson Structure Key: **P**-Problem Set Lesson, **M**-Modeling Cycle Lesson, **E**-Exploration Lesson, **S**-Socratic Lesson

Topic B: Linear Equations in Two Variables and Their Graphs

111

This work is derived from Eureka Math ™ and licensed by Great Minds. ©2015 Great Minds. eureka-math.org
G8-M4-TE-B3-1.3.0-07.2015

Lesson 10: A Critical Look at Proportional Relationships

Student Outcomes

- Students work with proportional relationships that involve average speed and constant speed in order to write a linear equation in two variables.
- Students use linear equations in two variables to answer questions about distance and time.

Classwork

Discussion/Examples 1–2 (25 minutes)

Example 1

- Consider the word problem below. We can do several things to answer this problem, but let's begin to organize our work using a table for time and distance:

  Scaffolding:
  It may be necessary to remind students of the relationship between distance traveled, rate, and time spent traveling at that rate.

  Example 1

  Paul walks 2 miles in 25 minutes. How many miles can Paul walk in 137.5 minutes?

Time (in minutes)	Distance (in miles)
25	2
50	4
75	6
100	8
125	10

As students answer the questions below, fill in the table.

- How many miles would Paul be able to walk in 50 minutes? Explain.
  - *Paul could walk 4 miles in 50 minutes because 50 minutes is twice the time we were given, so we can calculate twice the distance, which is 4.*
- How many miles would Paul be able to walk in 75 minutes? Explain.
  - *Paul could walk 6 miles in 75 minutes because 75 minutes is three times the number of minutes we were given, so we can calculate three times the distance, which is 6.*
- How many miles would Paul be able to walk in 100 minutes?
  - *He could walk 8 miles.*
- How many miles would he walk in 125 minutes?
  - *He could walk 10 miles.*
- How could we determine the number of miles Paul could walk in 137.5 minutes?

Provide students time to think about the answer to this question. They may likely say that they can write a proportion to figure it out. Allow them to share and demonstrate their solutions. Then, proceed with the discussion below, if necessary.

- Since the relationship between the distance Paul walks and the time it takes him to walk that distance is proportional, we let y represent the distance Paul walks in 137.5 minutes and write the following:

$$\frac{25}{2} = \frac{137.5}{y}$$
$$25y = 137.5(2)$$
$$25y = 275$$
$$y = 11$$

Therefore, Paul can walk 11 miles in 137.5 minutes.

- How many miles, y, can Paul walk in x minutes?

Provide students time to think about the answer to this question. Allow them to share their ideas, and then proceed with the discussion below, if necessary.

- We know for a fact that Paul can walk 2 miles in 25 minutes, so we can write the ratio $\frac{25}{2}$ as we did with the proportion. We can write another ratio for the number of miles, y, Paul walks in x minutes. It is $\frac{x}{y}$. For the same reason we could write the proportion before, we can write one now with these two ratios:

$$\frac{25}{2} = \frac{x}{y}.$$

Does this remind you of something we have done recently? Explain.

  □ This is a linear equation in disguise. All we need to do is multiply each numerator by the other fraction's denominator, and then we will have a linear equation.

$$25y = 2x$$

- Recall our original question: How many miles, y, can Paul walk in x minutes? We need to solve this equation for y.

$$25y = 2x$$
$$y = \frac{2}{25}x$$
$$y = 0.08x$$

Paul can walk $0.08x$ miles in x minutes. This equation will allow us to answer all kinds of questions about Paul with respect to any given number of minutes or miles.

- Let's go back to the table and look for $y = 0.08x$ or its equivalent $y = \frac{2}{25}x$. What do you notice?

Time (in minutes)	Distance (in miles)
25	2
50	4
75	6
100	8
125	10

  □ The fraction $\frac{2}{25}$ came from the first row in the table. It is the distance traveled divided by the time it took to travel that distance. It is also in between each row of the table. For example, the difference between 4 miles and 2 miles is 2, and the difference between the associated times 50 and 25 is 25. The pattern repeats throughout the table.

Show on the table the +2 between each distance interval and the +25 between each time interval. Remind students that they have done work like this before, specifically finding a unit rate for a proportional relationship. Make clear that the unit rate found in the table was exactly the same as the unit rate found using the proportion and that the unit rate is the rate at which Paul walks.

- Let's look at another problem where only a table is provided.

Time (in hours)	Distance (in miles)
3	123
6	246
9	369
12	492
	y

  We want to know how many miles, y, can be traveled in any number of hours x. Using our previous work, what should we do?
  □ We can write and solve a proportion that contains both x and y or use the table to help us determine the unit rate.

- How many miles, y, can be traveled in any number of hours x?
  □ Student work:

$$\frac{123}{3} = \frac{y}{x}$$
$$123x = 3y$$
$$\frac{123}{3}x = y$$
$$41x = y$$

- What does the equation $y = 41x$ mean?
  □ It means that the distance traveled, y, is equal to the rate of 41 multiplied by the number of hours x traveled at that rate.

A STORY OF RATIOS   Lesson 10   8•4

Example 2

The point of this problem is to make clear to students that constant rate must be assumed in order to write linear equations in two variables and to use those equations to answer questions about distance, time, and rate.

- Consider the following word problem: Alexxa walked from Grand Central Station (GCS) on 42nd Street to Penn Station on 7th Avenue. The total distance traveled was 1.1 miles. It took Alexxa 25 minutes to make the walk. How many miles did she walk in the first 10 minutes?

Give students a minute to think and/or work on the problem. Expect them to write a proportion and solve the problem. The next part of the discussion gets them to think about what is meant by "constant" speed or, rather, lack of it.

  - *She walked 0.44 miles. (Assuming students used a proportion to solve.)*

- Are you sure about your answer? How often do you walk at a constant speed? Notice the problem did not even mention that she was walking at the same rate throughout the entire 1.1 miles. What if you have more information about her walk: Alexxa walked from GCS along 42nd Street to an ATM 0.3 miles away in 8 minutes. It took her 2 minutes to get some money out of the machine. Do you think your answer is still correct?

  - *Probably not since we now know that she had to stop at the ATM.*

- Let's continue with Alexxa's walk: She reached the 7th Avenue junction 13 minutes after she left GCS, a distance of 0.6 miles. There, she met her friend Karen with whom she talked for 2 minutes. After leaving her friend, she finally got to Penn Station 25 minutes after her walk began.

- Is this a more realistic situation than believing that she walked the exact same speed throughout the entire trip? What other events typically occur during walks in the city?

  - *Stoplights at crosswalks, traffic, maybe a trip/fall, or running an errand*

- This is precisely the reason we need to take a critical look at what we call *proportional relationships* and constant speed, in general.

- The following table shows an accurate picture of Alexxa's walk:

Time (in minutes)	Distance Traveled (in miles)
0	0
8	0.3
10	0.3
13	0.6
15	0.6
25	1.1

With this information, we can answer the question. Alexxa walked *exactly* 0.3 miles in 10 minutes.

- Now that we have an idea of what could go wrong when we assume a person walks at a constant rate or that a proportion can give us the correct answer all of the time, let's define what is called *average speed*.

- Suppose a person walks a distance of d (miles) in a given time interval t (minutes). Then, the **average speed** in the given time interval is $\frac{d}{t}$ in miles per minute.

- With this definition, we can calculate Alexxa's average speed: The distance that Alexxa traveled divided by the time interval she walked is $\frac{1.1}{25}$ miles per minute.

Lesson 10:   A Critical Look at Proportional Relationships   115

- If we assume that someone can actually walk at the same average speed over *any* time interval, then we say that the person is walking at a *constant speed*.

  Suppose the average speed of a person is the *same* constant C for *any* given time interval. Then, we say that the person is walking at a **constant speed C**.

- If the original problem included information specifying constant speed, then we could write the following:

  Alexxa's average speed for 25 minutes is $\frac{1.1}{25}$.

  Let y represent the distance Alexxa walked in 10 minutes. Then, her average speed for 10 minutes is $\frac{y}{10}$.

  Since Alexxa is walking at a constant speed of C miles per minute, then we know that

  $$\frac{1.1}{25} = C, \quad \text{and} \quad \frac{y}{10} = C.$$

  Since both fractions are equal to C, then we can write

  $$\frac{1.1}{25} = \frac{y}{10}.$$

  With the assumption of constant speed, we now have a *proportional relationship*, which would make the answer you came up with in the beginning correct.

- We can go one step further and write a statement in general. If Alexxa walks y miles in x minutes, then

  $$\frac{y}{x} = C, \quad \text{and} \quad \frac{1.1}{25} = \frac{y}{x}.$$

  To find how many miles y Alexxa walks in x miles, we solve the equation for y:

  $$\frac{1.1}{25} = \frac{y}{x}$$

  $$25y = 1.1x$$

  $$\frac{25}{25}y = \frac{1.1}{25}x$$

  $$y = \frac{1.1}{25}x,$$

  where the last equation is an example of a linear equation in two variables x and y. With this general equation, we can find the distance y Alexxa walks in any given time x. Since we have more information about Alexxa's walk, where and when she stopped, we know that the equation cannot accurately predict the distance she walks after a certain number of minutes. To do so requires us to assume that she walks at a constant rate. This is an assumption we generally take for granted when solving problems about rate.

Exercises (5 minutes)

Students complete Exercises 1–2 independently or in pairs.

Exercises

1. Wesley walks at a constant speed from his house to school 1.5 miles away. It took him 25 minutes to get to school.

   a. What fraction represents his constant speed, C?

   $$\frac{1.5}{25} = C$$

   b. You want to know how many miles he has walked after 15 minutes. Let y represent the distance he traveled after 15 minutes of walking at the given constant speed. Write a fraction that represents the constant speed, C, in terms of y.

   $$\frac{y}{15} = C$$

   c. Write the fractions from parts (a) and (b) as a proportion, and solve to find how many miles Wesley walked after 15 minutes.

   $$\frac{1.5}{25} = \frac{y}{15}$$
   $$25y = 22.5$$
   $$\frac{25}{25}y = \frac{22.5}{25}$$
   $$y = 0.9$$

   Wesley walks 0.9 miles in 15 minutes.

   d. Let y be the distance in miles that Wesley traveled after x minutes. Write a linear equation in two variables that represents how many miles Wesley walked after x minutes.

   $$\frac{1.5}{25} = \frac{y}{x}$$
   $$25y = 1.5x$$
   $$\frac{25}{25}y = \frac{1.5}{25}x$$
   $$y = \frac{1.5}{25}x$$

2. Stefanie drove at a constant speed from her apartment to her friend's house 20 miles away. It took her 45 minutes to reach her destination.

   a. What fraction represents her constant speed, C?

   $$\frac{20}{45} = C$$

   b. What fraction represents constant speed, C, if it takes her x number of minutes to get halfway to her friend's house?

   $$\frac{10}{x} = C$$

Lesson 10: A Critical Look at Proportional Relationships

c. Write and solve a proportion using the fractions from parts (a) and (b) to determine how many minutes it takes her to get to the halfway point.

$$\frac{20}{45} = \frac{10}{x}$$
$$20x = 450$$
$$\frac{20}{20}x = \frac{450}{20}$$
$$x = 22.5$$

Stefanie gets halfway to her friend's house, 10 miles away, after 22.5 minutes.

d. Write a two-variable equation to represent how many miles Stefanie can drive over any time interval.

Let y represent the distance traveled over any time interval x. Then,

$$\frac{20}{45} = \frac{y}{x}$$
$$20x = 45y$$
$$\frac{20}{45}x = \frac{45}{45}y$$
$$\frac{4}{9}x = y.$$

Discussion (4 minutes)

- Consider the problem: Dave lives 15 miles from town A. He is driving at a constant speed of 50 miles per hour from his home away from (in the opposite direction of) the city. How far away is Dave from the town after x hours of driving?
- Since we know he is driving at a constant speed of 50 miles per hour, then we need to determine the distance he travels over a time interval.
- If we say that Dave is y miles from town A after driving x hours, how can we express the actual number of miles that Dave traveled?
  □ Dave is 15 miles from town A to begin with, so the total number of miles Dave traveled is $y - 15$.
- If Dave's average speed in x hours is $\frac{y - 15}{x}$, which is equal to a constant (i.e., his constant speed), then we have the equation

$$\frac{y - 15}{x} = 50.$$

- We want to know how many miles Dave is from town A, y, after driving for x hours. Solve this equation for y.
  □ Student work:

$$\frac{y - 15}{x} = 50$$
$$y - 15 = 50x$$
$$y - 15 + 15 = 50x + 15$$
$$y = 50x + 15$$

Lesson 10: A Critical Look at Proportional Relationships

- With this equation, $y = 50x + 15$, we can find the distance Dave is from town A for any given time x. How far away is Dave from town A after one hour?
  - If $x = 1$, then

    $y = 50(1) + 15$

    $y = 65$.

    Dave is 65 miles from town A after one hour.

Exercise 3 (4 minutes)

Students complete Exercise 3 independently or in pairs.

3. The equation that represents how many miles, y, Dave travels after x hours is $y = 50x + 15$. Use the equation to complete the table below.

x (hours)	Linear Equation: $y = 50x + 15$	y (miles)
1	$y = 50(1) + 15$	65
2	$y = 50(2) + 15$	115
3	$y = 50(3) + 15$	165
3.5	$y = 50(3.5) + 15$	190
4.1	$y = 50(4.1) + 15$	220

Closing (3 minutes)

Summarize, or ask students to summarize, the main points from the lesson:

- Average speed is found by taking the total distance traveled in a given time interval, divided by the time interval.
- If we assume the same average speed over any time interval, then we have constant speed, which can then be used to express a linear equation in two variables relating distance and time.
- We know how to use linear equations to answer questions about distance and time.
- We cannot assume that a problem can be solved using a proportion unless we know that the situation involves constant speed (or rate).

> **Lesson Summary**
>
> Average speed is found by taking the total distance traveled in a given time interval, divided by the time interval.
>
> If y is the total distance traveled in a given time interval x, then $\frac{y}{x}$ is the average speed.
>
> If we assume the same average speed over any time interval, then we have constant speed, which can then be used to express a linear equation in two variables relating distance and time.
>
> If $\frac{y}{x} = C$, where C is a constant, then you have constant speed.

Exit Ticket (4 minutes)

Lesson 10: A Critical Look at Proportional Relationships

Name _____ Date _____

Lesson 10: A Critical Look at Proportional Relationships

Exit Ticket

Alex skateboards at a constant speed from his house to school 3.8 miles away. It takes him 18 minutes.

a. What fraction represents his constant speed, C?

b. After school, Alex skateboards at the same constant speed to his friend's house. It takes him 10 minutes. Write the fraction that represents constant speed, C, if he travels a distance of y.

c. Write the fractions from parts (a) and (b) as a proportion, and solve to find out how many miles Alex's friend's house is from school. Round your answer to the tenths place.

A STORY OF RATIOS   Lesson 10   8•4

Exit Ticket Sample Solutions

> Alex skateboards at a constant speed from his house to school 3.8 miles away. It takes him 18 minutes.
>
> a. What fraction represents his constant speed, C?
>
> $$\frac{3.8}{18} = C$$
>
> b. After school, Alex skateboards at the same constant speed to his friend's house. It takes him 10 minutes. Write the fraction that represents constant speed, C, if he travels a distance of y.
>
> $$\frac{y}{10} = C$$
>
> c. Write the fractions from parts (a) and (b) as a proportion, and solve to find out how many miles Alex's friend's house is from school. Round your answer to the tenths place.
>
> $$\frac{3.8}{18} = \frac{y}{10}$$
> $$3.8(10) = 18y$$
> $$38 = 18y$$
> $$\frac{38}{18} = y$$
> $$2.1 \approx y$$
>
> Alex's friend lives about 2.1 miles from school.

Problem Set Sample Solutions

Students practice writing and solving proportions to solve constant speed problems. Students write two variable equations to represent situations, generally.

> 1. Eman walks from the store to her friend's house, 2 miles away. It takes her 35 minutes.
>
>   a. What fraction represents her constant speed, C?
>
> $$\frac{2}{35} = C$$
>
>   b. Write the fraction that represents her constant speed, C, if she walks y miles in 10 minutes.
>
> $$\frac{y}{10} = C$$
>
>   c. Write and solve a proportion using the fractions from parts (a) and (b) to determine how many miles she walks after 10 minutes. Round your answer to the hundredths place.
>
> $$\frac{2}{35} = \frac{y}{10}$$
> $$35y = 20$$
> $$\frac{35}{35}y = \frac{20}{35}$$
> $$y = 0.57142 \ldots$$
>
>   Eman walks about 0.57 miles after 10 minutes.

Lesson 10:   A Critical Look at Proportional Relationships

d. Write a two-variable equation to represent how many miles Eman can walk over any time interval.

Let y represent the distance Eman walks in x minutes.

$$\frac{2}{35} = \frac{y}{x}$$
$$35y = 2x$$
$$\frac{35}{35}y = \frac{2}{35}x$$
$$y = \frac{2}{35}x$$

2. Erika drives from school to soccer practice 1.3 miles away. It takes her 7 minutes.

   a. What fraction represents her constant speed, C?

   $$\frac{1.3}{7} = C$$

   b. What fraction represents her constant speed, C, if it takes her x minutes to drive exactly 1 mile?

   $$\frac{1}{x} = C$$

   c. Write and solve a proportion using the fractions from parts (a) and (b) to determine how much time it takes her to drive exactly 1 mile. Round your answer to the tenths place.

   $$\frac{1.3}{7} = \frac{1}{x}$$
   $$1.3x = 7$$
   $$\frac{1.3}{1.3}x = \frac{7}{1.3}$$
   $$x = 5.38461\ldots$$

   It takes Erika about 5.4 minutes to drive exactly 1 mile.

   d. Write a two-variable equation to represent how many miles Erika can drive over any time interval.

   Let y be the number of miles Erika travels in x minutes.

   $$\frac{1.3}{7} = \frac{y}{x}$$
   $$7y = 1.3x$$
   $$\frac{7}{7}y = \frac{1.3}{7}x$$
   $$y = \frac{1.3}{7}x$$

3. Darla drives at a constant speed of 45 miles per hour.

   a. If she drives for y miles and it takes her x hours, write the two-variable equation to represent the number of miles Darla can drive in x hours.

   $$\frac{y}{x} = 45$$
   $$y = 45x$$

b. Darla plans to drive to the market 14 miles from her house, then to the post office 3 miles from the market, and then return home, which is 15 miles from the post office. Assuming she drives at a constant speed the entire time, how long will it take her to run her errands and get back home? Round your answer to the hundredths place.

Altogether, Darla plans to drive 32 miles because $14 + 3 + 15 = 32$.

$$32 = 45x$$
$$\frac{32}{45} = \frac{45}{45}x$$
$$0.71111\ldots = x$$

It will take Darla about 0.71 hours to run her errands and get back home.

4. Aaron walks from his sister's house to his cousin's house, a distance of 4 miles, in 80 minutes. How far does he walk in 30 minutes?

   I cannot say for sure how far Aaron walks in 30 minutes because I do not know if he is walking at a constant speed. Maybe he stopped at his friend's house for 20 minutes.

5. Carlos walks 4 miles every night for exercise. It takes him exactly 63 minutes to finish his walk.

   a. Assuming he walks at a constant rate, write an equation that represents how many miles, y, Carlos can walk in x minutes.

   Since $\frac{4}{63} = C$ and $\frac{y}{x} = C$, then

   $$\frac{4}{63} = \frac{y}{x}$$
   $$63y = 4x$$
   $$\frac{63}{63}y = \frac{4}{63}x$$
   $$y = \frac{4}{63}x.$$

   b. Use your equation from part (a) to complete the table below. Use a calculator, and round all values to the hundredths place.

x (minutes)	Linear Equation: $y = \frac{4}{63}x$	y (miles)
15	$y = \frac{4}{63}(15)$	0.95
30	$y = \frac{4}{63}(30)$	1.90
40	$y = \frac{4}{63}(40)$	2.54
60	$y = \frac{4}{63}(60)$	3.81
75	$y = \frac{4}{63}(75)$	4.76

Lesson 10: A Critical Look at Proportional Relationships

A STORY OF RATIOS   Lesson 11   8•4

 Lesson 11: Constant Rate

Student Outcomes

- Students know the definition of constant rate in varied contexts as expressed using two variables where one is t representing a time interval.
- Students graph points on a coordinate plane related to constant rate problems.

Classwork

Example 1 (6 minutes)

Give students the first question below, and allow them time to work. Ask them to share their solutions with the class, and then proceed with the discussion, table, and graph to finish Example 1.

> **Example 1**
>
> Pauline mows a lawn at a constant rate. Suppose she mows a 35-square-foot lawn in 2.5 minutes. What area, in square feet, can she mow in 10 minutes? t minutes?

- What is Pauline's average rate in 2.5 minutes?
  - Pauline's average rate in 2.5 minutes is $\frac{35}{2.5}$ square feet per minute.
- What is Pauline's average rate in 10 minutes?
  - Let A represent the square feet of the area mowed in 10 minutes. Pauline's average rate in 10 minutes is $\frac{A}{10}$ square feet per minute.
- Let C be Pauline's constant rate in square feet per minute; then, $\frac{35}{2.5} = C$, and $\frac{A}{10} = C$. Therefore,

$$\frac{35}{2.5} = \frac{A}{10}$$
$$350 = 2.5A$$
$$\frac{350}{2.5} = \frac{2.5}{2.5}A$$
$$140 = A$$

Pauline mows 140 square feet of lawn in 10 minutes.

- If we let y represent the number of square feet Pauline can mow in t minutes, then Pauline's average rate in t minutes is $\frac{y}{t}$ square feet per minute.
- Write the two-variable equation that represents the area of lawn, y, Pauline can mow in t minutes.

124   Lesson 11:   Constant Rate

This work is derived from Eureka Math ™ and licensed by Great Minds. ©2015 Great Minds. eureka-math.org
G8-M4-TE-B3-1.3.0-07.2015

A STORY OF RATIOS   Lesson 11   8•4

$$\frac{35}{2.5} = \frac{y}{t}$$
$$2.5y = 35t$$
$$\frac{2.5}{2.5}y = \frac{35}{2.5}t$$
$$y = \frac{35}{2.5}t$$

MP.7

- What is the meaning of $\frac{35}{2.5}$ in the equation $y = \frac{35}{2.5}t$?
    - The number $\frac{35}{2.5}$ represents the constant rate at which Pauline can mow a lawn.
- We can organize the information in a table.

t (time in minutes)	Linear Equation: $y = \frac{35}{2.5}t$	y (area in square feet)
0	$y = \frac{35}{2.5}(0)$	0
1	$y = \frac{35}{2.5}(1)$	$\frac{35}{2.5} = 14$
2	$y = \frac{35}{2.5}(2)$	$\frac{70}{2.5} = 28$
3	$y = \frac{35}{2.5}(3)$	$\frac{105}{2.5} = 42$
4	$y = \frac{35}{2.5}(4)$	$\frac{140}{2.5} = 56$

- On a coordinate plane, we will let the x-axis represent time t, in minutes, and the y-axis represent the area of mowed lawn in square feet. Then we have the following graph.

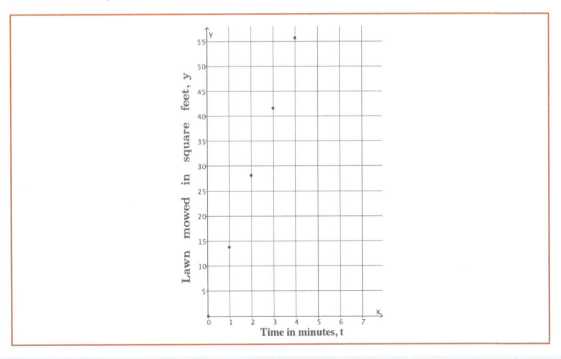

Lesson 11:   Constant Rate

A STORY OF RATIOS — Lesson 11 8•4

- Because Pauline mows at a constant rate, we would expect the square feet of mowed lawn to continue to rise as the time, in minutes, increases.

Concept Development (6 minutes)

- In the last lesson, we learned about average speed and constant speed. Constant speed problems are just a special case of a larger variety of problems known as constant rate problems. Some of these problems were topics in Grade 7, such as water pouring out of a faucet into a tub, painting a house, and mowing a lawn.
- First, we define the average rate:

  Suppose V gallons of water flow from a faucet in a given time interval t (minutes). Then, the *average rate* of water flow in the given time interval is $\frac{V}{t}$ in gallons per minute.

- Then, we define the constant rate:

  Suppose the average rate of water flow is the same constant C for *any* given time interval. Then we say that the water is flowing at a *constant rate*, C.

- Similarly, suppose A square feet of lawn are mowed in a given time interval t (minutes). Then, the *average rate* of lawn mowing in the given time interval is $\frac{A}{t}$ square feet per minute. If we assume that the average rate of lawn mowing is the same constant, C, for *any* given time interval, then we say that the lawn is mowed at a constant rate, C.

- Describe the average rate of painting a house.
    - Suppose A square feet of house are painted in a given time interval t (minutes). Then the average rate of house painting in the given time interval is $\frac{A}{t}$ square feet per minute.

- Describe the constant rate of painting a house.
    - If we assume that the average rate of house painting is the same constant, C, over any given time interval, then we say that the wall is painted at a constant rate, C.

- What is the difference between average rate and constant rate?
    - Average rate is the rate in which something can be done over a specific time interval. Constant rate assumes that the average rate is the same over any time interval.

- As you can see, the way we define average rate and constant rate for a given situation is very similar. In each case, a transcription of the given information leads to an expression in two variables.

Example 2 (8 minutes)

> **Example 2**
>
> Water flows at a constant rate out of a faucet. Suppose the volume of water that comes out in three minutes is 10.5 gallons. How many gallons of water come out of the faucet in t minutes?

- Write the linear equation that represents the volume of water, V, that comes out in t minutes.

126 Lesson 11: Constant Rate

> Let C represent the constant rate of water flow.
>
> $\dfrac{10.5}{3} = C$, and $\dfrac{V}{t} = C$; then, $\dfrac{10.5}{3} = \dfrac{V}{t}$.
>
> $$\dfrac{10.5}{3} = \dfrac{V}{t}$$
> $$3V = 10.5t$$
> $$\dfrac{3}{3}V = \dfrac{10.5}{3}t$$
> $$V = \dfrac{10.5}{3}t$$

- What is the meaning of the number $\dfrac{10.5}{3}$ in the equation $V = \dfrac{10.5}{3}t$?

  ▫ The number $\dfrac{10.5}{3}$ represents the constant rate at which water flows from a faucet.

- Using the linear equation $V = \dfrac{10.5}{3}t$, complete the table.

t (time in minutes)	Linear Equation: $V = \dfrac{10.5}{3}t$	V (in gallons)
0	$V = \dfrac{10.5}{3}(0)$	0
1	$V = \dfrac{10.5}{3}(1)$	$\dfrac{10.5}{3} = 3.5$
2	$V = \dfrac{10.5}{3}(2)$	$\dfrac{21}{3} = 7$
3	$V = \dfrac{10.5}{3}(3)$	$\dfrac{31.5}{3} = 10.5$
4	$V = \dfrac{10.5}{3}(4)$	$\dfrac{42}{3} = 14$

- On a coordinate plane, we will let the x-axis represent time t in minutes and the y-axis represent the volume of water. Graph the data from the table.

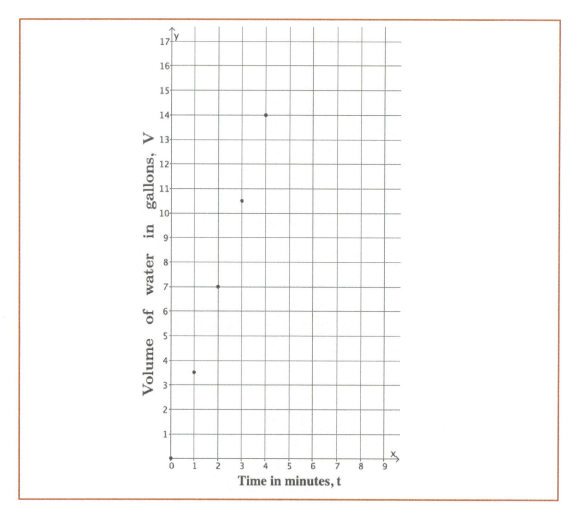

- Using the graph, about how many gallons of water do you think would flow after $1\frac{1}{2}$ minutes? Explain.
  - After $1\frac{1}{2}$ minutes, between $3\frac{1}{2}$ and 7 gallons of water will flow. Since the water is flowing at a constant rate, we can expect the volume of water to rise between 1 and 2 minutes. The number of gallons that flow after $1\frac{1}{2}$ minutes then would have to be between the number of gallons that flow out after 1 minute and 2 minutes.
- Using the graph, about how long would it take for 15 gallons of water to flow out of the faucet? Explain.
  - It would take between 4 and 5 minutes for 15 gallons of water to flow out of the faucet. It takes 4 minutes for 14 gallons to flow; therefore, it must take more than 4 minutes for 15 gallons to come out. It must take less than 5 minutes because $3\frac{1}{2}$ gallons flow out every minute.
- Graphing proportional relationships like these last two constant rate problems provides us more information than simply solving an equation and calculating one value. The graph provides information that is not so obvious in an equation.

Lesson 11

Exercises (15 minutes)

Students complete Exercises 1–3 independently.

Exercises

1. Juan types at a constant rate. He can type a full page of text in $3\frac{1}{2}$ minutes. We want to know how many pages, p, Juan can type after t minutes.

   a. Write the linear equation in two variables that represents the number of pages Juan types in any given time interval.

   Let C represent the constant rate that Juan types in pages per minute. Then,

   $\frac{1}{3.5} = C$, and $\frac{p}{t} = C$; therefore, $\frac{1}{3.5} = \frac{p}{t}$.

   $$\frac{1}{3.5} = \frac{p}{t}$$
   $$3.5p = t$$
   $$\frac{3.5}{3.5}p = \frac{1}{3.5}t$$
   $$p = \frac{1}{3.5}t$$

   b. Complete the table below. Use a calculator, and round your answers to the tenths place.

t (time in minutes)	Linear Equation: $p = \frac{1}{3.5}t$	p (pages typed)
0	$p = \frac{1}{3.5}(0)$	0
5	$p = \frac{1}{3.5}(5)$	$\frac{5}{3.5} \approx 1.4$
10	$p = \frac{1}{3.5}(10)$	$\frac{10}{3.5} \approx 2.9$
15	$p = \frac{1}{3.5}(15)$	$\frac{15}{3.5} \approx 4.3$
20	$p = \frac{1}{3.5}(20)$	$\frac{20}{3.5} \approx 5.7$

   c. Graph the data on a coordinate plane.

   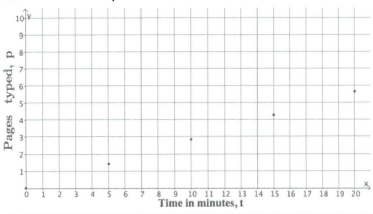

Lesson 11: Constant Rate

d. About how long would it take Juan to type a 5-page paper? Explain.

It would take him between 15 and 20 minutes. After 15 minutes, he will have typed 4.3 pages. In 20 minutes, he can type 5.7 pages. Since 5 pages is between 4.3 and 5.7, then it will take him between 15 and 20 minutes.

2. Emily paints at a constant rate. She can paint 32 square feet in 5 minutes. What area, A, in square feet, can she paint in t minutes?

   a. Write the linear equation in two variables that represents the number of square feet Emily can paint in any given time interval.

   Let C be the constant rate that Emily paints in square feet per minute. Then,

   $\frac{32}{5} = C$, and $\frac{A}{t} = C$; therefore, $\frac{32}{5} = \frac{A}{t}$.

   $$\frac{32}{5} = \frac{A}{t}$$
   $$5A = 32t$$
   $$\frac{5}{5}A = \frac{32}{5}t$$
   $$A = \frac{32}{5}t$$

   b. Complete the table below. Use a calculator, and round answers to the tenths place.

t (time in minutes)	Linear Equation: $A = \frac{32}{5}t$	A (area painted in square feet)
0	$A = \frac{32}{5}(0)$	0
1	$A = \frac{32}{5}(1)$	$\frac{32}{5} = 6.4$
2	$A = \frac{32}{5}(2)$	$\frac{64}{5} = 12.8$
3	$A = \frac{32}{5}(3)$	$\frac{96}{5} = 19.2$
4	$A = \frac{32}{5}(4)$	$\frac{128}{5} = 25.6$

Lesson 11: Constant Rate

c. Graph the data on a coordinate plane.

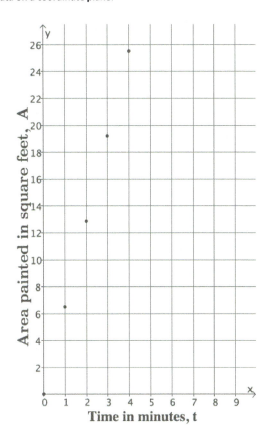

d. About how many square feet can Emily paint in $2\frac{1}{2}$ minutes? Explain.

Emily can paint between 12.8 and 19.2 square feet in $2\frac{1}{2}$ minutes. After 2 minutes, she paints 12.8 square feet, and after 3 minutes, she will have painted 19.2 square feet.

3. Joseph walks at a constant speed. He walked to a store that is one-half mile away in 6 minutes. How many miles, m, can he walk in t minutes?

   a. Write the linear equation in two variables that represents the number of miles Joseph can walk in any given time interval, t.

   Let C be the constant rate that Joseph walks in miles per minute. Then,

   $\frac{0.5}{6} = C$, *and* $\frac{m}{t} = C$; *therefore,* $\frac{0.5}{6} = \frac{m}{t}$.

   $$\frac{0.5}{6} = \frac{m}{t}$$
   $$6m = 0.5t$$
   $$\frac{6}{6}m = \frac{0.5}{6}t$$
   $$m = \frac{0.5}{6}t$$

Lesson 11: Constant Rate

b. Complete the table below. Use a calculator, and round answers to the tenths place.

t (time in minutes)	Linear Equation: $m = \frac{0.5}{6}t$	m (distance in miles)
0	$m = \frac{0.5}{6}(0)$	0
30	$m = \frac{0.5}{6}(30)$	$\frac{15}{6} = 2.5$
60	$m = \frac{0.5}{6}(60)$	$\frac{30}{6} = 5$
90	$m = \frac{0.5}{6}(90)$	$\frac{45}{6} = 7.5$
120	$m = \frac{0.5}{6}(120)$	$\frac{60}{6} = 10$

c. Graph the data on a coordinate plane.

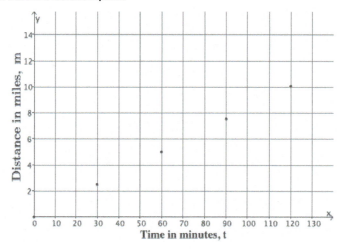

d. Joseph's friend lives 4 miles away from him. About how long would it take Joseph to walk to his friend's house? Explain.

It will take Joseph a little less than an hour to walk to his friend's house. Since it takes 30 minutes for him to walk 2.5 miles and 60 minutes to walk 5 miles, and 4 is closer to 5 than 2.5, it will take Joseph less than an hour to walk the 4 miles.

Closing (5 minutes)

Summarize, or ask students to summarize, the main points from the lesson:

- Constant rate problems appear in a variety of contexts like painting a house, typing, walking, or water flow.
- We can express the constant rate as a two-variable equation representing proportional change.
- We can graph the constant rate situation by completing a table to compute data points.

Lesson Summary

When constant rate is stated for a given problem, then you can express the situation as a two-variable equation. The equation can be used to complete a table of values that can then be graphed on a coordinate plane.

Exit Ticket (5 minutes)

Name _____   Date _____

Lesson 11: Constant Rate

Exit Ticket

Vicky reads at a constant rate. She can read 5 pages in 9 minutes. We want to know how many pages, p, Vicky can read after t minutes.

a. Write a linear equation in two variables that represents the number of pages Vicky reads in any given time interval.

b. Complete the table below. Use a calculator, and round answers to the tenths place.

t (time in minutes)	Linear Equation:	p (pages read)
0		
20		
40		
60		

c. About how long would it take Vicky to read 25 pages? Explain.

Exit Ticket Sample Solutions

Vicky reads at a constant rate. She can read 5 pages in 9 minutes. We want to know how many pages, p, Vicky can read after t minutes.

a. Write a linear equation in two variables that represents the number of pages Vicky reads in any given time interval.

Let C represent the constant rate that Vicky reads in pages per minute. Then,

$\frac{5}{9} = C$, and $\frac{p}{t} = C$; therefore, $\frac{5}{9} = \frac{p}{t}$.

$$\frac{5}{9} = \frac{p}{t}$$
$$9p = 5t$$
$$\frac{9}{9}p = \frac{5}{9}t$$
$$p = \frac{5}{9}t$$

b. Complete the table below. Use a calculator, and round answers to the tenths place.

t (time in minutes)	Linear Equation: $p = \frac{5}{9}t$	p (pages read)
0	$p = \frac{5}{9}(0)$	0
20	$p = \frac{5}{9}(20)$	$\frac{100}{9} \approx 11.1$
40	$p = \frac{5}{9}(40)$	$\frac{200}{9} \approx 22.2$
60	$p = \frac{5}{9}(60)$	$\frac{300}{9} \approx 33.3$

c. About how long would it take Vicky to read 25 pages? Explain.

It would take her a little over 40 minutes. After 40 minutes, she can read about 22.2 pages, and after 1 hour, she can read about 33.3 pages. Since 25 pages is between 22.2 and 33.3, it will take her between 40 and 60 minutes to read 25 pages.

Problem Set Sample Solutions

Students practice writing two-variable equations that represent a constant rate.

1. A train travels at a constant rate of 45 miles per hour.

   a. What is the distance, d, in miles, that the train travels in t hours?

   Let C be the constant rate the train travels. Then, $\frac{45}{1} = C$, and $\frac{d}{t} = C$; therefore, $\frac{45}{1} = \frac{d}{t}$.

   $$\frac{45}{1} = \frac{d}{t}$$
   $$d = 45t$$

Lesson 11: Constant Rate

b. How many miles will it travel in 2.5 hours?

$$d = 45(2.5)$$
$$= 112.5$$

The train will travel 112.5 miles in 2.5 hours.

2. Water is leaking from a faucet at a constant rate of $\frac{1}{3}$ gallons per minute.

   a. What is the amount of water, w, in gallons per minute, that is leaked from the faucet after t minutes?

   Let C be the constant rate the water leaks from the faucet in gallons per minute. Then,

   $\frac{\frac{1}{3}}{1} = C$, and $\frac{w}{t} = C$; therefore, $\frac{\frac{1}{3}}{1} = \frac{w}{t}$.

   $$\frac{\frac{1}{3}}{1} = \frac{w}{t}$$
   $$w = \frac{1}{3}t$$

   b. How much water is leaked after an hour?

   $$w = \frac{1}{3}t$$
   $$= \frac{1}{3}(60)$$
   $$= 20$$

   The faucet will leak 20 gallons in one hour.

3. A car can be assembled on an assembly line in 6 hours. Assume that the cars are assembled at a constant rate.

   a. How many cars, y, can be assembled in t hours?

   Let C be the constant rate the cars are assembled in cars per hour. Then,

   $\frac{1}{6} = C$, and $\frac{y}{t} = C$; therefore, $\frac{1}{6} = \frac{y}{t}$.

   $$\frac{1}{6} = \frac{y}{t}$$
   $$6y = t$$
   $$\frac{6}{6}y = \frac{1}{6}t$$
   $$y = \frac{1}{6}t$$

   b. How many cars can be assembled in a week?

   A week is $24 \times 7 = 168$ hours. So, $y = \frac{1}{6}(168) = 28$. Twenty-eight cars can be assembled in a week.

4. A copy machine makes copies at a constant rate. The machine can make 80 copies in $2\frac{1}{2}$ minutes.

   a. Write an equation to represent the number of copies, n, that can be made over any time interval in minutes, t.

   Let C be the constant rate that copies can be made in copies per minute. Then,

   $\frac{80}{2\frac{1}{2}} = C$, and $\frac{n}{t} = C$; therefore, $\frac{80}{2\frac{1}{2}} = \frac{n}{t}$.

   $$\frac{80}{2\frac{1}{2}} = \frac{n}{t}$$
   $$2\frac{1}{2}n = 80t$$
   $$\frac{5}{2}n = 80t$$
   $$\frac{2}{5} \cdot \frac{5}{2}n = \frac{2}{5} \cdot 80t$$
   $$n = 32t$$

   b. Complete the table below.

t (time in minutes)	Linear Equation: $n = 32t$	n (number of copies)
0	$n = 32(0)$	0
0.25	$n = 32(0.25)$	8
0.5	$n = 32(0.5)$	16
0.75	$n = 32(0.75)$	24
1	$n = 32(1)$	32

c. Graph the data on a coordinate plane.

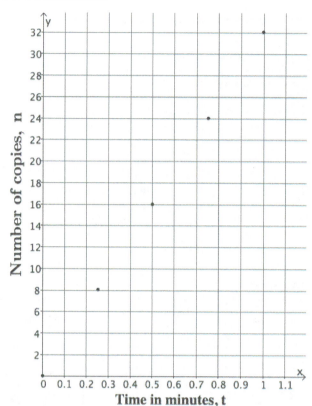

d. The copy machine runs for 20 seconds and then jams. About how many copies were made before the jam occurred? Explain.

Since 20 seconds is approximately 0.3 of a minute, then the number of copies made will be between 8 and 16 because 0.3 is between 0.25 and 0.5.

5. Connor runs at a constant rate. It takes him 34 minutes to run 4 miles.

   a. Write the linear equation in two variables that represents the number of miles Connor can run in any given time interval in minutes, t.

   Let C be the constant rate that Connor runs in miles per minute, and let m represent the number of miles he ran in t minutes. Then,

   $$\frac{4}{34} = C, \text{ and } \frac{m}{t} = C; \text{ therefore, } \frac{4}{34} = \frac{m}{t}.$$

   $$\frac{4}{34} = \frac{m}{t}$$
   $$34m = 4t$$
   $$\frac{34}{34}m = \frac{4}{34}t$$
   $$m = \frac{4}{34}t$$
   $$m = \frac{2}{17}t$$

b. Complete the table below. Use a calculator, and round answers to the tenths place.

t (time in minutes)	Linear Equation: $m = \frac{2}{17}t$	m (distance in miles)
0	$m = \frac{2}{17}(0)$	0
15	$m = \frac{2}{17}(15)$	$\frac{30}{17} \approx 1.8$
30	$m = \frac{2}{17}(30)$	$\frac{60}{17} \approx 3.5$
45	$m = \frac{2}{17}(45)$	$\frac{90}{17} \approx 5.3$
60	$m = \frac{2}{17}(60)$	$\frac{120}{17} \approx 7.1$

c. Graph the data on a coordinate plane.

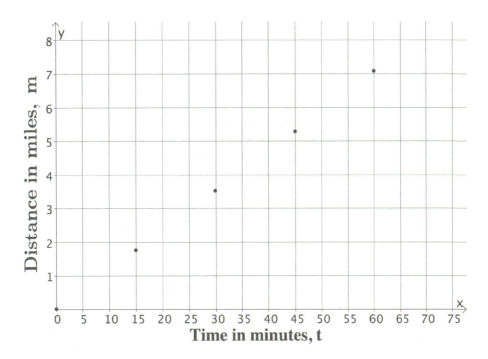

d. Connor ran for 40 minutes before tripping and spraining his ankle. About how many miles did he run before he had to stop? Explain.

Since Connor ran for 40 minutes, he ran more than 3.5 miles but less than 5.3 miles. Since 40 is between 30 and 45, then we can use those reference points to make an estimate of how many miles he ran in 40 minutes, probably about 5 miles.

Lesson 11: Constant Rate

A STORY OF RATIOS   Lesson 12   8•4

 # Lesson 12: Linear Equations in Two Variables

Student Outcomes

- Students use a table to find solutions to a given linear equation and plot the solutions on a coordinate plane.

Lesson Notes

In this lesson, students find solutions to a linear equation in two variables using a table and then plot the solutions as points on the coordinate plane. Students need graph paper in order to complete the Exercises and the Problem Set.

Classwork

Opening Exercise (5 minutes)

Students complete the Opening Exercise independently in preparation for the discussion about standard form and the solutions that follow.

Opening Exercise

Emily tells you that she scored 32 points in a basketball game. Write down all the possible ways she could have scored 32 with only two- and three-point baskets. Use the table below to organize your work.

Number of Two-Pointers	Number of Three-Pointers
16	0
13	2
10	4
7	6
4	8
1	10

Let x be the number of two-pointers and y be the number of three-pointers that Emily scored. Write an equation to represent the situation.

$$2x + 3y = 32$$

Discussion (10 minutes)

- An equation in the form of $ax + by = c$ is called a *linear equation in two variables*, where a, b, and c are constants, and at least one of a and b are not zero. In this lesson, neither a nor b will be equal to zero. In the Opening Exercise, what equation did you write to represent Emily's score at the basketball game?
  - $2x + 3y = 32$
- The equation $2x + 3y = 32$ is an example of a linear equation in two variables.

- An equation of this form, $ax + by = c$, is also referred to as an equation in *standard form*. Is the equation you wrote in the Opening Exercise in standard form?
  - *Yes. It is in the same form as $ax + by = c$.*
- In the equation $ax + by = c$, the symbols a, b, and c are constants. What, then, are x and y?
  - *The symbols x and y are numbers. Since they are not constants, it means they are unknown numbers, typically called variables, in the equation $ax + by = c$.*
- For example, $-50x + y = 15$ is a linear equation in x and y. As you can easily see, not just *any* pair of numbers x and y will make the equation true. Consider $x = 1$ and $y = 2$. Does it make the equation true?
  - *No, because $-50(1) + 2 = -50 + 2 = -48 \neq 15$.*
- What pairs of numbers did you find that worked for Emily's basketball score? Did just any pair of numbers work? Explain.
  - *Students should identify the pairs of numbers in the table of the Opening Exercise. No, not just any pair of numbers worked. For example, I couldn't say that Emily scored 15 two-pointers and 1 three-pointer because that would mean she scored 33 points in the game, and she only scored 32 points.*
- A *solution* to the linear equation in two variables is an ordered pair of numbers (x, y) so that x and y makes the equation a true statement. The pairs of numbers that you wrote in the table for Emily are solutions to the equation $2x + 3y = 32$ because they are pairs of numbers that make the equation true. The question becomes, how do we find an unlimited number of solutions to a given linear equation?
  - *Guess numbers until you find a pair that makes the equation true.*
- A strategy that will help us find solutions to a linear equation in two variables is as follows: We fix a number for x. That means we pick any number we want and call it x. Since we know how to solve a linear equation in one variable, then we solve for y. The number we picked for x and the number we get when we solve for y is the ordered pair (x, y), which is a solution to the two-variable linear equation.
- For example, let $x = 5$. Then, in the equation $-50x + y = 15$, we have
$$-50(5) + y = 15$$
$$-250 + y = 15$$
$$-250 + 250 + y = 15 + 250$$
$$y = 265.$$
Therefore, $(5, 265)$ is a solution to the equation $-50x + y = 15$.
- Similarly, we can fix a number for y and solve for x. Let $y = 10$; then
$$-50x + 10 = 15$$
$$-50x + 10 - 10 = 15 - 10$$
$$-50x = 5$$
$$\frac{-50}{-50}x = \frac{5}{-50}$$
$$x = -\frac{1}{10}.$$
Therefore, $\left(-\frac{1}{10}, 10\right)$ is a solution to the equation $-50x + y = 15$.

Ask students to provide a number for x or y and demonstrate how to find a solution. This can be done more than once in order to prove to students that they can find a solution no matter which number they choose to fix for x or y. Once they are convinced, allow them to work on the Exploratory Challenge.

Lesson 12: Linear Equations in Two Variables

A STORY OF RATIOS — Lesson 12 8•4

Exploratory Challenge/Exercises (20 minutes)

Students can work independently or in pairs to complete the exercises. Every few minutes, have students share their tables and graphs with the class. Make suggestions to students as they work as to which values for x or y they could choose. For example, in Exercises 1 and 2, small numbers would ease the mental math work. Exercise 3 may be made easier if they choose a number for y and solve for x. Exercise 4 can be made easier if students choose values for x that are multiples of 5. While making suggestions, ask students why the suggestions would make the work easier.

Exploratory Challenge/Exercises

1. Find five solutions for the linear equation $x + y = 3$, and plot the solutions as points on a coordinate plane.

x	Linear Equation: $x + y = 3$	y
1	$1 + y = 3$	2
2	$2 + y = 3$	1
3	$3 + y = 3$	0
4	$4 + y = 3$	-1
5	$5 + y = 3$	-2

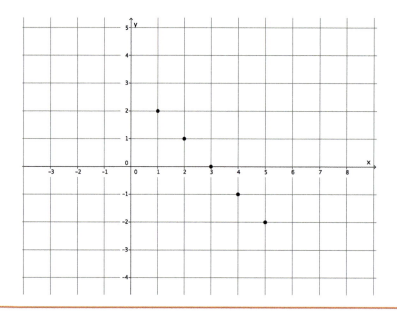

Lesson 12: Linear Equations in Two Variables

2. Find five solutions for the linear equation $2x - y = 10$, and plot the solutions as points on a coordinate plane.

x	Linear Equation: $2x - y = 10$	y
1	$2(1) - y = 10$ $2 - y = 10$ $2 - 2 - y = 10 - 2$ $-y = 8$ $y = -8$	-8
2	$2(2) - y = 10$ $4 - y = 10$ $4 - 4 - y = 10 - 4$ $-y = 6$ $y = -6$	-6
3	$2(3) - y = 10$ $6 - y = 10$ $6 - 6 - y = 10 - 6$ $-y = 4$ $y = -4$	-4
4	$2(4) - y = 10$ $8 - y = 10$ $8 - 8 - y = 10 - 8$ $-y = 2$ $y = -2$	-2
5	$2(5) - y = 10$ $10 - y = 10$ $y = 0$	0

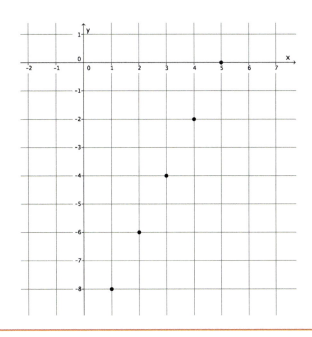

3. Find five solutions for the linear equation $x + 5y = 21$, and plot the solutions as points on a coordinate plane.

x	Linear Equation: $x + 5y = 21$	y
16	$x + 5(1) = 21$ $x + 5 = 21$ $x = 16$	1
11	$x + 5(2) = 21$ $x + 10 = 21$ $x = 11$	2
6	$x + 5(3) = 21$ $x + 15 = 21$ $x = 6$	3
1	$x + 5(4) = 21$ $x + 20 = 21$ $x = 1$	4
−4	$x + 5(5) = 21$ $x + 25 = 21$ $x = -4$	5

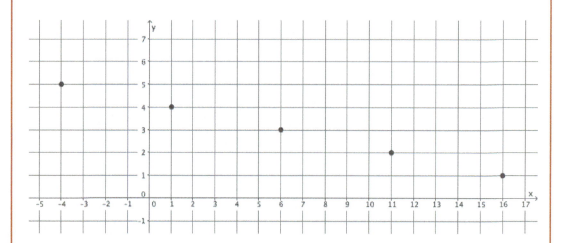

4. Consider the linear equation $\frac{2}{5}x + y = 11$.

   a. Will you choose to fix values for x or y? Explain.

   If I fix values for x, it will make the computations easier. Solving for y can be done in one step.

   b. Are there specific numbers that would make your computational work easier? Explain.

   Values for x that are multiples of 5 will make the computations easier. When I multiply $\frac{2}{5}$ by a multiple of 5, I will get an integer.

Lesson 12: Linear Equations in Two Variables

c. Find five solutions to the linear equation $\frac{2}{5}x + y = 11$, and plot the solutions as points on a coordinate plane.

x	Linear Equation: $\frac{2}{5}x + y = 11$	y
5	$\frac{2}{5}(5) + y = 11$ $2 + y = 11$ $y = 9$	9
10	$\frac{2}{5}(10) + y = 11$ $4 + y = 11$ $y = 7$	7
15	$\frac{2}{5}(15) + y = 11$ $6 + y = 11$ $y = 5$	5
20	$\frac{2}{5}(20) + y = 11$ $8 + y = 11$ $y = 3$	3
25	$\frac{2}{5}(25) + y = 11$ $10 + y = 11$ $y = 1$	1

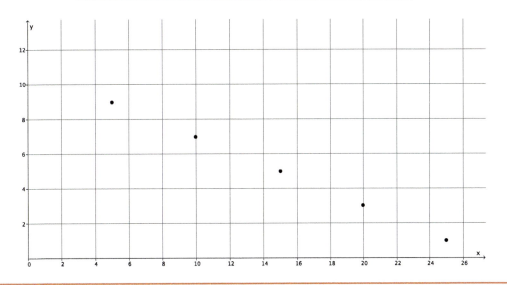

5. At the store, you see that you can buy a bag of candy for $2 and a drink for $1. Assume you have a total of $35 to spend. You are feeling generous and want to buy some snacks for you and your friends.

   a. Write an equation in standard form to represent the number of bags of candy, x, and the number of drinks, y, that you can buy with $35.

   $$2x + y = 35$$

   b. Find five solutions to the linear equation from part (a), and plot the solutions as points on a coordinate plane.

x	Linear Equation: $2x + y = 35$	y
4	$2(4) + y = 35$ $8 + y = 35$ $y = 27$	27
5	$2(5) + y = 35$ $10 + y = 35$ $y = 25$	25
8	$2(8) + y = 35$ $16 + y = 35$ $y = 19$	19
10	$2(10) + y = 35$ $20 + y = 35$ $y = 15$	15
15	$2(15) + y = 35$ $30 + y = 35$ $y = 5$	5

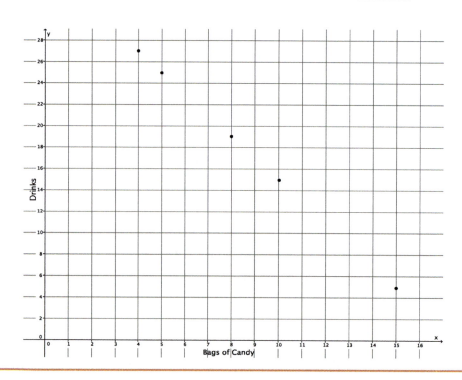

Closing (5 minutes)

Summarize, or ask students to summarize, the main points from the lesson:

- A two-variable equation in the form of $ax + by = c$ is known as a *linear equation in standard form*.
- A solution to a linear equation in two variables is an ordered pair (x, y) that makes the given equation true.
- We can find solutions by fixing a number for x or y and then solving for the other variable. Our work can be made easier by thinking about the computations we will need to make before fixing a number for x or y. For example, if x has a coefficient of $\frac{1}{3}$, we should select values for x that are multiples of 3.

Lesson Summary

A linear equation in two-variables x and y is in standard form if it is the form $ax + by = c$ for numbers a, b, and c, where a and b are both not zero. The numbers a, b, and c are called constants.

A solution to a linear equation in two variables is the ordered pair (x, y) that makes the given equation true. Solutions can be found by fixing a number for x and solving for y or fixing a number for y and solving for x.

Exit Ticket (5 minutes)

Lesson 12: Linear Equations in Two Variables

Name _____ Date _____

Lesson 12: Linear Equations in Two Variables

Exit Ticket

1. Is the point $(1, 3)$ a solution to the linear equation $5x - 9y = 32$? Explain.

2. Find three solutions for the linear equation $4x - 3y = 1$, and plot the solutions as points on a coordinate plane.

x	Linear Equation: $4x - 3y = 1$	y

Exit Ticket Sample Solutions

1. Is the point $(1, 3)$ a solution to the linear equation $5x - 9y = 32$? Explain.

   No, $(1, 3)$ is not a solution to $5x - 9y = 32$ because $5(1) - 9(3) = 5 - 27 = -22$, and $-22 \neq 32$.

2. Find three solutions for the linear equation $4x - 3y = 1$, and plot the solutions as points on a coordinate plane.

x	Linear Equation: $4x - 3y = 1$	y
1	$4(1) - 3y = 1$ $4 - 3y = 1$ $-3y = -3$ $y = 1$	1
4	$4x - 3(5) = 1$ $4x - 15 = 1$ $4x = 16$ $x = 4$	5
7	$4(7) - 3y = 1$ $28 - 3y = 1$ $-3y = -27$ $y = 9$	9

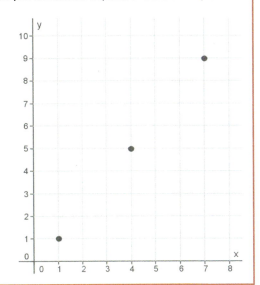

Problem Set Sample Solutions

Students practice finding and graphing solutions for linear equations that are in standard form.

1. Consider the linear equation $x - \frac{3}{2}y = -2$.

   a. Will you choose to fix values for x or y? Explain.

   If I fix values for y, it will make the computations easier. Solving for x can be done in one step.

   b. Are there specific numbers that would make your computational work easier? Explain.

   Values for y that are multiples of 2 will make the computations easier. When I multiply $\frac{3}{2}$ by a multiple of 2, I will get a whole number.

Lesson 12: Linear Equations in Two Variables

c. Find five solutions to the linear equation $x - \frac{3}{2}y = -2$, and plot the solutions as points on a coordinate plane.

x	Linear Equation: $x - \frac{3}{2}y = -2$	y
1	$x - \frac{3}{2}(2) = -2$ $x - 3 = -2$ $x - 3 + 3 = -2 + 3$ $x = 1$	2
4	$x - \frac{3}{2}(4) = -2$ $x - 6 = -2$ $x - 6 + 6 = -2 + 6$ $x = 4$	4
7	$x - \frac{3}{2}(6) = -2$ $x - 9 = -2$ $x - 9 + 9 = -2 + 9$ $x = 7$	6
10	$x - \frac{3}{2}(8) = -2$ $x - 12 = -2$ $x - 12 + 12 = -2 + 12$ $x = 10$	8
13	$x - \frac{3}{2}(10) = -2$ $x - 15 = -2$ $x - 15 + 15 = -2 + 15$ $x = 13$	10

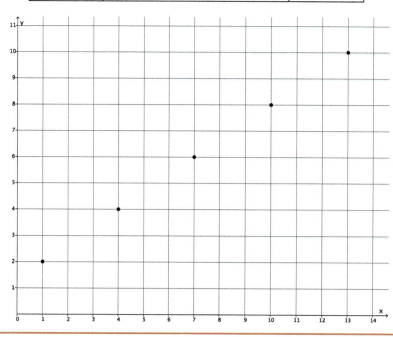

Lesson 12: Linear Equations in Two Variables

2. Find five solutions for the linear equation $\frac{1}{3}x + y = 12$, and plot the solutions as points on a coordinate plane.

x	Linear Equation: $\frac{1}{3}x + y = 12$	y
3	$\frac{1}{3}(3) + y = 12$ $1 + y = 12$ $y = 11$	11
6	$\frac{1}{3}(6) + y = 12$ $2 + y = 12$ $y = 10$	10
9	$\frac{1}{3}(9) + y = 12$ $3 + y = 12$ $y = 9$	9
12	$\frac{1}{3}(12) + y = 12$ $4 + y = 12$ $y = 8$	8
15	$\frac{1}{3}(15) + y = 12$ $5 + y = 12$ $y = 7$	7

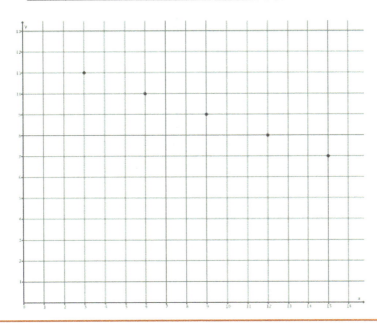

3. Find five solutions for the linear equation $-x + \frac{3}{4}y = -6$, and plot the solutions as points on a coordinate plane.

x	Linear Equation: $-x + \frac{3}{4}y = -6$	y
9	$-x + \frac{3}{4}(4) = -6$ $-x + 3 = -6$ $-x + x + 3 = -6 + x$ $3 = -6 + x$ $3 + 6 = -6 + 6 + x$ $9 = x$	4
12	$-x + \frac{3}{4}(8) = -6$ $-x + 6 = -6$ $-x + x + 6 = -6 + x$ $6 = -6 + x$ $6 + 6 = -6 + 6 + x$ $12 = x$	8
15	$-x + \frac{3}{4}(12) = -6$ $-x + 9 = -6$ $-x + x + 9 = -6 + x$ $9 = -6 + x$ $9 + 6 = -6 + 6 + x$ $15 = x$	12
18	$-x + \frac{3}{4}(16) = -6$ $-x + 12 = -6$ $-x + x + 12 = -6 + x$ $12 = -6 + x$ $12 + 6 = -6 + 6 + x$ $18 = x$	16
21	$-x + \frac{3}{4}(20) = -6$ $-x + 15 = -6$ $-x + x + 15 = -6 + x$ $15 = -6 + x$ $15 + 6 = -6 + 6 + x$ $21 = x$	20

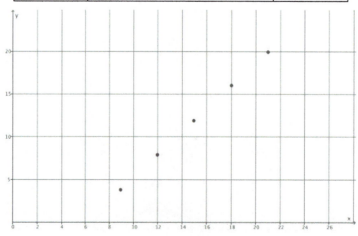

Lesson 12: Linear Equations in Two Variables

4. Find five solutions for the linear equation $2x + y = 5$, and plot the solutions as points on a coordinate plane.

x	Linear Equation: $2x + y = 5$	y
1	$2(1) + y = 5$ $2 + y = 5$ $y = 3$	3
2	$2(2) + y = 5$ $4 + y = 5$ $y = 1$	1
3	$2(3) + y = 5$ $6 + y = 5$ $y = -1$	-1
4	$2(4) + y = 5$ $8 + y = 5$ $y = -3$	-3
5	$2(5) + y = 5$ $10 + y = 5$ $y = -5$	-5

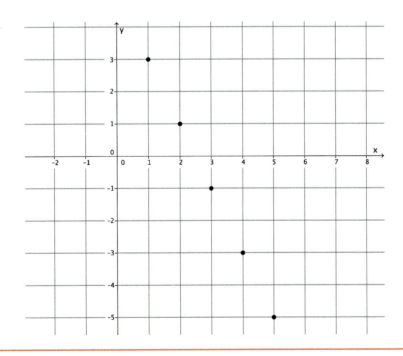

Lesson 12: Linear Equations in Two Variables

5. Find five solutions for the linear equation $3x - 5y = 15$, and plot the solutions as points on a coordinate plane.

x	Linear Equation: $3x - 5y = 15$	y
$\dfrac{20}{3}$	$3x - 5(1) = 15$ $3x - 5 = 15$ $3x - 5 + 5 = 15 + 5$ $3x = 20$ $\dfrac{3}{3}x = \dfrac{20}{3}$ $x = \dfrac{20}{3}$	1
$\dfrac{25}{3}$	$3x - 5(2) = 15$ $3x - 10 = 15$ $3x - 10 + 10 = 15 + 10$ $3x = 25$ $\dfrac{3}{3}x = \dfrac{25}{3}$ $x = \dfrac{25}{3}$	2
10	$3x - 5(3) = 15$ $3x - 15 = 15$ $3x - 15 + 15 = 15 + 15$ $3x = 30$ $x = 10$	3
$\dfrac{35}{3}$	$3x - 5(4) = 15$ $3x - 20 = 15$ $3x - 20 + 20 = 15 + 20$ $3x = 35$ $\dfrac{3}{3}x = \dfrac{35}{3}$ $x = \dfrac{35}{3}$	4
$\dfrac{40}{3}$	$3x - 5(5) = 15$ $3x - 25 = 15$ $3x - 25 + 25 = 15 + 25$ $3x = 40$ $\dfrac{3}{3}x = \dfrac{40}{3}$ $x = \dfrac{40}{3}$	5

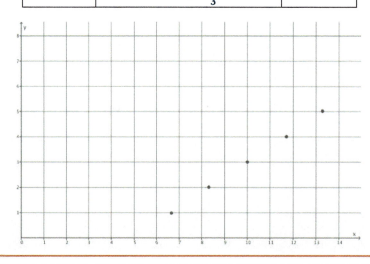

Lesson 12: Linear Equations in Two Variables

Lesson 13: The Graph of a Linear Equation in Two Variables

Student Outcomes

- Students predict the shape of a graph of a linear equation by finding and plotting solutions on a coordinate plane.
- Students informally explain why the graph of a linear equation is not curved in terms of solutions to the given linear equation.

Classwork

Discussion (20 minutes)

- In the last lesson, we saw that the solutions of a linear equation in two variables can be plotted on a coordinate plane as points. The collection of all points (x, y) in the coordinate plane so that each (x, y) is a solution of $ax + by = c$ is called the *graph* of $ax + by = c$.
- Do you think it is possible to plot *all* of the solutions of a linear equation on a coordinate plane?
  - *No, it is not possible. There are an infinite number of values we can use to fix one of the variables.*
- For that reason, we cannot draw the graph of a linear equation. What we can do is plot a few points of an equation and make predictions about what the graph should look like.
- Let's find five solutions to the linear equation $x + y = 6$ and plot the points on a coordinate plane. Name a solution.

As students provide solutions (samples provided below), organize them in an x-y table, as shown. It is most likely that students give whole number solutions only. Accept them for now.

x	y
0	6
1	5
2	4
3	3
4	2

- Now let's plot these points of the graph of $x + y = 6$ on a coordinate plane.

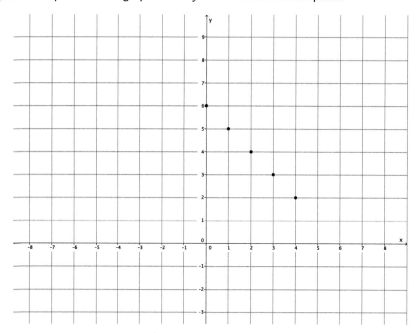

- Can you predict the shape of the graph of this linear equation based on just the five points we have so far?
  - *It looks like the points lie on a line.*
- Yes, at this point, it looks like the graph of the equation is a line, but for all we know, there can be some curved parts between some of these points.

For all we know, the graph of $x + y = 6$ could be the following curve. Notice that this curve passes through the selected five points.

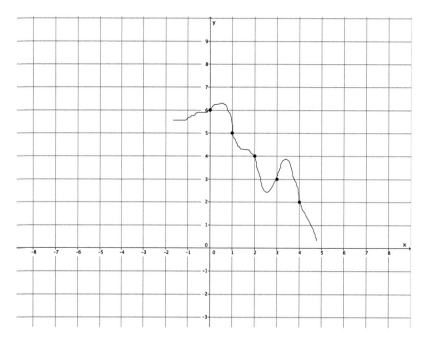

- The only thing we can do at this point is find more solutions that would show what the graph looks like between the existing points. That means we will have to look at some points with coordinates that are fractions. Name a solution that will plot as a point between the points we already have on the graph.

Add to the x-y table. Sample solutions are provided below.

x	y
0	6
1	5
2	4
3	3
4	2
$\frac{1}{2}$	$5\frac{1}{2}$
$1\frac{1}{2}$	$4\frac{1}{2}$
$2\frac{1}{2}$	$3\frac{1}{2}$
$3\frac{1}{2}$	$2\frac{1}{2}$
$4\frac{1}{2}$	$1\frac{1}{2}$

- Now let's add these points of the graph of $x + y = 6$ on our coordinate plane.

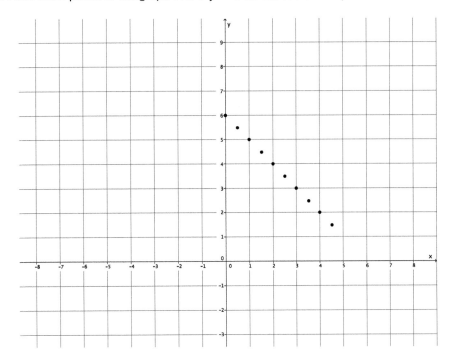

- Are you convinced that the graph of the linear equation $x + y = 6$ is a line? You shouldn't be! What if it looked like this?

Again, draw curves between each of the points on the graph to show what the graph could look like. Make sure to have curves in Quadrants II and IV to illustrate the need to find points that fit (or do not fit) the pattern of the graph when x- and y-values are negative.

- Now it is time to find more solutions. This time, we need to come up with solutions where the x-value is negative or the y-value is negative.

Add to the x-y table. Sample solutions are provided below.

x	y
0	6
1	5
2	4
3	3
4	2
$\frac{1}{2}$	$5\frac{1}{2}$
$1\frac{1}{2}$	$4\frac{1}{2}$
$2\frac{1}{2}$	$3\frac{1}{2}$
$3\frac{1}{2}$	$2\frac{1}{2}$
$4\frac{1}{2}$	$1\frac{1}{2}$
-1	7
-2	8
-3	9
7	-1
8	-2

- Now we have 15 solutions to the equation $x + y = 6$. Are you convinced that the graph of this equation is a line?

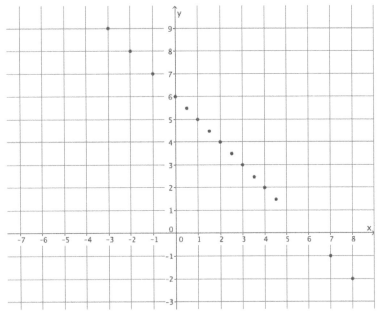

  - Students should say that they are still unsure; even though it looks like the graph is forming a line, there is a possibility for a curve between some of the existing points.

- Based on the look of the graph, does the point $\left(\frac{17}{3}, \frac{1}{3}\right)$ belong to the graph of the linear equation $x + y = 6$? Why or why not?
  - It looks like $\left(\frac{17}{3}, \frac{1}{3}\right)$ should be on the graph of the linear equation because it looks like the point would follow the pattern of the rest of the points (i.e., be on the line).

- Just by looking at the graph, we cannot be sure. We can predict that $\left(\frac{17}{3}, \frac{1}{3}\right)$ is on the graph of $x + y = 6$ because it looks like it should be. What can we do to find out for sure?
  - Since each point on the graph represents a solution to the equation and, if $\frac{17}{3} + \frac{1}{3} = 6$, which it does, then we can say for sure that $\left(\frac{17}{3}, \frac{1}{3}\right)$ is a point on the graph of $x + y = 6$.

- Just by looking at the graph, would you say that $\left(-2\frac{3}{7}, 4\right)$ is a point on the graph of $x + y = 6$?
  - Based on the graph, it does not look like the point $\left(-2\frac{3}{7}, 4\right)$ would be on the graph of $x + y = 6$.

- How do you know that the graph does not curve down at that point? How can you be sure?
  - I would have to see if $\left(-2\frac{3}{7}, 4\right)$ is a solution to $x + y = 6$; it is not a solution because $-2\frac{3}{7} + 4 = 1\frac{4}{7}$, not 6. Therefore, the point $\left(-2\frac{3}{7}, 4\right)$ is not a solution to the equation.

- At this point, we can predict that the graph of this linear equation is a line. Does that mean that the graph of every linear equation is a line? Might there be some linear equations so that the graphs of those linear equations are not lines? For now, our only method of proving or disproving our prediction is plotting solutions on a coordinate plane. The more we learn about linear equations, the better we will be able to answer the questions just asked.

A STORY OF RATIOS   Lesson 13   8•4

Exercises (15 minutes)

Students need graph paper to complete Exercises 1–2. Students work independently on Exercises 1–2 for the first 10 minutes. Then, they share their solutions with their partners and plot more points on their graphs. As students work, verify through discussion that they are choosing a variety of rational numbers to get a good idea of what the graph of the linear equation looks like. Exercise 6 is an optional exercise because it challenges students to come up with an equation that does not graph as a line.

Exercises

1. Find at least ten solutions to the linear equation $3x + y = -8$, and plot the points on a coordinate plane.

x	Linear Equation: $3x + y = -8$	y
1	$3(1) + y = -8$ $3 + y = -8$ $y = -11$	-11
$1\frac{1}{2}$	$3\left(1\frac{1}{2}\right) + y = -8$ $4\frac{1}{2} + y = -8$ $y = -12\frac{1}{2}$	$-12\frac{1}{2}$
2	$3(2) + y = -8$ $6 + y = -8$ $y = -14$	-14
3	$3(3) + y = -8$ $9 + y = -8$ $y = -17$	-17
$3\frac{1}{2}$	$3\left(3\frac{1}{2}\right) + y = -8$ $10\frac{1}{2} + y = -8$ $y = -18\frac{1}{2}$	$-18\frac{1}{2}$
4	$3(4) + y = -8$ $12 + y = -8$ $y = -20$	-20
-1	$3(-1) + y = -8$ $-3 + y = -8$ $y = -5$	-5
-2	$3(-2) + y = -8$ $-6 + y = -8$ $y = -2$	-2
-3	$3(-3) + y = -8$ $-9 + y = -8$ $y = 1$	1
-4	$3(-4) + y = -8$ $-12 + y = -8$ $y = 4$	4

Lesson 13:   The Graph of a Linear Equation in Two Variables

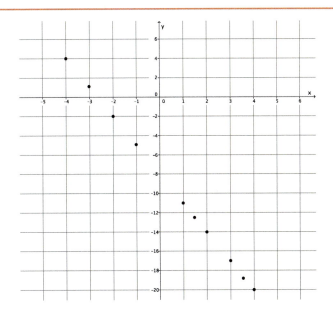

What shape is the graph of the linear equation taking?

The graph appears to be the shape of a line.

2. Find at least ten solutions to the linear equation $x - 5y = 11$, and plot the points on a coordinate plane.

x	Linear Equation: $x - 5y = 11$	y
$13\frac{1}{2}$	$x - 5\left(\frac{1}{2}\right) = 11$ $x - 2\frac{1}{2} = 11$ $x = 13\frac{1}{2}$	$\frac{1}{2}$
16	$x - 5(1) = 11$ $x - 5 = 11$ $x = 16$	1
$18\frac{1}{2}$	$x - 5\left(1\frac{1}{2}\right) = 11$ $x - 7\frac{1}{2} = 11$ $x = 18\frac{1}{2}$	$1\frac{1}{2}$
21	$x - 5(2) = 11$ $x - 10 = 11$ $x = 21$	2
$23\frac{1}{2}$	$x - 5\left(2\frac{1}{2}\right) = 11$ $x - 12\frac{1}{2} = 11$ $x = 23\frac{1}{2}$	$2\frac{1}{2}$

Lesson 13: The Graph of a Linear Equation in Two Variables

26	$x - 5(3) = 11$ $x - 15 = 11$ $x = 26$	3
6	$x - 5(-1) = 11$ $x + 5 = 11$ $x = 6$	-1
1	$x - 5(-2) = 11$ $x + 10 = 11$ $x = 1$	-2
-4	$x - 5(-3) = 11$ $x + 15 = 11$ $x = -4$	-3
-9	$x - 5(-4) = 11$ $x + 20 = 11$ $x = -9$	-4

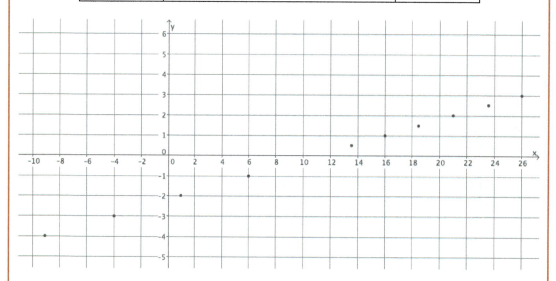

What shape is the graph of the linear equation taking?

The graph appears to be the shape of a line.

3. Compare the solutions you found in Exercise 1 with a partner. Add the partner's solutions to your graph. Is the prediction you made about the shape of the graph still true? Explain.

   Yes. With the additional points, the graph still appears to be the shape of a line.

4. Compare the solutions you found in Exercise 2 with a partner. Add the partner's solutions to your graph. Is the prediction you made about the shape of the graph still true? Explain.

   Yes. With the additional points, the graph still appears to be the shape of a line.

A STORY OF RATIOS   Lesson 13   8•4

5. Joey predicts that the graph of $-x + 2y = 3$ will look like the graph shown below. Do you agree? Explain why or why not.

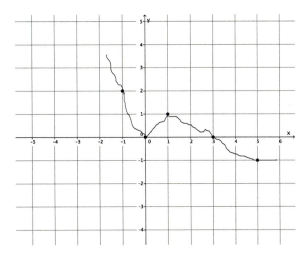

No, I do not agree with Joey. The graph that Joey drew contains the point $(0, 0)$. If $(0, 0)$ is on the graph of the linear equation, then it will be a solution to the equation; however, it is not. Therefore, the point cannot be on the graph of the equation, which means Joey's prediction is incorrect.

6. We have looked at some equations that appear to be lines. Can you write an equation that has solutions that do not form a line? Try to come up with one, and prove your assertion on the coordinate plane.

Answers will vary. Any nonlinear equation that students write will graph as something other than a line. For example, the graph of $y = x^2$ or the graph of $y = x^3$ will not be a line.

Closing (5 minutes)

Summarize, or ask students to summarize, the main points from the lesson:

- All of the graphs of linear equations we have done so far appear to take the shape of a line.
- We can show whether or not a point is on the graph of an equation by checking to see if it is a solution to the equation.

> **Lesson Summary**
>
> One way to determine if a given point is on the graph of a linear equation is by checking to see if it is a solution to the equation. Note that all graphs of linear equations appear to be lines.

Exit Ticket (5 minutes)

Lesson 13:   The Graph of a Linear Equation in Two Variables   163

Lesson 13: The Graph of a Linear Equation in Two Variables

Exit Ticket

1. Ethan found solutions to the linear equation $3x - y = 8$ and graphed them. What shape is the graph of the linear equation taking?

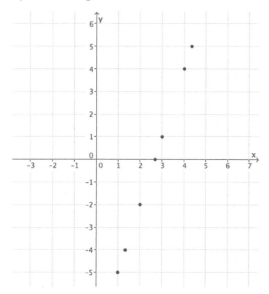

2. Could the following points be on the graph of $-x + 2y = 5$?

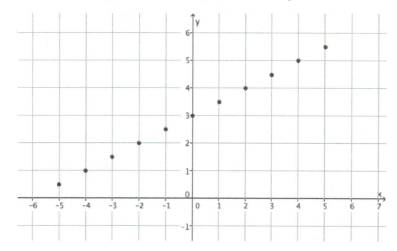

Exit Ticket Sample Solutions

1. Ethan found solutions to the linear equation $3x - y = 8$ and graphed them. What shape is the graph of the linear equation taking?

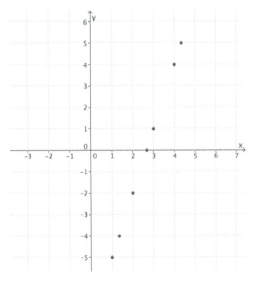

   It appears to take the shape of a line.

2. Could the following points be on the graph of $-x + 2y = 5$?

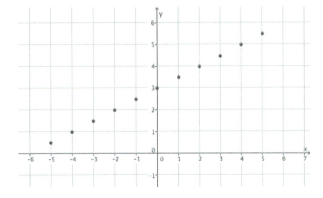

   Students may have chosen any point to make the claim that this is not the graph of the equation $-x + 2y = 5$.

   Although the graph appears to be a line, the graph contains the point $(0, 3)$. The point $(0, 3)$ is not a solution to the linear equation; therefore, this is not the graph of $-x + 2y = 5$.

   Note to teacher: Accept any point as not being a solution to the linear equation.

Problem Set Sample Solutions

In Problem 1, students graph linear equations by plotting points that represent solutions. For that reason, they need graph paper. Students informally explain why the graph of a linear equation is not curved by showing that a point on the curve is not a solution to the linear equation.

1. Find at least ten solutions to the linear equation $\frac{1}{2}x + y = 5$, and plot the points on a coordinate plane.

x	Linear Equation: $\frac{1}{2}x + y = 5$	y
0	$\frac{1}{2}(0) + y = 5$ $0 + y = 5$ $y = 5$	5
1	$\frac{1}{2}(1) + y = 5$ $\frac{1}{2} + y = 5$ $y = 4\frac{1}{2}$	$4\frac{1}{2}$
$1\frac{1}{2}$	$\frac{1}{2}\left(1\frac{1}{2}\right) + y = 5$ $\frac{3}{4} + y = 5$ $y = 4\frac{1}{4}$	$4\frac{1}{4}$
2	$\frac{1}{2}(2) + y = 5$ $1 + y = 5$ $y = 4$	4
3	$\frac{1}{2}(3) + y = 5$ $1\frac{1}{2} + y = 5$ $y = 3\frac{1}{2}$	$3\frac{1}{2}$
-1	$\frac{1}{2}(-1) + y = 5$ $-\frac{1}{2} + y = 5$ $y = 5\frac{1}{2}$	$5\frac{1}{2}$
$-1\frac{1}{2}$	$\frac{1}{2}\left(-1\frac{1}{2}\right) + y = 5$ $-\frac{3}{4} + y = 5$ $y = 5\frac{3}{4}$	$5\frac{3}{4}$
-2	$\frac{1}{2}(-2) + y = 5$ $-1 + y = 5$ $y = 6$	6

-3	$\frac{1}{2}(-3) + y = 5$ $-\frac{3}{2} + y = 5$ $y = 6\frac{1}{2}$	$6\frac{1}{2}$
$-3\frac{1}{2}$	$\frac{1}{2}\left(-3\frac{1}{2}\right) + y = 5$ $-\frac{7}{4} + y = 5$ $y = 6\frac{3}{4}$	$6\frac{3}{4}$

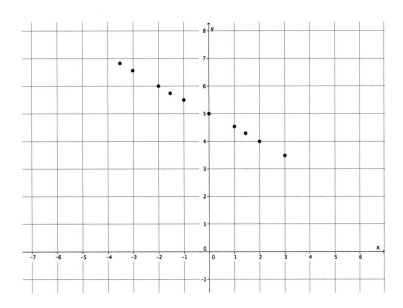

What shape is the graph of the linear equation taking?

The graph appears to be the shape of a line.

2. Can the following points be on the graph of the equation $x - y = 0$? Explain.

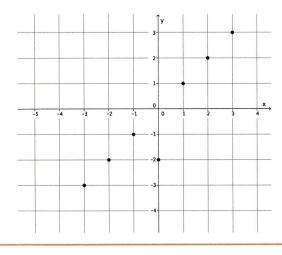

The graph shown contains the point $(0, -2)$. If $(0, -2)$ is on the graph of the linear equation, then it will be a solution to the equation. It is not; therefore, the point cannot be on the graph of the equation, which means the graph shown cannot be the graph of the equation $x - y = 0$.

Lesson 13: The Graph of a Linear Equation in Two Variables

3. Can the following points be on the graph of the equation $x + 2y = 2$? Explain.

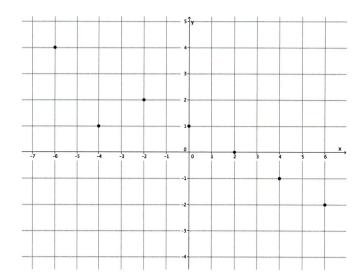

The graph shown contains the point $(-4, 1)$. If $(-4, 1)$ is on the graph of the linear equation, then it will be a solution to the equation. It is not; therefore, the point cannot be on the graph of the equation, which means the graph shown cannot be the graph of the equation $x + 2y = 2$.

4. Can the following points be on the graph of the equation $x - y = 7$? Explain.

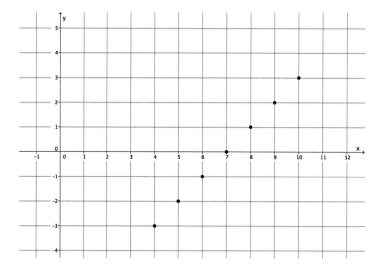

Yes, because each point on the graph represents a solution to the linear equation $x - y = 7$.

5. Can the following points be on the graph of the equation $x + y = 2$? Explain.

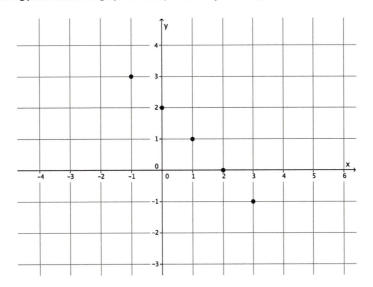

Yes, because each point on the graph represents a solution to the linear equation $x + y = 2$.

6. Can the following points be on the graph of the equation $2x - y = 9$? Explain.

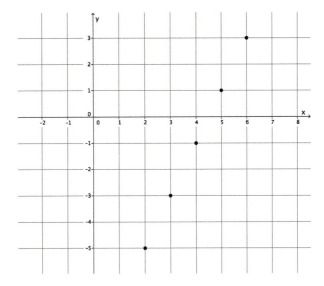

Yes, because each point on the graph represents a solution to the linear equation $2x - y = 9$.

7. Can the following points be on the graph of the equation $x - y = 1$? Explain.

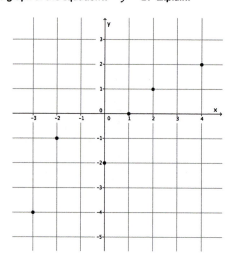

The graph shown contains the point $(-2, -1)$. If $(-2, -1)$ is on the graph of the linear equation, then it will be a solution to the equation. It is not; therefore, the point cannot be on the graph of the equation, which means the graph shown cannot be the graph of the equation $x - y = 1$.

A STORY OF RATIOS

Lesson 14 8•4

Lesson 14: The Graph of a Linear Equation—Horizontal and Vertical Lines

Student Outcomes

- Students graph linear equations in standard form, $ax + by = c$ (a or $b = 0$), that produce a horizontal or a vertical line.

Lesson Notes

The goal of this lesson is for students to know that the graph of an equation in the form of $x = c$ or $y = c$, where c is a constant, is the graph of a vertical or horizontal line, respectively. In order to show this, the lesson begins with linear equations in two variables, $ax + by = c$, where one of the coefficients of x or y is equal to zero. The reason behind this is that students know an ordered pair (x, y) is a point on a coordinate plane, as well as a solution to the equation $ax + by = c$. Frequently, when students see an equation in the form of $x = c$ or $y = c$, they think of it as just a number, not a point or a line. To avoid this, the approach to graphs of horizontal and vertical lines is embedded with what students already know about points and solutions to linear equations in two variables. In this lesson, students begin by graphing and exploring the connection between $ax + by = c$ and $x = c$ in the first three exercises, and then a discussion follows to solidify the concept that the graph of the linear equation $x = c$ is a vertical line. Similar exercises precede the discussion for the graph of $y = c$ as a horizontal line.

Classwork

Exercises 1–3 (5 minutes)

Students complete Exercises 1–3 independently or in pairs in preparation for the discussion that follows. Students need graph paper to complete the exercises.

Exercises

1. Find at least four solutions to graph the linear equation $1x + 2y = 5$.

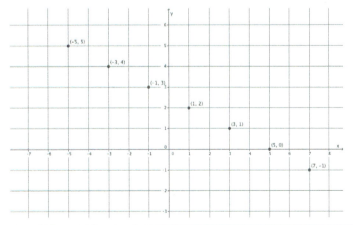

2. Find at least four solutions to graph the linear equation $1x + 0y = 5$.

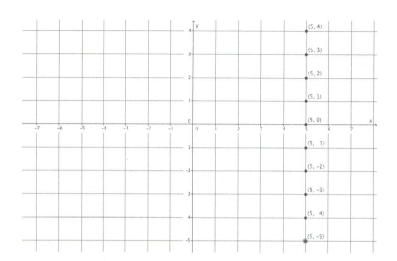

3. What was different about the equations in Exercises 1 and 2? What effect did this change have on the graph?

   In the first equation, the coefficient of y was 2. In the second equation, the coefficient of y was 0. The graph changed from being slanted to a vertical line.

Discussion (14 minutes)

- From Lesson 13, we can say that the graph of a linear equation in two variables looks like a line. We want to be able to prove that the graph is a line, not just predict. For that reason, we will begin with two special cases of linear equations in two variables.

  Given a linear equation in two variables $ax + by = c$, where a, b, and c are constants, we will look first at Case 1, where $a = 1$ and $b = 0$. (In the past, we have said that a and $b \neq 0$). To do some investigating, let's say that $c = 5$. Then we have the following equation:

  $$1 \cdot x + 0 \cdot y = 5.$$

- For the linear equation $1 \cdot x + 0 \cdot y = 5$, we want to find solutions, (x, y), to plot as points on a coordinate plane. How have we found solutions in prior lessons?
  - *We find solutions by picking a number for x or y and then solve for the other variable. The numbers (x, y) are a solution to the equation.*

- What happens if we pick 7 for x? Explain.
  - *If we pick 7 for x, then we get*

  $$1 \cdot 7 + 0 \cdot y = 5$$
  $$7 + 0 \cdot y = 5$$
  $$0 \cdot y = -2$$
  $$0 \neq -2.$$

  If we replace x with 7, then we get an untrue statement. Therefore, $(7, y)$ is not a solution to this linear equation.

Lesson 14: The Graph of a Linear Equation—Horizontal and Vertical Lines

- What happens if we pick 7 for y? Explain.
  - *If we pick 7 for y, then we get*
  $$1 \cdot x + 0 \cdot 7 = 5$$
  $$x + 0 = 5$$
  $$x = 5.$$
  If we replace y with 7, then we see that $x = 5$. Therefore, $(5, 7)$ is a solution to this linear equation.

- What happens if we pick -3 for y? Explain.
  - *If we pick -3 for y, then we get*
  $$1 \cdot x + 0 \cdot (-3) = 5$$
  $$x + 0 = 5$$
  $$x = 5.$$
  If we replace y with -3, then we see that $x = 5$. Therefore, $(5, -3)$ is a solution to this linear equation.

- What happens if we pick $\frac{1}{2}$ for y? Explain.
  - *If we pick $\frac{1}{2}$ for y, then we get*
  $$1 \cdot x + 0 \cdot \frac{1}{2} = 5$$
  $$x + 0 = 5$$
  $$x = 5.$$
  If we replace y with $\frac{1}{2}$, then we see that $x = 5$. Therefore, $\left(5, \frac{1}{2}\right)$ is a solution to this linear equation.

- What do you notice about the x-value each time we pick a number for y?
  - *Each time we pick a number for y, we keep getting $x = 5$.*

- Look at the equation again. Can we show that x must always be equal to 5?
  - *Yes. If we transform the equation, we see that $x = 5$:*
  $$1 \cdot x + 0 \cdot y = 5$$
  $$x + 0 = 5$$
  $$x = 5.$$

- What does that mean for our y-values? Which number will produce a solution where $x = 5$?
  - *Our y-values can be any number. No matter which number we pick for y, the value for x will always be equal to 5.*

- Let's graph the solutions we found and a few more.

Lesson 14: The Graph of a Linear Equation—Horizontal and Vertical Lines

Add points to the graph as a result of student responses to "If y is a number, then what value is x?" The graph should look similar to what is shown below.

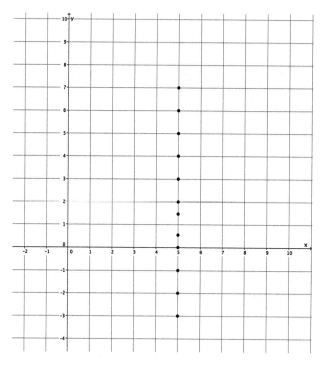

Focus in on one unit, between 2 and 3, for example, and explain that all of the fractional values for y between 2 and 3 produce a solution where $x = 5$.

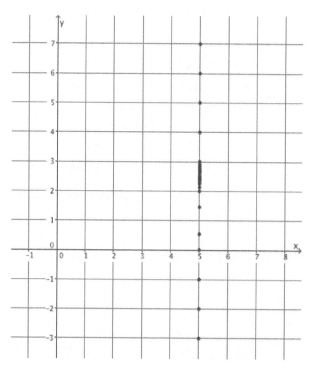

- If this process is repeated between all integers, then the result is the vertical line $x = 5$, as shown in blue.

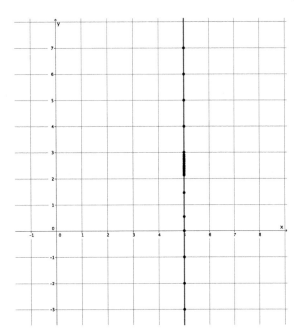

- The graph of the equation $ax + by = c$, where $a = 1$, $b = 0$, and $c = 5$, is the vertical line passing through point $(5, 0)$.
- The above situation is not unique. That is, in the equation $ax + by = c$, we chose the value for c to be 5. The same reasoning can be used for any value of c. If we chose c to be 6, what do you think the graph would look like?
    - *The graph would probably be a vertical line passing through the point $(6, 0)$.*
- If we chose c to be $-\frac{1}{2}$, what do you think the graph would look like?
    - *The graph would probably be a vertical line passing through the point $\left(-\frac{1}{2}, 0\right)$.*
- Notice that the equation $1 \cdot x + 0 \cdot y = c$ is equivalent to $x = c$. Therefore, we can make the following conclusion in the form of a theorem:

    THEOREM: The graph of $x = c$ is the vertical line passing through $(c, 0)$, where c is a constant.

- To prove this theorem, we first want to show that the graph of $x = c$ will intersect the x-axis. To show that the graph of $x = c$ intersects the x-axis, we actually have to show that the graph of $x = c$ is not parallel to the x-axis because if it were parallel, it would not intersect the x-axis. Therefore, if we can show that $x = c$ is not parallel to the x-axis, then it must intersect it at some point. How do we know that the graph of $x = c$ is not parallel to the x-axis?
    - *We know it is not parallel to the x-axis because it intersects the x-axis at $(c, 0)$.*
- Then the graph of $x = c$ must be parallel to the y-axis. This is because of how we define/set up the coordinate plane. The plane comprises horizontal lines parallel to the x-axis and vertical lines parallel to the y-axis. Since $x = c$ is not parallel to the x-axis, then it must be parallel to the y-axis.

- Now we need to show that $x = c$ is the graph of the only line that is parallel to the y-axis going through the point $(c, 0)$. How do we show that there is only one line parallel to the y-axis passing through a specific point?
  - That goes back to what we learned about basic rigid motions. If we rotate a line around a center $180°$ (in this particular case, we would rotate around the point $(2.5, 0)$), then we will get an image of the line parallel to the line itself. There exists only one line parallel to a given line that goes through a specific point.
- For that reason, we know that the graph of $x = c$ is the vertical line that passes through $(c, 0)$.

Exercises 4–6 (3 minutes)

Students complete Exercises 4–6 independently. Students need graph paper to complete the exercises.

4. Graph the linear equation $x = -2$.

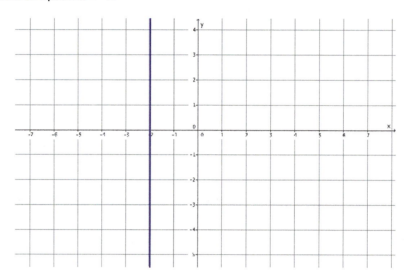

5. Graph the linear equation $x = 3$.

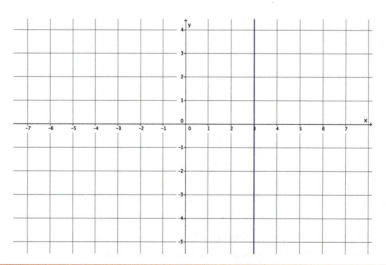

Lesson 14: The Graph of a Linear Equation—Horizontal and Vertical Lines

6. What will the graph of $x = 0$ look like?

   The graph of $x = 0$ will look like a vertical line that goes through the point $(0,0)$. It will be the same as the y-axis.

Exercises 7–9 (5 minutes)

Students complete Exercises 7–9 independently or in pairs in preparation for the discussion that follows. Students need graph paper to complete the exercises.

7. Find at least four solutions to graph the linear equation $2x + 1y = 2$.

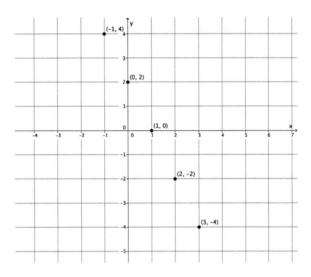

8. Find at least four solutions to graph the linear equation $0x + 1y = 2$.

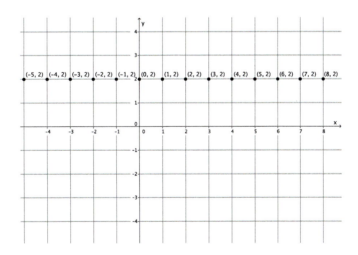

9. What was different about the equations in Exercises 7 and 8? What effect did this change have on the graph?

   In the first equation, the coefficient of x was 2. In the second equation, the coefficient of x was 0. The graph changed from being a slanted line to a horizontal line.

A STORY OF RATIOS — Lesson 14   8•4

Discussion (8 minutes)

- Now for Case 2. We need to look at the graph of the linear equation in two variables $ax + by = c$, where $a = 0$, $b = 1$, and c is a constant.
- Let's pick a number for c, as we did for Case 1. Let $c = 2$. Then we have the equation $0 \cdot x + 1 \cdot y = 2$.
- Now let's find a few solutions for this linear equation. What happens if we pick 7 for y? Explain.
  - *If we pick 7 for y, then we get*
  $$0 \cdot x + 1 \cdot 7 = 2$$
  $$0 + 7 = 2$$
  $$7 \neq 2.$$

  If we replace y with 7, then we get an untrue statement. Therefore, $(x, 7)$ is not a solution to this linear equation.
- If $x = 5$, what value does y have? Explain.
  - *The value of y is 2 because*
  $$0 \cdot 5 + 1 \cdot y = 2$$
  $$0 + y = 2$$
  $$y = 2.$$

  Therefore, $(5, 2)$ is a solution to the linear equation.
- If $x = -5$, what value does y have? Explain.
  - *The value of y is 2 because*
  $$0 \cdot (-5) + 1 \cdot y = 2$$
  $$0 + y = 2$$
  $$y = 2.$$

  Therefore, $(-5, 2)$ is a solution to the linear equation.
- If $x = \frac{1}{2}$, what value does y have? Explain.
  - *The value of y is 2 because*
  $$0 \cdot \frac{1}{2} + 1 \cdot y = 2$$
  $$0 + y = 2$$
  $$y = 2.$$

  Therefore, $\left(\frac{1}{2}, 2\right)$ is a solution to the linear equation.
- Do you see a similar pattern emerging for this linear equation? Explain.
  - *Yes. No matter which value we pick for x, the y-value is always 2. Therefore, the graph of $0 \cdot x + 1 \cdot y = 2$ is equivalent to $y = 2$.*
- What do you think the graph of $y = 2$ will look like?
  - *Students may predict that the graph of $y = 2$ is the horizontal line that goes through the point $(0, 2)$.*

Lesson 14: The Graph of a Linear Equation—Horizontal and Vertical Lines

As before, place the solutions found as points of a graph. Add points to the graph as a result of student responses to "If x is a number, then what value is y?" The graph should look something like what is shown below.

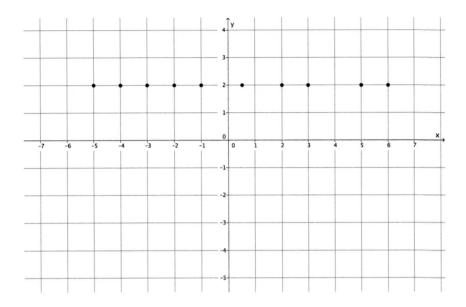

Again, focus in on one unit, between 5 and 6, for example, and explain that all of the fractional values for x between 5 and 6 produce a solution where $y = 2$.

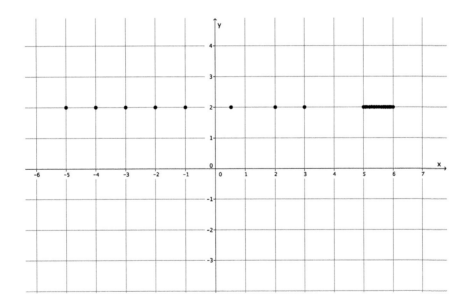

- If this process is repeated between all integers, then the result is the horizontal line $y = 2$, as shown in blue.

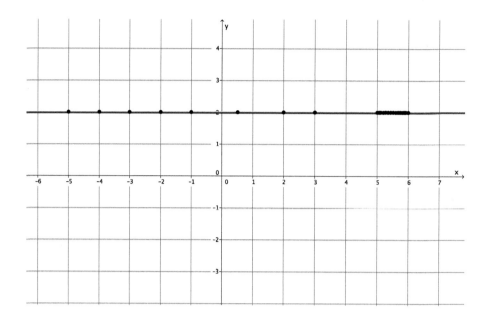

- The graph of the equation $ax + by = c$, where $a = 0$, $b = 1$, and $c = 2$, is the horizontal line passing through point $(0, 2)$.
- The above situation is not unique. As before in the equation $ax + by = c$, we chose the value for c to be 2. The same reasoning can be used for any value c. If we chose c to be 6, what do you think the graph would look like?
  - The graph would probably be a horizontal line passing through the point $(0, 6)$.
- If we chose c to be $-\frac{1}{2}$, what do you think the graph would look like?
  - The graph would probably be a horizontal line passing through the point $\left(0, -\frac{1}{2}\right)$.
- We can generalize this to a theorem:

THEOREM: The graph of $y = c$ is the horizontal line passing through the point $(0, c)$.

- We can also say that there is only one line with the equation $y = c$ whose graph is parallel to the x-axis that goes through the point $(0, c)$.
- The proofs of these statements are similar to the proofs for vertical lines.

Exercises 10–12 (3 minutes)

Students complete Exercises 10–12 independently. Students need graph paper to complete the exercises.

10. Graph the linear equation $y = -2$.

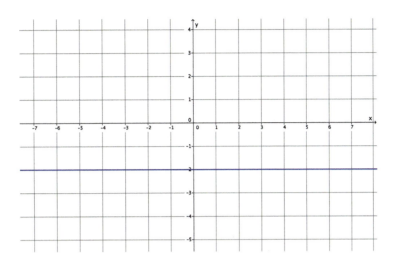

11. Graph the linear equation $y = 3$.

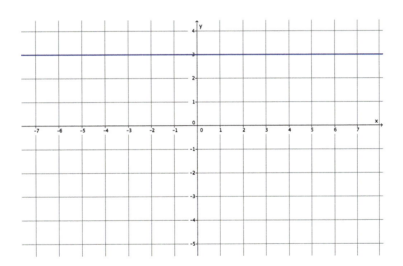

12. What will the graph of $y = 0$ look like?

    The graph of $y = 0$ will look like a horizontal line that goes through the point $(0, 0)$. It will be the same as the x-axis.

Closing (3 minutes)

Summarize, or ask students to summarize, the main points from the lesson:

- The graph of the linear equation in two variables $ax + by = c$, where $a = 1$ and $b = 0$, is the graph of the equation $x = c$. The graph of $x = c$ is the vertical line that passes through the point $(c, 0)$.
- The graph of the linear equation in two variables $ax + by = c$, where $a = 0$ and $b = 1$, is the graph of the equation $y = c$. The graph of $y = c$ is the horizontal line that passes through the point $(0, c)$.

Lesson Summary

In a coordinate plane with perpendicular x- and y-axes, a *vertical line* is either the y-axis or any other line parallel to the y-axis. The graph of the linear equation in two variables $ax + by = c$, where $a = 1$ and $b = 0$, is the graph of the equation $x = c$. The graph of $x = c$ is the vertical line that passes through the point $(c, 0)$.

In a coordinate plane with perpendicular x- and y-axes, a *horizontal line* is either the x-axis or any other line parallel to the x-axis. The graph of the linear equation in two variables $ax + by = c$, where $a = 0$ and $b = 1$, is the graph of the equation $y = c$. The graph of $y = c$ is the horizontal line that passes through the point $(0, c)$.

Exit Ticket (4 minutes)

Name _____ Date _____

Lesson 14: The Graph of a Linear Equation—Horizontal and Vertical Lines

Exit Ticket

1. Graph the linear equation $ax + by = c$, where $a = 0$, $b = 1$, and $c = 1.5$.

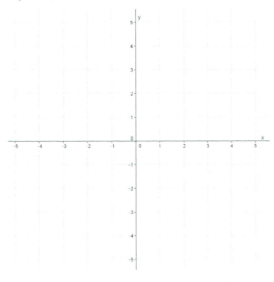

2. Graph the linear equation $ax + by = c$, where $a = 1$, $b = 0$, and $c = -\frac{5}{2}$.

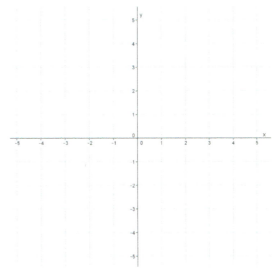

3. What linear equation represents the graph of the line that coincides with the x-axis?

4. What linear equation represents the graph of the line that coincides with the y-axis?

Exit Ticket Sample Solutions

1. Graph the linear equation $ax + by = c$, where $a = 0$, $b = 1$, and $c = 1.5$.

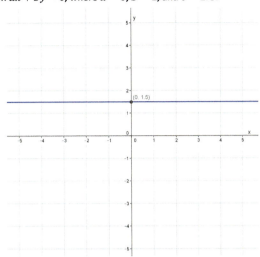

2. Graph the linear equation $ax + by = c$, where $a = 1$, $b = 0$, and $c = -\frac{5}{2}$.

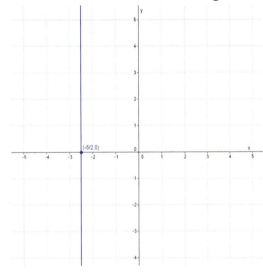

3. What linear equation represents the graph of the line that coincides with the x-axis?

   $y = 0$

4. What linear equation represents the graph of the line that coincides with the y-axis?

   $x = 0$

Problem Set Sample Solutions

Students need graph paper to complete the Problem Set.

1. Graph the two-variable linear equation $ax + by = c$, where $a = 0$, $b = 1$, and $c = -4$.

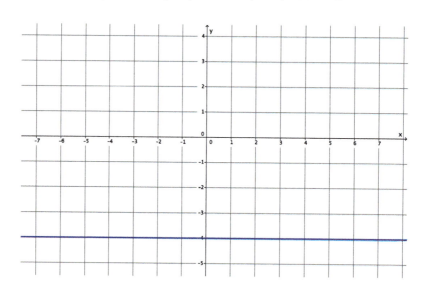

2. Graph the two-variable linear equation $ax + by = c$, where $a = 1$, $b = 0$, and $c = 9$.

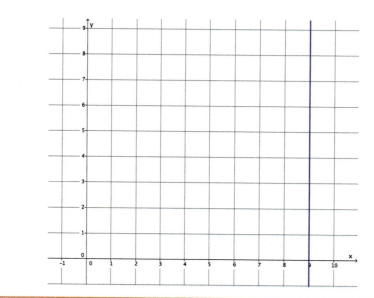

3. Graph the linear equation $y = 7$.

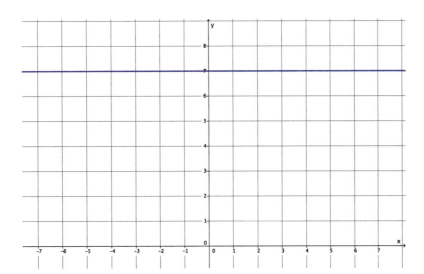

4. Graph the linear equation $x = 1$.

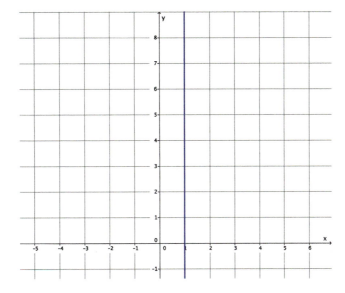

5. Explain why the graph of a linear equation in the form of $y = c$ is the horizontal line, parallel to the x-axis passing through the point $(0, c)$.

   The graph of $y = c$ passes through the point $(0, c)$, which means the graph of $y = c$ cannot be parallel to the y-axis because the graph intersects it. For that reason, the graph of $y = c$ must be the horizontal line parallel to the x-axis.

6. Explain why there is only one line with the equation $y = c$ that passes through the point $(0, c)$.

   There can only be one line parallel to another that goes through a given point. Since the graph of $y = c$ is parallel to the x-axis and it goes through the point $(0, c)$, then it must be the only line that does. Therefore, there is only one line that is the graph of the equation $y = c$ that passes through $(0, c)$.

Name _____ Date _____

1. Write and solve each of the following linear equations.

   a. Ofelia has a certain amount of money. If she spends $12, then she has $\frac{1}{5}$ of the original amount left. How much money did Ofelia have originally?

   b. Three consecutive integers have a sum of 234. What are the three integers?

   c. Gil is reading a book that has 276 pages. He already read some of it last week. He plans to read 20 pages tomorrow. By then, he will be $\frac{2}{3}$ of the way through the book. How many pages did Gil read last week?

2.

   a. Without solving, identify whether each of the following equations has a unique solution, no solution, or infinitely many solutions.

   i. $3x + 5 = -2$

   ii. $6(x - 11) = 15 - 4x$

   iii. $12x + 9 = 8x + 1 + 4x$

   iv. $2(x - 3) = 10x - 6 - 8x$

   v. $5x + 6 = 5x - 4$

   b. Solve the following equation for a number x. Verify that your solution is correct.

   $$-15 = 8x + 1$$

   c. Solve the following equation for a number x. Verify that your solution is correct.

   $$7(2x + 5) = 4x - 9 - x$$

3.

a. Parker paid $4.50 for three pounds of gummy candy. Assuming each pound of gummy candy costs the same amount, complete the table of values representing the cost of gummy candy in pounds.

Gummy Candy in Pounds (x)	1	2	3	4	5	6	7	8	9
Cost in Dollars (y)			4.50						

b. Graph the data on the coordinate plane.

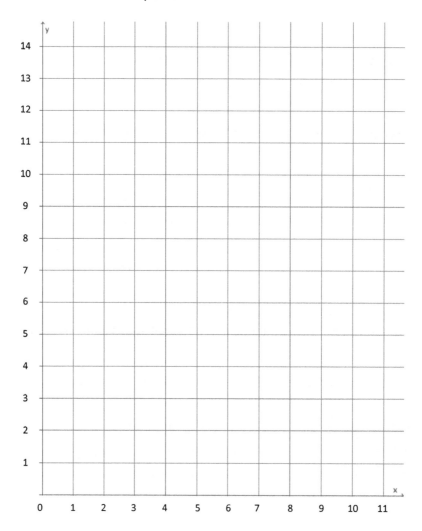

c. On the same day, Parker's friend, Peggy, was charged $5 for $1\frac{1}{2}$ lb. of gummy candy. Explain in terms of the graph why this must be a mistake.

A STORY OF RATIOS — Mid-Module Assessment Task 8•4

A Progression Toward Mastery					
Assessment Task Item		STEP 1 Missing or incorrect answer and little evidence of reasoning or application of mathematics to solve the problem.	STEP 2 Missing or incorrect answer but evidence of some reasoning or application of mathematics to solve the problem.	STEP 3 A correct answer with some evidence of reasoning or application of mathematics to solve the problem, OR an incorrect answer with substantial evidence of solid reasoning or application of mathematics to solve the problem.	STEP 4 A correct answer supported by substantial evidence of solid reasoning or application of mathematics to solve the problem.
1	a 8.EE.C.7b	Student makes no attempt to solve the problem or leaves the problem blank. OR Student may or may not have identified the variable.	Student does not set up an equation (i.e., guesses the answer). OR Student may or may not have identified the variable.	Student may or may not have set up the correct equation. OR Student may or may not have identified the variable. OR Student makes calculation errors.	Student identifies the variable as, "Let x be the amount of money Ofelia had," or something similar. AND Student sets up a correct equation, $x - 12 = \frac{1}{5}x$ or other equivalent version. AND Student solves for the variable correctly, $x = 15$.
	b 8.EE.C.7b	Student makes no attempt to solve the problem or leaves the problem blank. OR Student may or may not have identified the variable.	Student does not set up an equation (i.e., guesses the answer). OR Student may or may not have identified the variable. OR Student makes calculation errors. OR Student only answers part of the question, stating, for example, that the first number is 77, but does not give all three numbers.	Student attempts to set up an equation but may have set up an incorrect equation. OR Student may or may not have identified the variable. OR Student makes calculation errors. OR Student only answers part of the question, stating, for example, that the first number is 77, but does not give all three numbers.	Student identifies the variable as, "Let x be the first integer." AND Student sets up a correct equation, $3x + 3 = 234$ or other equivalent version. AND Student solves the equation correctly and identifies all three numbers correctly (i.e., 77, 78, and 79).

Module 4: Linear Equations

	c 8.EE.C.7b	Student makes no attempt to solve the problem or leaves the problem blank. OR Student may or may not have identified the variable.	Student does not set up an equation (i.e., guesses the answer). OR Student may or may not have identified the variable.	Student attempts to set up an equation but may have set up an incorrect equation. OR Student may or may not have identified the variable. OR Student makes calculation errors leading to an incorrect answer.	Student identifies the variable as, "Let x be the number of pages Gil read last week," or something similar. AND Student sets up a correct equation, $x + 20 = 184$ or other equivalent version. AND Student solves for the number of pages Gil read last week as 164 pages.
2	a 8.EE.C.7a	Student makes no attempt to determine the type of solution or leaves the problem blank. OR Student determines 0 of the solution types correctly. OR Student may have attempted to determine the solutions by solving.	Student determines 1–2 of the solution types correctly. OR Student may have attempted to determine the solutions by solving.	Student determines 3–5 of the solution types correctly. OR Student may have attempted to determine the solutions by solving.	Student determines 5 of the solution types correctly. Equations 1 and 2 have unique solutions, equation 3 has no solution, equation 4 has infinitely many solutions, and equation 5 has no solution. AND Student determines the solutions by observation only.
	b 8.EE.C.7b	Student makes no attempt to solve the problem or leaves the problem blank.	Student uses properties of equality incorrectly, (e.g., subtracts 1 from just one side of the equation or divides by 8 on just one side of the equation), leading to an incorrect solution.	Student correctly uses properties of rational numbers to solve the equation but makes a computational error leading to an incorrect solution. For example, student may have subtracted 1 from each side of the equation, but $-15 - 1$ led to an incorrect answer. Student may or may not have verified the answer.	Student correctly uses properties of rational numbers to solve the equation (i.e., finds $x = -2$). There is evidence that student verifies the solution.
	c 8.EE.C.7b	Student makes no attempt to solve the problem or leaves the problem blank.	Student uses the distributive property incorrectly on both sides of the equation (e.g., $7(2x + 5) = 14x + 5$ or $4x - x = 4$), leading to an incorrect solution.	Student uses the distributive property correctly on one or both sides of the equation but makes a computational error leading to an incorrect solution. Student may or may not have verified the answer.	Student uses the distributive property correctly on both sides of the equation leading to a correct solution (i.e., $x = -4$). There is evidence that student verifies the solution.

Module 4: Linear Equations

3	a 8.EE.B.5	Student makes no attempt to complete the table or uses completely random numbers in the blanks.	Student completes the table incorrectly but only because of a simple computational error in finding the cost of one pound of candy, leading to all other parts being incorrect.	Student completes 6–7 parts of the table correctly. A computational error leads to 1–2 parts being incorrect.	Student completes all 8 parts of the table correctly. (See the table below for the correct answers.)
			Gummy Candy in Pounds (x)	1	2	3	4	5	6	7	8	9
			Cost in Dollars (y)	1.50	3.00	4.50	6.00	7.50	9.00	10.50	12.00	13.50
	b 8.EE.B.5	Student makes no attempt to put the data on the graph, or points are graphed randomly.	Student plots data points on the graph but misplaces a few points. OR Student inverses the data (i.e., plots points according to (y, x) instead of (x, y)).	Student plots 6–8 data points correctly according to the data in the table.	Student plots all 9 data points correctly according to the data in the table.
	c 8.EE.B.5	Student leaves the problem blank.	Student performs a computation to prove the mistake. Little or no reference to the graph is made in the argument.	Student makes a weak argument as to why $(1.5, 5)$ could not be correct. Student may have connected the dots on the graph to show $(1.5, 5)$ could not be correct.	Student makes a convincing argument as to why the point $(1.5, 5)$ could not be correct. Student references the relationship being proportional and/or predicts that all points should fall into a line based on the existing pattern of points on the graph.

Module 4: Linear Equations

A STORY OF RATIOS   Mid-Module Assessment Task   8•4

Name _____   Date _____

1. Write and solve each of the following linear equations.

   a. Ofelia has a certain amount of money. If she spends $12, then she has $\frac{1}{5}$ of the original amount left. How much money did Ofelia have originally?

   LET x BE THE AMOUNT OF MONEY OFELIA HAD

   $x - 12 = \frac{1}{5}x$

   $x - \frac{1}{5}x - 12 + 12 = \frac{1}{5}x - \frac{1}{5}x + 12$

   $\frac{4}{5}x = 12$

   $x = 12 \cdot \frac{5}{4} = \frac{60}{4}$

   OFELIA HAD $15.00 ORIGINALLY.

   b. Three consecutive integers have a sum of 234. What are the three integers?

   LET x BE THE FIRST INTEGER.

   $x + x+1 + x+2 = 234$

   $3x + 3 = 234$

   $3x + 3 - 3 = 234 - 3$

   $3x = 231$

   $x = 77$

   THE INTEGERS ARE 77, 78, AND 79.

   c. Gil is reading a book that has 276 pages. He already read some of it last week. He plans to read 20 pages tomorrow. By then, he will be $\frac{2}{3}$ of the way through the book. How many pages did Gil read last week?

   LET x BE THE NUMBER OF PAGES GIL READ LAST WEEK.

   $x + 20 = \frac{2}{3}(276)$

   $x + 20 = 184$

   $x + 20 - 20 = 184 - 20$

   $x = 164$

   GIL READ 164 PAGES LAST WEEK.

Module 4: Linear Equations

2.

  a. Without solving, identify whether each of the following equations has a unique solution, no solution, or infinitely many solutions.

   i. $3x + 5 = -2$ UNIQUE

   ii. $6(x - 11) = 15 - 4x$ UNIQUE

   iii. $12x + 9 = 8x + 1 + 4x$ NO SOLUTION

   iv. $2(x - 3) = 10x - 6 - 8x$ INFINITELY MANY SOLUTIONS

   v. $5x + 6 = 5x - 4$ NO SOLUTION

  b. Solve the following equation for a number x. Verify that your solution is correct.

   $-15 = 8x + 1$
   $-1 \quad\quad -1$
   $\overline{-16 = 8x}$
   $\dfrac{-16}{8} = \dfrac{8x}{8}$
   $-2 = x$

   $-15 = 8(-2) + 1$
   $-15 = -16 + 1$
   $-15 = -15$

  c. Solve the following equation for a number x. Verify that your solution is correct.

   $7(2x + 5) = 4x - 9 - x$
   $14x + 35 = 4x - x - 9$
   $14x + 35 = 3x - 9$
   $14x - 3x + 35 = 3x - 3x - 9$
   $11x + 35 = -9$
   $11x + 35 - 35 = -9 - 35$
   $11x = -44$
   $x = -4$

   $7(2(-4) + 5) = 4(-4) - 9 - (-4)$
   $7(-8 + 5) = -16 - 9 + 4$
   $7(-3) = -25 + 4$
   $-21 = -21$

3.

a. Parker paid $4.50 for three pounds of gummy candy. Assuming each pound of gummy candy costs the same amount, complete the table of values representing the cost of gummy candy in pounds.

Gummy Candy in Pounds (x)	1	2	3	4	5	6	7	8	9
Cost in Dollars (y)	1.50	3.00	4.50	6.00	7.50	9.00	10.50	12.00	13.50

b. Graph the data on the coordinate plane.

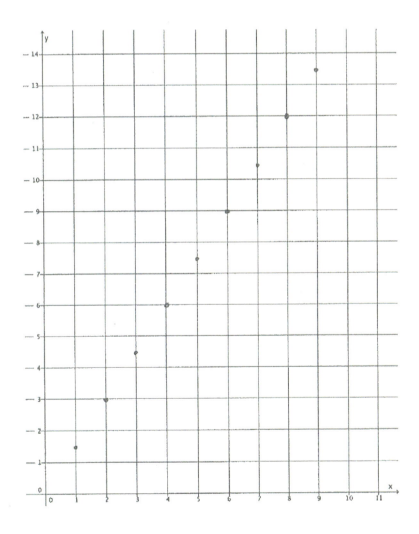

c. On the same day, Parker's friend, Peggy, was charged $5 for $1\frac{1}{2}$ lb. of gummy candy. Explain in terms of the graph why this must be a mistake.

Even though 1½ pounds of candy isn't a point on the graph, it is reasonable to believe it will fall in line with the other points. The cost of 1½ pounds of candy does not fit the pattern.

A STORY OF RATIOS

Mathematics Curriculum

GRADE 8 • MODULE 4

Topic C
Slope and Equations of Lines

8.EE.B.5, 8.EE.B.6

Focus Standards:	8.EE.B.5	Graph proportional relationships, interpreting the unit rate as the slope of the graph. Compare two different proportional relationships represented in different ways. *For example, compare a distance-time graph to a distance-time equation to determine which of two moving objects has greater speed.*
	8.EE.B.6	Use similar triangles to explain why the slope m is the same between any two distinct points on a non-vertical line in the coordinate plane; derive the equation $y = mx$ for a line through the origin and the equation $y = mx + b$ for a line intercepting the vertical axis at b.
Instructional Days:	9	
Lesson 15:	The Slope of a Non-Vertical Line (P)[1]	
Lesson 16:	The Computation of the Slope of a Non-Vertical Line (S)	
Lesson 17:	The Line Joining Two Distinct Points of the Graph $y = mx + b$ Has Slope m (S)	
Lesson 18:	There Is Only One Line Passing Through a Given Point with a Given Slope (P)	
Lesson 19:	The Graph of a Linear Equation in Two Variables Is a Line (S)	
Lesson 20:	Every Line Is a Graph of a Linear Equation (P)	
Lesson 21:	Some Facts About Graphs of Linear Equations in Two Variables (P)	
Lesson 22:	Constant Rates Revisited (P)	
Lesson 23:	The Defining Equation of a Line (E)	

Topic C begins with students examining the slope of non-vertical lines. Students relate what they know about unit rate in terms of the slope of the graph of a line (**8.EE.B.5**). In Lesson 16, students learn the formula for computing slope between any two points. Students reason that any two points on the same line can be used to determine slope because of what they know about similar triangles (**8.EE.B.6**). In Lesson 17, students transform the standard form of an equation into slope-intercept form. Further, students learn that the slope of a line joining any two distinct points is the graph of a linear equation with slope, m. In Lesson 18, students investigate the concept of uniqueness of a line and recognize that if two lines have the same slope and a common point, the two lines are the same.

---

[1]Lesson Structure Key: **P**-Problem Set Lesson, **M**-Modeling Cycle Lesson, **E**-Exploration Lesson, **S**-Socratic Lesson

Topic C: Slope and Equations of Lines

Lessons 19 and 20 prove to students that the graph of a linear equation is a line and that a line is a graph of a linear equation. In Lesson 21, students learn that the y-intercept is the location on the coordinate plane where the graph of a linear equation crosses the y-axis. Also in this lesson, students learn to write the equation of a line given the slope and a point. In Lesson 22, constant rate problems are revisited. Students learn that any constant rate problem can be described by a linear equation in two variables where the slope of the graph is the constant rate (i.e., rate of change). Lesson 22 also presents students with two proportional relationships expressed in different ways. Given a graph and an equation, students must use what they know about slope to determine which of the two has a greater rate of change. Lesson 23 introduces students to the symbolic representation of two linear equations that would graph as the same line.

Lesson 15: The Slope of a Non-Vertical Line

Student Outcomes

- Students know slope is a number that describes the steepness or slant of a line.
- Students interpret the unit rate as the slope of a graph.

Lesson Notes

In Lesson 13, some predictions were made about what the graph of a linear equation would look like. In all cases, it was predicted that the graph of a linear equation in two variables would be a line. In Lesson 14, students learned that the graph of the linear equation $x = c$ is the vertical line passing through the point $(c, 0)$, and the graph of the linear equation $y = c$ is the horizontal line passing through the point $(0, c)$.

We would like to prove that our predictions are true: The graph of a linear equation in two variables is a line. Before doing so, some tools are needed:

1. A number must be defined for each non-vertical line that can be used to measure the steepness or slant of the line. Once defined, this number is called the *slope* of the line. (Later, when a linear equation is being used to represent a function, it corresponds to the *rate of change* of the function.)

Rate of change is terminology that is used in later lessons in the context of linear functions. In this first exposure, slope, is characterized as a number that describes the *steepness* or *slant* of a line. In this lesson, students make observations about the steepness of a line. Further, students give directions about how to get from one point on the line to another point on the line. This leads students to the conclusion that the units in the directions have the same ratio. Students then compare ratios between graphs and describe lines as steeper or flatter.

2. It must be shown that *any* two points on a non-vertical line can be used to find the slope of the line.
3. It must be shown that the line joining two points on a graph of a linear equation of the form $y = mx + b$ has the slope m.
4. It must be shown that there is only one line passing through a given point with a given slope.

These tools are developed over the next few lessons. It is recommended that students are made aware of the four-part plan to achieve the goal of proving that the graph of a linear equation in two variables is a line. These parts are referenced in the next few lessons to help students make sense of the problem and persevere in solving it. In this lesson, students look specifically at what is meant by the terms *steepness* and *slant* by defining the slope of a line. This lesson defines slope in the familiar context of unit rate; that is, slope is defined when the horizontal distance between two points is fixed at one. Defining slope this way solidifies the understanding that the unit rate is the slope of a graph. Further, students see that the number that describes slope is the distance between the y-coordinates, leading to the general slope formula. This approach to slope owes a debt to the 2013 MPDI notes of Prof. Hung Hsi-Wu. It appears here with the professor's full permission.

A STORY OF RATIOS — Lesson 15 — 8•4

Classwork

Opening Exercise (8 minutes)

To develop a conceptual understanding of slope, have students complete the Opening Exercise where they make informal observations about the steepness of lines. Model for students how to answer the questions with the first pair of graphs. Then, have students work independently or in pairs to describe how to move from one point to another on a line in terms of units up or down and units right or left. Students also compare the ratios of their descriptions and relate the ratios to the steepness or flatness of the graph of the line.

Examine each pair of graphs, and answer the questions that follow.

Opening Exercise

Graph A   Graph B

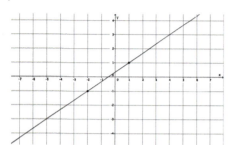 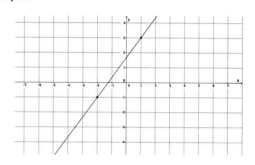

a. Which graph is steeper?

It looks like Graph B is steeper.

b. Write directions that explain how to move from one point on the graph to the other for both Graph A and Graph B.

For Graph A, move 2 units up and 3 units right. For Graph B, move 4 units up and 3 units right.

c. Write the directions from part (b) as ratios, and then compare the ratios. How does this relate to which graph was steeper in part (a)?

$\frac{2}{3} < \frac{4}{3}$. Graph B was steeper and had the greater ratio.

Lesson 15: The Slope of a Non-Vertical Line

Pair 1:

Graph A **Graph B**

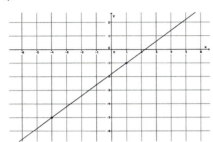 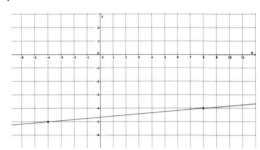

a. Which graph is steeper?

It looks like Graph A is steeper.

b. Write directions that explain how to move from one point on the graph to the other for both Graph A and Graph B.

For Graph A, move 4 units up and 5 units right. For Graph B, move 1 unit up and 12 units right.

c. Write the directions from part (b) as ratios, and then compare the ratios. How does this relate to which graph was steeper in part (a)?

$\frac{4}{5} > \frac{1}{12}$. Graph A was steeper and had the greater ratio.

Pair 2:

Graph A **Graph B**

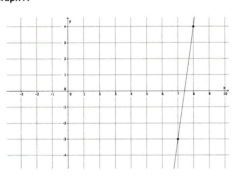 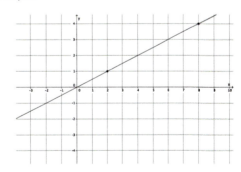

a. Which graph is steeper?

It looks like Graph A is steeper.

b. Write directions that explain how to move from one point on the graph to the other for both Graph A and Graph B.

For Graph A, move 7 units up and 1 unit right. For Graph B, move 3 units up and 6 units right.

c. Write the directions from part (b) as ratios, and then compare the ratios. How does this relate to which graph was steeper in part (a)?

$\frac{7}{1} > \frac{3}{6}$. Graph A was steeper and had the greater ratio.

Lesson 15: The Slope of a Non-Vertical Line

Pair 3:

Graph A

Graph B

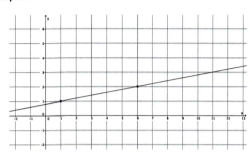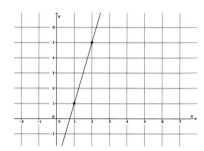

a. Which graph is steeper?

It looks like Graph B is steeper.

b. Write directions that explain how to move from one point on the graph to the other for both Graph A and Graph B.

For Graph A, move 1 unit up and 5 units right. For Graph B, move 4 units up and 1 unit right.

c. Write the directions from part (b) as ratios, and then compare the ratios. How does this relate to which graph was steeper in part (a)?

$\frac{1}{5} < \frac{4}{1}$. Graph B was steeper and had the greater ratio.

Pair 4:

Graph A

Graph B

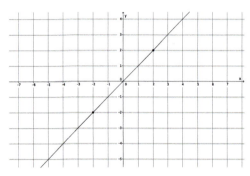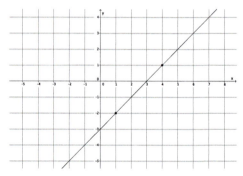

a. Which graph is steeper?

They look about the same steepness.

b. Write directions that explain how to move from one point on the graph to the other for both Graph A and Graph B.

For Graph A, move 4 units up and 4 units right. For Graph B, move 3 units up and 3 units right.

c. Write the directions from part (b) as ratios, and then compare the ratios. How does this relate to which graph was steeper in part (a)?

$\frac{4}{4} = \frac{3}{3}$. The graphs have equal ratios, which may explain why they look like the same steepness.

Example 1 (2 minutes)

- Putting horizontal lines off to the side for a moment, there are two other types of non-vertical lines. There are those that are *left-to-right inclining,* as in the graph of ℓ_1, and those that are *left-to-right declining*, as in the graph of ℓ_2. Both are shown below.

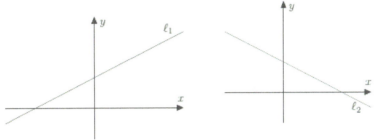

- We want to use a number to describe the amount of steepness or slant that each line has. The definition should be written in such a way that a horizontal line has a slope of 0, that is, no steepness and no slant.
- We begin by stating that lines in the coordinate plane that are *left-to-right inclining* are said to have a positive slope, and lines in the coordinate plane that are *left-to-right declining* are said to have a negative slope. We will discuss this more in a moment.

Example 2 (6 minutes)

- Now let's look more closely at finding the number that will be the slope of a line. Suppose a non-vertical line l is given in the coordinate plane. We let P be the point on line l that goes through the origin.

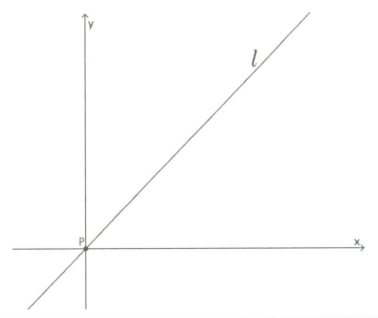

Lesson 15: The Slope of a Non-Vertical Line

- Next, we locate a point Q, exactly one unit to the right of point P.

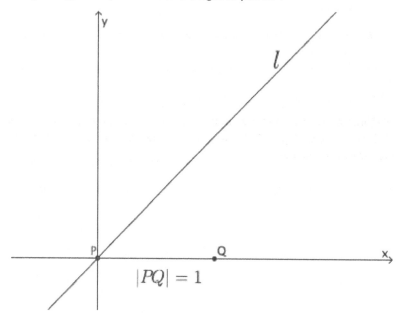

- Next, we draw a line, l_1, through point Q that is parallel to the y-axis. The point of intersection of line l and l_1 will be named point R.

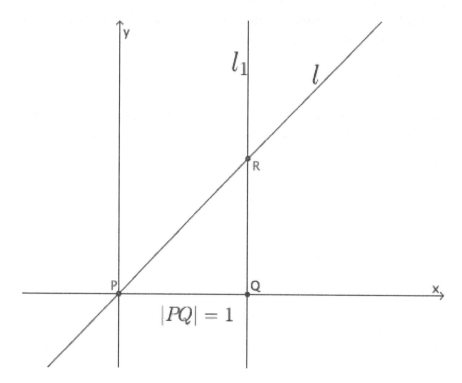

- Then, the slope of line l is the number, m, associated with the y-coordinate of point R.

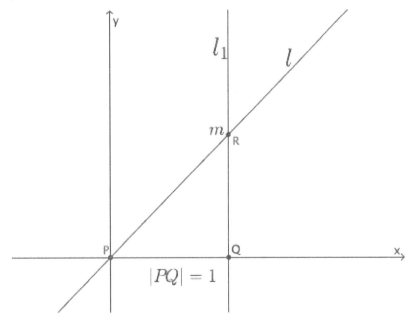

- The question remains, what is that number? Think of l_1 as a number line with point Q as zero; however many units R is from Q is the number we are looking for, the slope m. Another way of thinking about this is through our basic rigid motion translation. We know that translation preserves lengths of segments. If we translate everything in the plane one unit to the left so that point Q maps onto the origin, then the segment QR will coincide with the y-axis, and the y-coordinate of point R is the slope of the line.

If needed, students can use a transparency to complete the exercises in this lesson. Consider tracing the graph of the line and points P, Q, R onto a transparency. Demonstrate for students the translation along vector $\overrightarrow{QP}$ so that point Q is at the origin. Make clear that the translation moves point R to the y-axis, which is why the y-coordinate of point $R(0, m)$ is the number that represents the slope of the line, m.

- Let's look at an example with specific numbers. We have the same situation as just described. We have translated everything in the plane one unit to the left so that point Q maps onto the origin, and the segment QR coincides with the y-axis. What is the slope of this line?

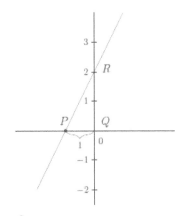

   □ The slope of the line is 2, or $m = 2$.

Lesson 15: The Slope of a Non-Vertical Line

- This explains why the slope of lines that are *left-to-right inclining* is positive. When we consider the number line, the point R is at positive 2; therefore, this line has a positive slope.

Example 3 (4 minutes)

- Suppose a non-vertical line is given in the coordinate plane. As before, we mark a point P on the line and go one unit to the right of P and mark point Q.

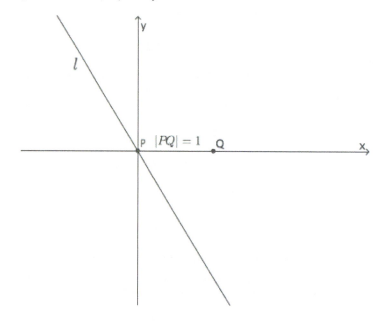

- Then, we draw a line, l_1, through point Q that is parallel to the y-axis.

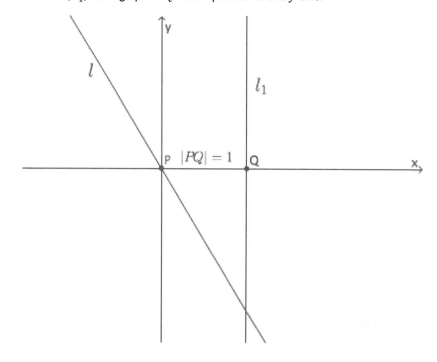

Lesson 15: The Slope of a Non-Vertical Line

- We mark the intersection of lines l and l_1 as point R. Again, recall that we consider the line l_1 to be a vertical number line where point Q is at zero. Then the number associated with the y-coordinate of point R is the slope of the line.

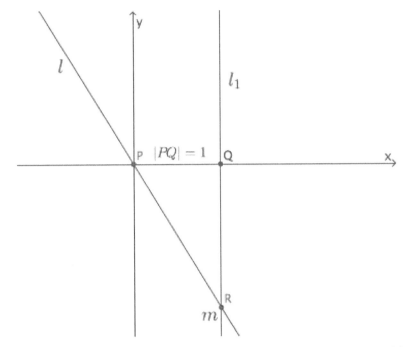

If needed, students can use a transparency to complete the exercises in this lesson. Again, consider tracing the graph of the line and points P, Q, R onto a transparency. Demonstrate for students the translation along vector $\overrightarrow{QP}$ so that point Q is at the origin. Make clear that the translation moves point R to the y-axis, which is why the y-coordinate of point $R(0, m)$ is the number that represents the slope of the line, m.

- Let's look at this example with specific numbers. We have the same situation as just described. We have translated everything in the plane one unit to the left so that point Q maps onto the origin, and the segment QR coincides with the y-axis. What is the slope of this line?

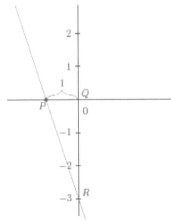

  ○ The slope of the line is -3, or $m = -3$.

- This explains why the slope of lines that are *left-to-right declining* is negative. When we consider the number line, the point R is at negative 3; therefore, this line has a negative slope.

Example 4 (4 minutes)

- Now we have a line l that does not go through the origin of the graph.

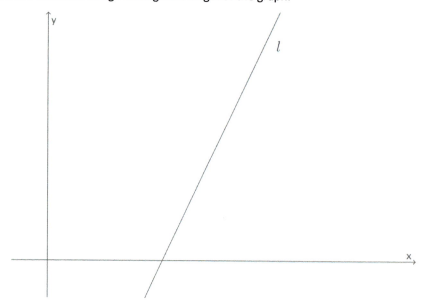

- Our process for finding slope changes only slightly. We will mark the point P at *any* location on the line l. Other than that, the work remains the same.

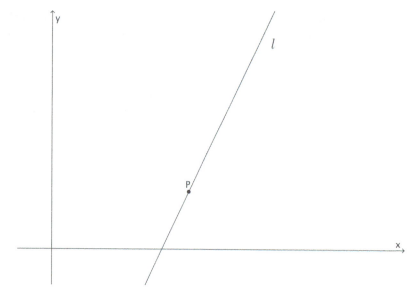

- We will go one unit to the right of P and mark point Q and then draw a line through Q parallel to the y-axis. We mark the intersection of lines l and l_1 as point R. Again, recall that we consider the line l_1 to be a vertical number line where point Q is at zero. Then the number associated with the y-coordinate of point R is the slope of the line.

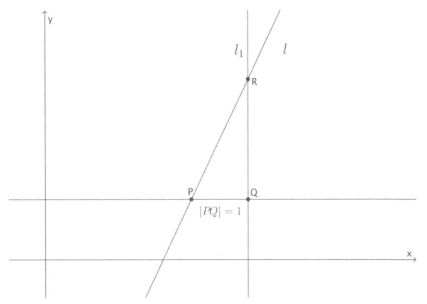

- Just as before, we translate so that point Q maps onto the origin, and the segment QR coincides with the y-axis.

If needed, students can use a transparency to complete the exercises in this lesson. Again, consider tracing the graph of the line and points P, Q, R onto a transparency. Demonstrate for students the translation so that point Q is at the origin (along a vector from Q to the origin). Make clear that the translation moves point R to the y-axis, which is why the y-coordinate of point $R(0, m)$ is the number that represents the slope of the line, m.

- What is the slope of this line?

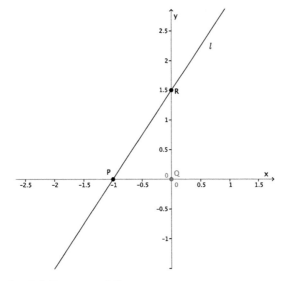

   □ The slope of the line is 1.5, or $m = 1.5$.

A STORY OF RATIOS   Lesson 15   8•4

- In general, we describe slope as an integer or a fraction. Given that we know the y-coordinate of R is 1.5, how can we express that number as a fraction?
  - The slope of the line is $\frac{3}{2}$, or $m = \frac{3}{2}$.

Exercises (5 minutes)

Students complete Exercises 1–6 independently. The exercises are referenced in the discussion that follows.

Exercises

Use your transparency to find the slope of each line if needed.

1. What is the slope of this non-vertical line?

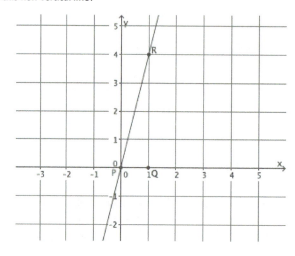

The slope of this line is 4, $m = 4$.

2. What is the slope of this non-vertical line?

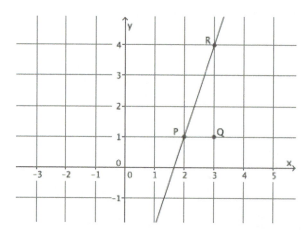

The slope of this line is 3, $m = 3$.

Lesson 15: The Slope of a Non-Vertical Line

3. Which of the lines in Exercises 1 and 2 is steeper? Compare the slopes of each of the lines. Is there a relationship between steepness and slope?

   The graph in Exercise 1 seems steeper. The slopes are 4 and 3. It seems like the greater the slope, the steeper the line.

4. What is the slope of this non-vertical line?

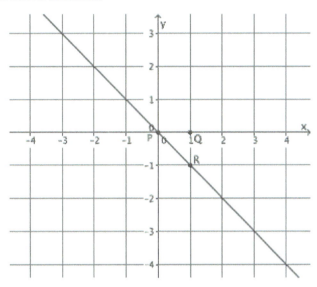

   The slope of this line is -1, $m = -1$.

5. What is the slope of this non-vertical line?

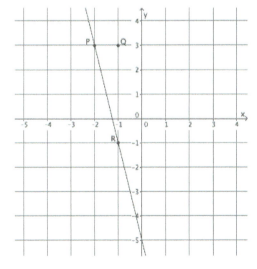

   The slope of this line is -4, $m = -4$.

A STORY OF RATIOS — Lesson 15 — 8•4

6. What is the slope of this non-vertical line?

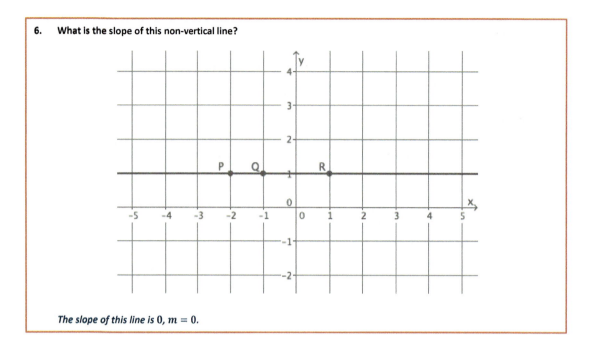

The slope of this line is 0, $m = 0$.

Discussion (6 minutes)

- When we began, we said that we wanted to define slope in a way so that a horizontal line would have a slope of 0 because a horizontal line has no steepness or slant. What did you notice in Exercise 6?
    - *The slope of the horizontal line was zero, like we wanted it to be.*
- In Exercise 3, you were asked to compare the steepness of the graphs of two lines and then compare their slopes. What did you notice?
    - *The steeper the line, the greater the number that describes the slope. For example, Exercise 1 had the steeper line and the greater slope.*
- Does this same relationship exist for lines with negative slopes? Look specifically at Exercises 4 and 5.

Provide students a minute or two to look back at Exercises 4 and 5. They should draw the conclusion that the absolute value of the slopes of lines that are *left-to-right* declining determines which is steeper. Use the points below to bring this fact to light.

- - *A similar relationship exists. The line in Exercise 5 was steeper with a slope of -4. The line in Exercise 4 had a slope of -1. Based on our previous reasoning, we would say that because $-1 > -4$, the line with a slope of -1 would have more steepness, but this is not the case.*
- We want to generalize the idea of steepness. When the slopes are positive, we expect the line with greater steepness to have the greater slope. When the slopes are negative, it is actually the smaller number that has more steepness. Is there a way that we can compare the slopes so that our reasoning is consistent? We want to say that a line is steeper than another when the number that describes the slope is larger than the other. How can we describe that mathematically, specifically when the slopes are negative?
    - *We can compare just the absolute value of the slopes. That way, we can say that the steeper line will be the slope with the greater absolute value.*

Lesson 15: The Slope of a Non-Vertical Line

Lesson 15

Example 5 (4 minutes)

- Let's take another look at one of the proportional relationships that we graphed in Lesson 11. Here is the problem and the work that we did.

    Pauline mows a lawn at a constant rate. Suppose she mowed a 35-square-foot lawn in 2.5 minutes.

t (time in minutes)	Linear Equation: $y = \dfrac{35}{2.5}t$	y (area in square feet)
0	$y = \dfrac{35}{2.5}(0)$	0
1	$y = \dfrac{35}{2.5}(1)$	$\dfrac{35}{2.5} = 14$
2	$y = \dfrac{35}{2.5}(2)$	$\dfrac{70}{2.5} = 28$
3	$y = \dfrac{35}{2.5}(3)$	$\dfrac{105}{2.5} = 42$
4	$y = \dfrac{35}{2.5}(4)$	$\dfrac{140}{2.5} = 56$

    Now, if we assume that the points we plot on the coordinate plane make a line, and the origin of the graph is point P, then we have the following graph.

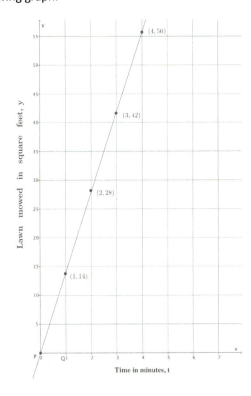

Lesson 15: The Slope of a Non-Vertical Line

- What is the slope of this line? Explain.
  - One unit to the right of point P is the point Q. That makes the point $(1, 14)$ the location of point R. Therefore, the slope of this line is 14, $m = 14$.
- What is the unit rate of mowing the lawn?
  - Pauline's unit rate of mowing a lawn is 14 square feet per 1 minute.
- When we graph proportional relationships, the unit rate is interpreted as the slope of the graph of the line, which is why slope is referred to as the rate of change.

Closing (3 minutes)

Summarize, or ask students to summarize, the main points from the lesson:

- We know that slope is a number that describes the steepness of a line.
- We know that lines that are left-to-right inclining have a positive slope, and lines that are left-to-right declining have a negative slope.
- We can find the slope of a line by looking at the graph's unit rate.

Lesson Summary

Slope is a number that can be used to describe the steepness of a line in a coordinate plane. The slope of a line is often represented by the symbol m.

Lines in a coordinate plane that are *left-to-right inclining* have a positive slope, as shown below.

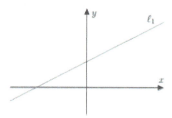

Lines in a coordinate plane that are *left-to-right declining* have a negative slope, as shown below.

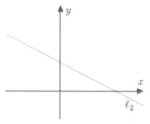

Determine the slope of a line when the horizontal distance between points is fixed at 1 by translating point Q to the origin of the graph and then identifying the y-coordinate of point R; by definition, that number is the slope of the line.

Lesson 15: The Slope of a Non-Vertical Line

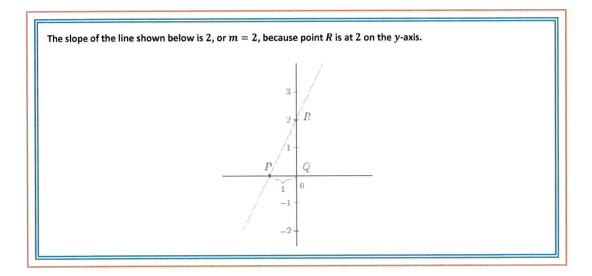

The slope of the line shown below is 2, or $m = 2$, because point R is at 2 on the y-axis.

Exit Ticket (3 minutes)

Lesson 15: The Slope of a Non-Vertical Line

A STORY OF RATIOS

Lesson 15

Name _____ Date _____

Lesson 15: The Slope of a Non-Vertical Line

Exit Ticket

1. What is the slope of this non-vertical line? Use your transparency if needed.

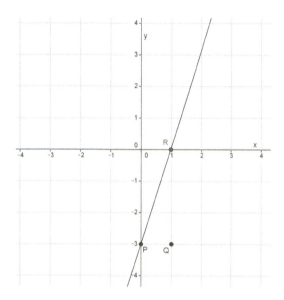

2. What is the slope of this non-vertical line? Use your transparency if needed.

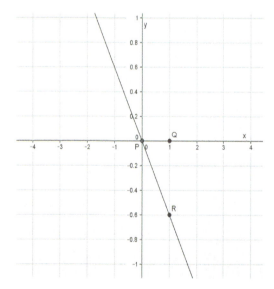

Exit Ticket Sample Solutions

1. What is the slope of this non-vertical line? Use your transparency if needed.

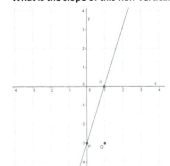

The slope of the line is 3, $m = 3$.

2. What is the slope of this non-vertical line? Use your transparency if needed.

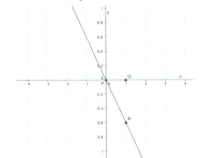

The slope of the line is -0.6, which is equal to $-\frac{3}{5}$, $m = -\frac{3}{5}$.

Problem Set Sample Solutions

Students practice identifying lines as having positive or negative slope. Students interpret the unit rate of a graph as the slope of the graph.

1. Does the graph of the line shown below have a positive or negative slope? Explain.

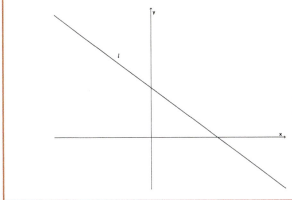

The graph of this line has a negative slope. First of all, it is left-to-right declining, which is an indication of negative slope. Also, if we were to mark a point P and a point Q one unit to the right of P and then draw a line parallel to the y-axis through Q, then the intersection of the two lines would be below Q, making the number that represents slope negative.

Lesson 15: The Slope of a Non-Vertical Line

2. Does the graph of the line shown below have a positive or negative slope? Explain.

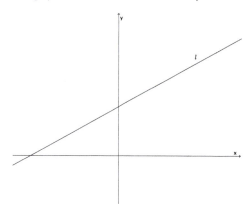

The graph of this line has a positive slope. First of all, it is left-to-right inclining, which is an indication of positive slope. Also, if we were to mark a point P and a point Q one unit to the right of P and then draw a line parallel to the y-axis through Q, then the intersection of the two lines would be above Q, making the number that represents slope positive.

3. What is the slope of this non-vertical line? Use your transparency if needed.

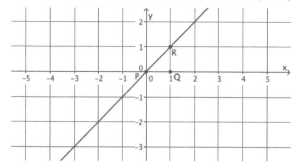

The slope of this line is 1, $m = 1$.

4. What is the slope of this non-vertical line? Use your transparency if needed.

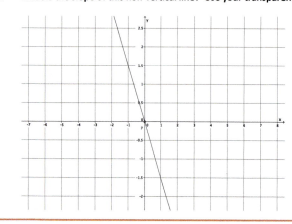

The slope of this line is $-\frac{3}{2}$, $m = -\frac{3}{2}$.

Lesson 15: The Slope of a Non-Vertical Line

5. What is the slope of this non-vertical line? Use your transparency if needed.

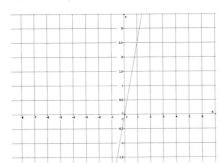

The slope of this line is $\dfrac{5}{2}$, $m = \dfrac{5}{2}$.

6. What is the slope of this non-vertical line? Use your transparency if needed.

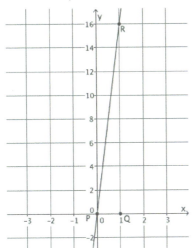

The slope of this line is 16, $m = 16$.

7. What is the slope of this non-vertical line? Use your transparency if needed.

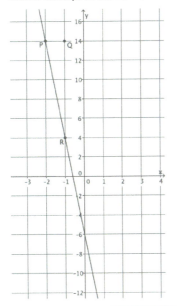

The slope of this line is -10, $m = -10$.

8. What is the slope of this non-vertical line? Use your transparency if needed.

The slope of this line is -5, $m = -5$.

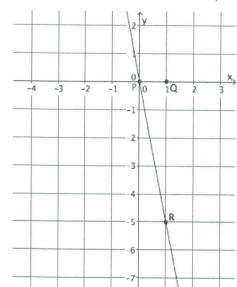

9. What is the slope of this non-vertical line? Use your transparency if needed.

The slope of this line is 2, $m = 2$.

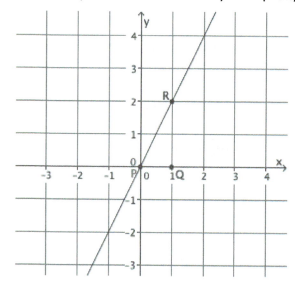

Lesson 15: The Slope of a Non-Vertical Line

10. What is the slope of this non-vertical line? Use your transparency if needed.

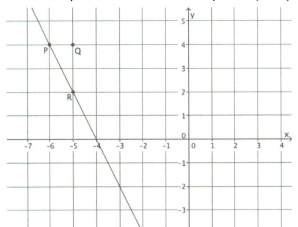

The slope of this line is -2, $m = -2$.

11. What is the slope of this non-vertical line? Use your transparency if needed.

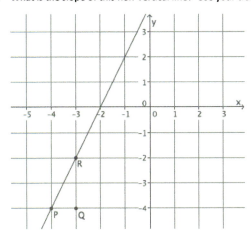

The slope of this line is 2, $m = 2$.

12. What is the slope of this non-vertical line? Use your transparency if needed.

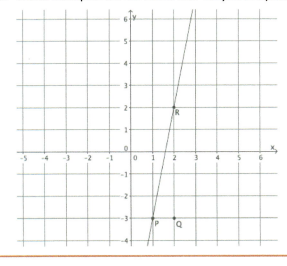

The slope of this line is 5, $m = 5$.

Lesson 15: The Slope of a Non-Vertical Line

13. What is the slope of this non-vertical line? Use your transparency if needed.

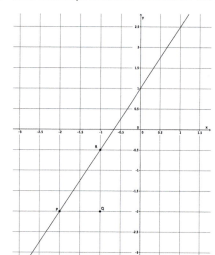

The slope of this line is $\frac{3}{2}$, $m = \frac{3}{2}$.

14. What is the slope of this non-vertical line? Use your transparency if needed.

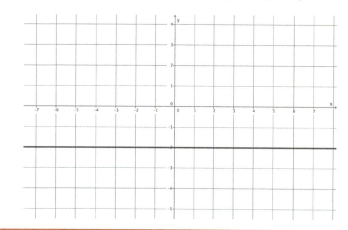

The slope of this line is 0, $m = 0$.

In Lesson 11, you did the work below involving constant rate problems. Use the table and the graphs provided to answer the questions that follow.

15. Suppose the volume of water that comes out in three minutes is 10.5 gallons.

t (time in minutes)	Linear Equation: $V = \dfrac{10.5}{3} t$	V (in gallons)
0	$V = \dfrac{10.5}{3}(0)$	0
1	$V = \dfrac{10.5}{3}(1)$	$\dfrac{10.5}{3} = 3.5$
2	$V = \dfrac{10.5}{3}(2)$	$\dfrac{21}{3} = 7$
3	$V = \dfrac{10.5}{3}(3)$	$\dfrac{31.5}{3} = 10.5$
4	$V = \dfrac{10.5}{3}(4)$	$\dfrac{42}{3} = 14$

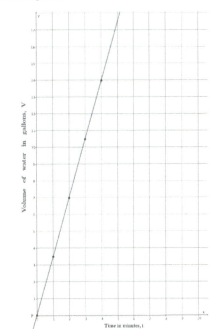

a. How many gallons of water flow out of the faucet per minute? In other words, what is the unit rate of water flow?

The unit rate of water flow is 3.5 gallons per minute.

b. Assume that the graph of the situation is a line, as shown in the graph. What is the slope of the line?

The slope of the line is 3.5, $m = 3.5$.

16. Emily paints at a constant rate. She can paint 32 square feet in five minutes.

t (time in minutes)	Linear Equation: $A = \dfrac{32}{5}t$	A (area painted in square feet)
0	$A = \dfrac{32}{5}(0)$	0
1	$A = \dfrac{32}{5}(1)$	$\dfrac{32}{5} = 6.4$
2	$A = \dfrac{32}{5}(2)$	$\dfrac{64}{5} = 12.8$
3	$A = \dfrac{32}{5}(3)$	$\dfrac{96}{5} = 19.2$
4	$A = \dfrac{32}{5}(4)$	$\dfrac{128}{5} = 25.6$

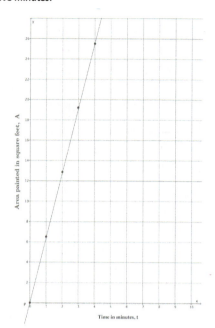

a. How many square feet can Emily paint in one minute? In other words, what is her unit rate of painting?

The unit rate at which Emily paints is 6.4 square feet per minute.

b. Assume that the graph of the situation is a line, as shown in the graph. What is the slope of the line?

The slope of the line is 6.4, $m = 6.4$.

17. A copy machine makes copies at a constant rate. The machine can make 80 copies in $2\frac{1}{2}$ minutes.

t (time in minutes)	Linear Equation: $n = 32t$	n (number of copies)
0	$n = 32(0)$	0
0.25	$n = 32(0.25)$	8
0.5	$n = 32(0.5)$	16
0.75	$n = 32(0.75)$	24
1	$n = 32(1)$	32

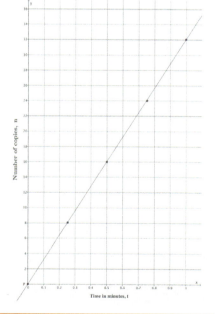

a. How many copies can the machine make each minute? In other words, what is the unit rate of the copy machine?

The unit rate of the copy machine is 32 copies per minute.

b. Assume that the graph of the situation is a line, as shown in the graph. What is the slope of the line?

The slope of the line is 32, $m = 32$.

Lesson 15: The Slope of a Non-Vertical Line

Lesson 16: The Computation of the Slope of a Non-Vertical Line

Student Outcomes

- Students use similar triangles to explain why the slope m is the same between any two distinct points on a non-vertical line in the coordinate plane.
- Students use the slope formula to compute the slope of a non-vertical line.

Lesson Notes

Throughout the lesson, the phrase *rate of change* is used in addition to *slope*. The goal is for students to know that these are references to the same thing with respect to the graph of a line. At this point in students' learning, the phrases *rate of change* and *slope* are interchangeable, but it could be said that rate of change refers to constant rate problems where time is involved, i.e., rate of change over time. In Module 5, when students work with nonlinear functions, they learn that the rate of change is not always constant as it is with linear equations and linear functions.

The points $P(p_1, p_2)$ and $R(r_1, r_2)$ are used throughout the lesson in order to make clear that students are looking at two distinct points. Using points P and R should decrease the chance of confusion compared to using the traditional (x_1, y_1) and (x_2, y_2). When considering what this this looks like in the formula, it should be clear that distinguishing the points by using different letters clarifies for students that they, in fact, have two distinct points. It is immediately recognizable that $m = \frac{r_2 - p_2}{p_1 - r_1}$ is written incorrectly compared to the traditional way of seeing the slope formula. Further, there should be less mixing up of the coordinates in the formula when it is presented with P and R.

There are several ways of representing the slope formula, each of which has merit.

$$m = \frac{p_2 - r_2}{p_1 - r_1}, \quad m = \frac{r_2 - p_2}{r_1 - p_1}, \quad m = \frac{\text{rise}}{\text{run}}, \quad m = \frac{\text{difference in } y\text{-values}}{\text{difference in } x\text{-values}}$$

Please make clear to students throughout this and subsequent lessons that no matter how slope is represented, it should result in the same value for a given line.

A STORY OF RATIOS   Lesson 16 8•4

Classwork

Example 1 (1 minute)

This example requires students to find the slope of a line where the horizontal distance between two points with integer coordinates is fixed at 1.

Example 1

Using what you learned in the last lesson, determine the slope of the line with the following graph.

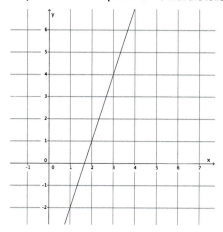

The slope of the line is 3.

Example 2 (1 minute)

This example requires students to find the slope of a line where the horizontal distance between two points with integer coordinates is fixed at 1.

Example 2

Using what you learned in the last lesson, determine the slope of the line with the following graph.

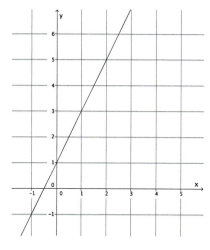

The slope of this line is 2.

---

228   Lesson 16:   The Computation of the Slope of a Non-Vertical Line

Example 3 (3 minutes)

This example requires students to find the slope of a line where the horizontal distance can be fixed at one, but determining the slope is difficult because it is not an integer. The point of this example is to make it clear to students that they need to develop a strategy that allows them to determine the slope of a line no matter what the horizontal distance is between the two points that are selected.

> **Example 3**
>
> What is different about this line compared to the last two examples?
>
>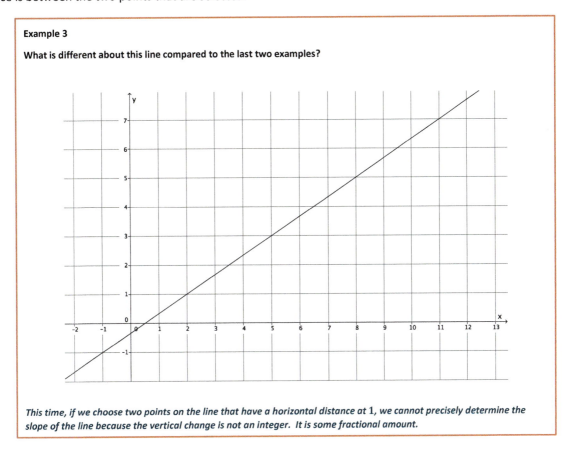
>
> This time, if we choose two points on the line that have a horizontal distance at 1, we cannot precisely determine the slope of the line because the vertical change is not an integer. It is some fractional amount.

- Make a conjecture about how you could find the slope of this line.

Have students write their conjectures and share their ideas about how to find the slope of the line in this example; then, continue with the Discussion that follows.

Discussion (10 minutes)

- In the last lesson, we found a number that described the slope or rate of change of a line. In each instance, we were looking at a special case of slope because the horizontal distance between the two points used to determine the slope, P and Q, was always 1. Since the horizontal distance was 1, the difference between the y-coordinates of points Q and R was equal to the slope or rate of change. For example, in the following graph, we thought of point Q as zero on a vertical number line and noted how many units point R was from point Q.

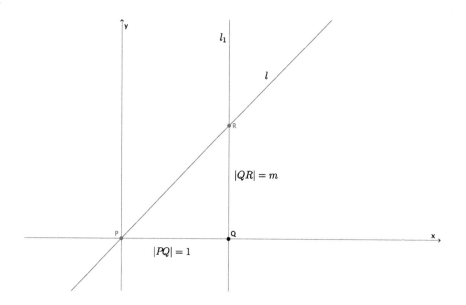

- Also in the last lesson, we found that the unit rate of a problem was equal to the slope. Using that knowledge, we can say that the slope or rate of change of a line $m = \dfrac{|QR|}{|PQ|}$.

- Now the task is to determine the rate of change of a non-vertical line when the distance between points P and Q is a number other than 1. We can use what we know already to guide our thinking.

- Let's take a closer look at Example 2. There are several points on the line with integer coordinates that we could use to help us determine the slope of the line. Recall that we want to select points with integer coordinates because our calculation of slope will be simpler. In each instance, from one point to the next, we have a horizontal distance of 1 unit noted by the red segment and the difference in the y-values between the two points, which is a distance of 2, noted by the blue segments. When we compare the change in the y-values to the change in the x-values, or more explicitly, when we compare the height of the slope triangle to the base of the slope triangle, we have a ratio of $2:1$ with a value of $\frac{2}{1}$ or just 2, which is equal to the slope of the line.

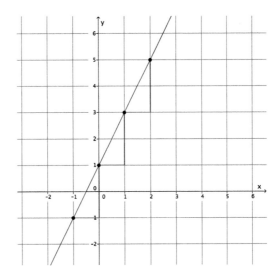

Each of the "slope triangles" shown have values of their ratios equal to $\frac{2}{1}$. Using the same line, let's look at a different pair of "slope triangles."

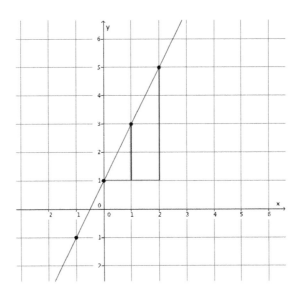

Lesson 16: The Computation of the Slope of a Non-Vertical Line

- What is the ratio of the larger slope triangle?
  - The value of the ratio of the larger slope triangle is $\frac{4}{2}$.
- What do you notice about the ratio of the smaller slope triangle and the ratio of the larger slope triangle?
  - The values of the ratios are equivalent: $\frac{2}{1} = \frac{4}{2} = 2$.
- We have worked with triangles before where the ratios of corresponding sides were equal. We called them *similar* triangles. Are the slope triangles in this diagram similar? How do you know?
  - Yes. The triangles are similar by the AA criterion. Each triangle has a right angle (at the intersection of the blue and red segments), and both triangles have a common angle (the angle formed by the red segment and the line itself).
- When we have similar triangles, we know that the ratios of corresponding side lengths must be equal. That is the reason that both of the slope triangles result in the same number for slope. Notice that we still got the correct number for the slope of the line even though the points chosen did not have a horizontal distance of 1. We can now find the slope of a line given any two points on the line. The horizontal distance between the points does not have to be 1.

Acknowledge any students who may have written or shared this strategy for finding slope from their work with Example 3.

- Now let's look again at Example 3. We did not have a strategy for finding slope before, but we do now. What do you think the slope of this line is? Explain.

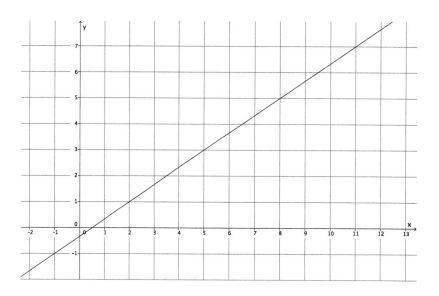

  - The slope of this line is $\frac{2}{3}$.

Ask students to share their work and explanations with the class. Specifically, have them show the slope triangle they used to determine the slope of the line. Select several students to share their work; ideally, students will pick different points and different slope triangles. Whether they do or not, have a discussion similar to the previous one that demonstrates that all slope triangles that could be drawn are similar and that the ratios of corresponding sides are equal.

Lesson 16

Exercise (4 minutes)

Students complete the Exercise independently.

> **Exercise**
>
> Let's investigate concretely to see if the claim that we can find slope between any two points is true.
>
>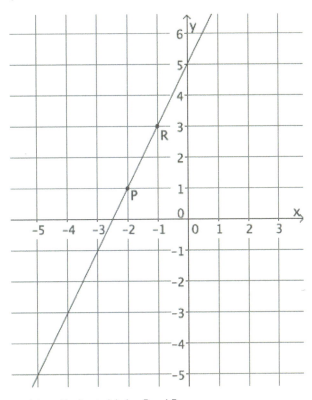
>
> a. Select any two points on the line to label as P and R.
>
>   *Sample points are selected on the graph.*
>
> b. Identify the coordinates of points P and R.
>
>   *Sample points are labeled on the graph.*
>
> c. Find the slope of the line using as many different points as you can. Identify your points, and show your work below.
>
>   *Points selected by students will vary, but the slope should always equal 2. Students could choose to use points $(0, 5)$, $(-1, 3)$, $(-2, 1)$, $(-3, -1)$, $(-4, -3)$, and $(-5, -5)$.*

Lesson 16: The Computation of the Slope of a Non-Vertical Line

Discussion (10 minutes)

- We want to show that the slope of a non-vertical line l can be found using any two points P and R on the line.
- Suppose we have point $P(p_1, p_2)$, where p_1 is the x-coordinate of point P, and p_2 is the y-coordinate of point P. Also, suppose we have points $Q(q_1, q_2)$ and $R(r_1, r_2)$.

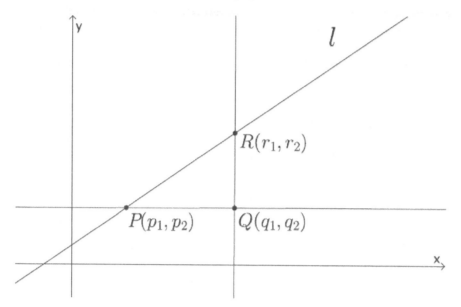

- Then, we claim that the slope m of line l is

$$m = \frac{|QR|}{|PQ|}.$$

- From the last lesson, we found the length of segment QR by looking at the y-coordinate. Without having to translate, we can find the length of segment QR by finding the difference between the y-coordinates of points R and Q (vertical distance). Similarly, we can find the length of the segment PQ by finding the difference between the x-coordinates of P and Q (horizontal distance). We claim

$$m = \frac{|QR|}{|PQ|} = \frac{(q_2 - r_2)}{(p_1 - q_1)}.$$

Lesson 16: The Computation of the Slope of a Non-Vertical Line

- We would like to remove any reference to the coordinates of Q, as it is not a point on the line. We can do this by looking more closely at the coordinates of point Q. Consider the following concrete example.

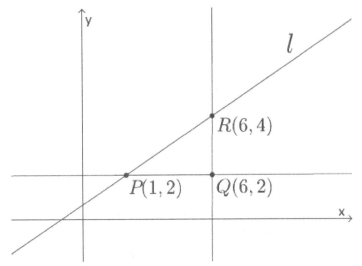

- What do you notice about the y-coordinates of points P and Q?
  - The y-coordinates of points P and Q are the same: 2.
- That means that $q_2 = p_2$. What do you notice about the x-coordinates of points R and Q?
  - The x-coordinates of points R and Q are the same: 6.
- That means that $q_1 = r_1$. Then, by substitution:
$$m = \frac{|QR|}{|PQ|} = \frac{(q_2 - r_2)}{(p_1 - q_1)} = \frac{(p_2 - r_2)}{(p_1 - r_1)}.$$
- Then, we claim that the slope can be calculated regardless of the choice of points. Also, we have discovered something called "the slope formula." With the formula for slope, or rate of change, $m = \frac{(p_2 - r_2)}{(p_1 - r_1)}$, the slope of a line can be found using any two points P and R on the line!

Ask students to translate the slope formula into words, and provide them with the traditional ways of describing slope. For example, students may say the slope of a line is the "height of the slope triangle over the base of the slope triangle" or "the difference in the y-coordinates over the difference in the x-coordinates." Tell students that slope can be referred to as "rise over run" as well.

Discussion (3 minutes)

Show that the formula to calculate slope is true for horizontal lines.

- Suppose we are given a horizontal line. Based on our work in the last lesson, what do we expect the slope of this line to be?

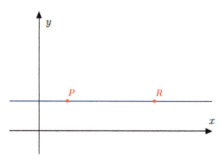

    □ The slope should be zero because if we go one unit to the right of P and then identify the vertical difference between that point and point R, there is no difference. Therefore, the slope is zero.

- As before, the coordinates of points P and R are represented as $P(p_1, p_2)$ and $R(r_1, r_2)$. Since this is a horizontal line, what do we know about the y-coordinates of each point?
    □ Horizontal lines are graphs of linear equations in the form of $y = c$, where the y-value does not change. Therefore, $p_2 = r_2$.
- By the slope formula:

$$m = \frac{(p_2 - r_2)}{(p_1 - r_1)} = \frac{0}{p_1 - r_1} = 0.$$

The slope of the horizontal line is zero, as expected, regardless of the value of the horizontal change.

Discussion (7 minutes)

- Now for the general case. We want to show that we can choose any two points P and R to find the slope, not just a point like R', where we have fixed the horizontal distance at 1. Consider the diagram below.

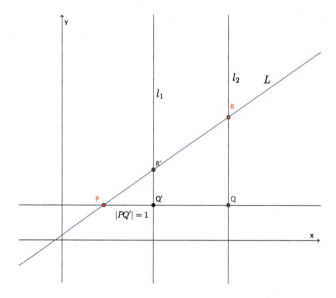

- Now we have a situation where point Q is an unknown distance from point P. We know that if $\triangle PQ'R'$ is similar to $\triangle PQR$, then the ratio of the corresponding sides will be equal, and the ratios are equal to the slope of the line L. Are $\triangle PQ'R'$ and $\triangle PQR$ similar? Explain.
  - Yes, the triangles are similar, i.e., $\triangle PQ'R' \sim \triangle PQR$. Both triangles have a common angle, $\angle RPQ$, and both triangles have a right angle, $\angle R'Q'P$ and $\angle RQP$. By the AA criterion, $\triangle PQ'R' \sim \triangle PQR$.

- Now what we want to do is find a way to express this information in a formula. Because we have similar triangles, we know the following:
$$\frac{|R'Q'|}{|RQ|} = \frac{|PQ'|}{|PQ|} = \frac{|PR'|}{|PR|} = r.$$

- Based on our previous knowledge, we know that $|R'Q'| = m$, and $|PQ'| = 1$. By substitution, we have
$$\frac{m}{|RQ|} = \frac{1}{|PQ|},$$
which is equivalent to
$$\frac{m}{1} = \frac{|RQ|}{|PQ|}$$
$$m = \frac{|RQ|}{|PQ|}.$$

- We also know from our work earlier that $|RQ| = p_2 - r_2$, and $|PQ| = p_1 - r_1$. By substitution, we have
$$m = \frac{p_2 - r_2}{p_1 - r_1}.$$

The slope of a line can be computed using any two points!

Closing (3 minutes)

Summarize, or ask students to summarize, the main points from the lesson:

- We know that the slope of a line can be calculated using any two points on the line because of what we know about similar triangles.
- Slope is referred to as the difference in y-values compared to the difference in x-values, or as the height compared to the base of the slope triangle, or as rise over run.
- We know that the formula to calculate slope is $m = \frac{p_2 - r_2}{p_1 - r_1}$, where (p_1, p_2) and (r_1, r_2) are two points on the line.

Lesson 16: The Computation of the Slope of a Non-Vertical Line

Lesson Summary

The slope of a line can be calculated using *any* two points on the same line because the slope triangles formed are similar, and corresponding sides will be equal in ratio.

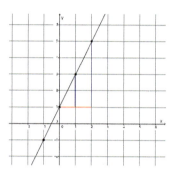

The *slope* of a non-vertical line in a coordinate plane that passes through two different points is the number given by the difference in y-coordinates of those points divided by the difference in the corresponding x-coordinates. For two points $P(p_1, p_2)$ and $R(r_1, r_2)$ on the line where $p_1 \neq r_1$, the slope of the line m can be computed by the formula

$$m = \frac{p_2 - r_2}{p_1 - r_1}.$$

The slope of a vertical line is not defined.

Exit Ticket (3 minutes)

Name _____ Date _____

Lesson 16: The Computation of the Slope of a Non-Vertical Line

Exit Ticket

Find the rate of change of the line by completing parts (a) and (b).

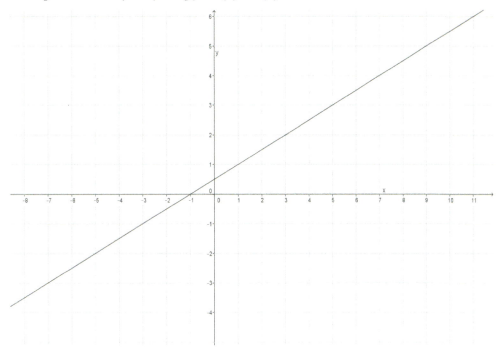

a. Select any two points on the line to label as P and R. Name their coordinates.

b. Compute the rate of change of the line.

Exit Ticket Sample Solutions

Find the rate of change of the line by completing parts (a) and (b).

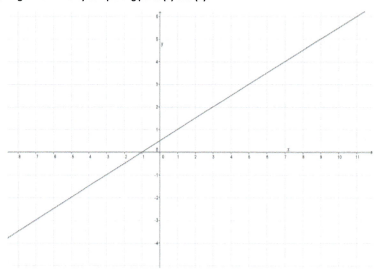

a. Select any two points on the line to label as P and R. Name their coordinates.

Answers will vary. Other points on the graph may have been chosen.

$P(-1, 0)$ and $R(5, 3)$

b. Compute the rate of change of the line.

$$m = \frac{(p_2 - r_2)}{(p_1 - r_1)}$$
$$= \frac{0 - 3}{-1 - 5}$$
$$= \frac{-3}{-6}$$
$$= \frac{1}{2}$$

Lesson 16: The Computation of the Slope of a Non-Vertical Line

A STORY OF RATIOS   Lesson 16   8•4

Problem Set Sample Solutions

Students practice finding slope between any two points on a line. Students also see that $m = \frac{p_2 - r_2}{p_1 - r_1}$ yields the same result as $m = \frac{r_2 - p_2}{r_1 - p_1}$.

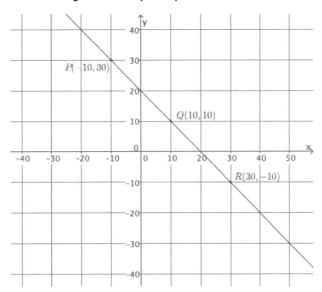

1. Calculate the slope of the line using two different pairs of points.

$$m = \frac{p_2 - r_2}{p_1 - r_1}$$
$$= \frac{30 - (-10)}{-10 - 30}$$
$$= \frac{40}{-40}$$
$$= -\frac{1}{1}$$
$$= -1$$

$$m = \frac{q_2 - r_2}{q_1 - r_1}$$
$$= \frac{10 - (-10)}{10 - 30}$$
$$= \frac{20}{-20}$$
$$= -1$$

Lesson 16:   The Computation of the Slope of a Non-Vertical Line

2. Calculate the slope of the line using two different pairs of points.

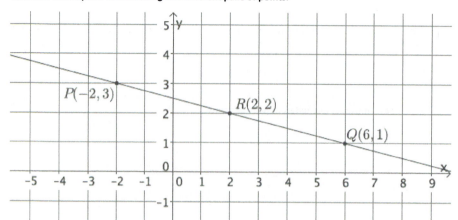

$$m = \frac{p_2 - r_2}{p_1 - r_1}$$
$$= \frac{3 - 2}{-2 - 2}$$
$$= \frac{1}{-4}$$
$$= -\frac{1}{4}$$

$$m = \frac{q_2 - r_2}{q_1 - r_1}$$
$$= \frac{1 - 2}{6 - 2}$$
$$= \frac{-1}{4}$$
$$= -\frac{1}{4}$$

3. Calculate the slope of the line using two different pairs of points.

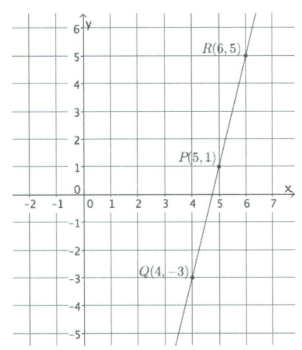

$m = \dfrac{p_2 - r_2}{p_1 - r_1}$

$= \dfrac{1 - 5}{5 - 6}$

$= \dfrac{-4}{-1}$

$= \dfrac{4}{1}$

$= 4$

$m = \dfrac{q_2 - r_2}{q_1 - r_1}$

$= \dfrac{-3 - 5}{4 - 6}$

$= \dfrac{-8}{-2}$

$= 4$

4. Calculate the slope of the line using two different pairs of points.

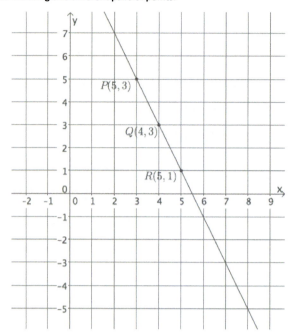

$m = \dfrac{p_2 - r_2}{p_1 - r_1}$

$= \dfrac{5 - 1}{3 - 5}$

$= \dfrac{4}{-2}$

$= -\dfrac{2}{1}$

$= -2$

$m = \dfrac{q_2 - r_2}{q_1 - r_1}$

$= \dfrac{3 - 1}{4 - 5}$

$= \dfrac{2}{-1}$

$= -2$

5. Calculate the slope of the line using two different pairs of points.

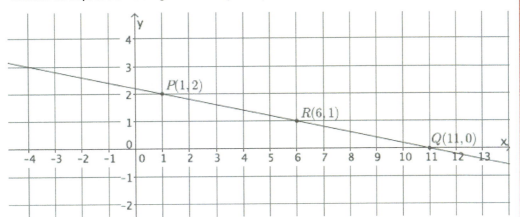

$$m = \frac{p_2 - r_2}{p_1 - r_1}$$
$$= \frac{2 - 1}{1 - 6}$$
$$= \frac{1}{-5}$$
$$= -\frac{1}{5}$$

$$m = \frac{q_2 - r_2}{q_1 - r_1}$$
$$= \frac{0 - 1}{11 - 6}$$
$$= \frac{-1}{5}$$
$$= -\frac{1}{5}$$

6. Calculate the slope of the line using two different pairs of points.

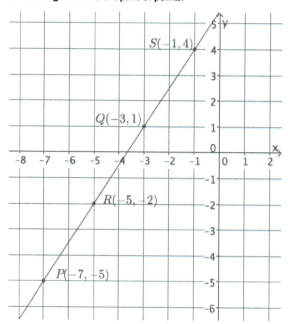

a. Select any two points on the line to compute the slope.

$$m = \frac{p_2 - r_2}{p_1 - r_1}$$
$$= \frac{-5 - (-2)}{-7 - (-5)}$$
$$= \frac{-3}{-2}$$
$$= \frac{3}{2}$$

b. Select two different points on the line to calculate the slope.

Let the two new points be $(-3, 1)$ and $(-1, 4)$.

$$m = \frac{q_2 - s_2}{q_1 - s_1}$$
$$= \frac{1 - 4}{-3 - (-1)}$$
$$= \frac{-3}{-2}$$
$$= \frac{3}{2}$$

c. What do you notice about your answers in parts (a) and (b)? Explain.

The slopes are equal in parts (a) and (b). This is true because of what we know about similar triangles. The slope triangle that is drawn between the two points selected in part (a) is similar to the slope triangle that is drawn between the two points in part (b) by the AA criterion. Then, because the corresponding sides of similar triangles are equal in ratio, the slopes are equal.

7. Calculate the slope of the line in the graph below.

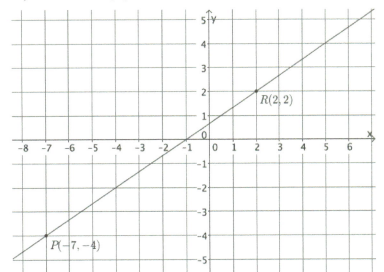

$$m = \frac{p_2 - r_2}{p_1 - r_1}$$
$$= \frac{-4 - 2}{-7 - 2}$$
$$= \frac{-6}{-9}$$
$$= \frac{2}{3}$$

8. Your teacher tells you that a line goes through the points $\left(-6, \frac{1}{2}\right)$ and $(-4, 3)$.

   a. Calculate the slope of this line.

   $$m = \frac{p_2 - r_2}{p_1 - r_1}$$
   $$= \frac{\frac{1}{2} - 3}{-6 - (-4)}$$
   $$= \frac{-\frac{5}{2}}{-2}$$
   $$= \frac{\frac{5}{2}}{2}$$
   $$= \frac{5}{2} \div 2$$
   $$= \frac{5}{2} \times \frac{1}{2}$$
   $$= \frac{5}{4}$$

   b. Do you think the slope will be the same if the order of the points is reversed? Verify by calculating the slope, and explain your result.

   The slope should be the same because we are joining the same two points.

   $$m = \frac{p_2 - r_2}{p_1 - r_1}$$
   $$= \frac{3 - \frac{1}{2}}{-4 - (-6)}$$
   $$= \frac{\frac{5}{2}}{2}$$
   $$= \frac{5}{4}$$

   Since the slope of a line can be computed using any two points on the same line, it makes sense that it does not matter which point we name as P and which point we name as R.

9. Use the graph to complete parts (a)–(c).

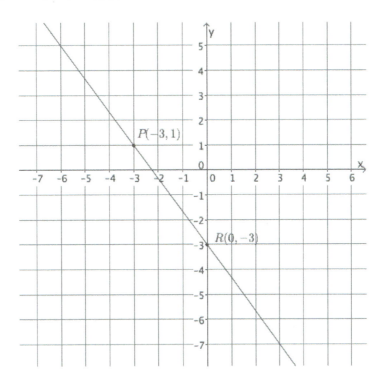

a. Select any two points on the line to calculate the slope.

$$m = \frac{p_2 - r_2}{p_1 - r_1}$$
$$= \frac{1 - (-3)}{-3 - 0}$$
$$= \frac{4}{-3}$$
$$= -\frac{4}{3}$$

b. Compute the slope again, this time reversing the order of the coordinates.

$$m = \frac{(r_2 - p_2)}{(r_1 - p_1)}$$
$$= \frac{-3 - 1}{0 - (-3)}$$
$$= \frac{-4}{3}$$
$$= -\frac{4}{3}$$

c. What do you notice about the slopes you computed in parts (a) and (b)?

The slopes are equal.

Lesson 16: The Computation of the Slope of a Non-Vertical Line

d. Why do you think $m = \frac{(p_2-r_2)}{(p_1-r_1)} = \frac{(r_2-p_2)}{(r_1-p_1)}$?

If I multiply the first fraction by $\frac{-1}{-1}$, then I get the second fraction:

$$\frac{-1}{-1} \times \left(\frac{(p_2-r_2)}{(p_1-r_1)}\right) = \frac{(r_2-p_2)}{(r_1-p_1)}.$$

I can do the same thing to the second fraction to obtain the first:

$$\frac{-1}{-1} \times \left(\frac{(r_2-p_2)}{(r_1-p_1)}\right) = \frac{(p_2-r_2)}{(p_1-r_1)}.$$

Also, since I know that I can find the slope between any two points, it should not matter which point I pick first.

10. Each of the lines in the lesson was non-vertical. Consider the slope of a vertical line, $x = 2$. Select two points on the line to calculate slope. Based on your answer, why do you think the topic of slope focuses only on non-vertical lines?

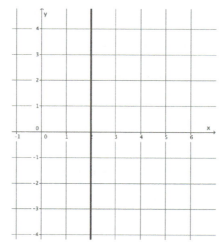

Students can use any points on the line $x = 2$ to determine that the slope is undefined. The computation of slope using the formula leads to a fraction with zero as its denominator. Since the slope of a vertical line is undefined, there is no need to focus on them.

Challenge:

11. A certain line has a slope of $\frac{1}{2}$. Name two points that may be on the line.

Answers will vary. Accept any answers that have a difference in y-values equal to 1 and a difference of x-values equal to 2. Points $(6, 4)$ and $(4, 3)$ may be on the line, for example.

A STORY OF RATIOS   Lesson 17   8•4

Lesson 17: The Line Joining Two Distinct Points of the Graph $y = mx + b$ Has Slope m

Student Outcomes

- Students show that the slope of a line joining any two distinct points of the graph of $y = mx + b$ has slope m.
- Students transform the standard form of an equation into $y = -\frac{a}{b}x + \frac{c}{b}$.

Lesson Notes

In the previous lesson, it was determined that slope can be calculated using any two points on the same line. In this lesson, students are shown that equations of the form $y = mx$ and $y = mx + b$ generate lines with slope m. Students need graph paper to complete some of the Exercises and Problem Set items.

Classwork

Exercises 1–3 (8 minutes)

Students work independently to complete Exercises 1–3.

Exercises

1. Find at least three solutions to the equation $y = 2x$, and graph the solutions as points on the coordinate plane. Connect the points to make a line. Find the slope of the line.

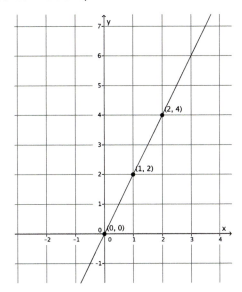

The slope of the line is 2; $m = 2$.

Lesson 17: The Line Joining Two Distinct Points of the Graph $y = mx + b$ Has Slope m 251

2. Find at least three solutions to the equation $y = 3x - 1$, and graph the solutions as points on the coordinate plane. Connect the points to make a line. Find the slope of the line.

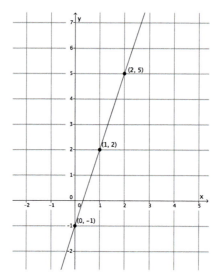

The slope of the line is 3; $m = 3$.

3. Find at least three solutions to the equation $y = 3x + 1$, and graph the solutions as points on the coordinate plane. Connect the points to make a line. Find the slope of the line.

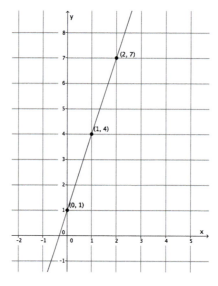

The slope of the line is 3; $m = 3$.

Discussion (12 minutes)

Recall the goal from Lesson 15: to prove that the graph of a linear equation is a line. To do so, some tools were needed, specifically two related to slope. Now that facts are known about slope, the focus is on showing that the line that joins two distinct points is a linear equation with slope m.

- We know from our previous work with slope that when the horizontal distance between two points is fixed at one, then the slope of the line is the difference in the y-coordinates. We also know that when the horizontal distance is not fixed at one, we can find the slope of the line using any two points because the ratio of corresponding sides of similar triangles will be equal. We can put these two facts together to prove that the graph of the line $y = mx$ has slope m. Consider the diagram below.

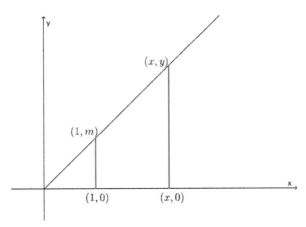

- Examine the diagram, and think of how we could prove that $\frac{y}{m} = \frac{x}{1}$.

Provide students time to work independently, and then provide time for them to discuss in pairs a possible proof of $\frac{y}{m} = \frac{x}{1}$. If necessary, use the four bullet points below to guide students' thinking.

- Do we have similar triangles? Explain.
  - Yes. Each of the triangles has a common angle at the origin, and each triangle has a right angle. By the AA criterion, these triangles are similar.
- What is the slope of the line? Explain.
  - The slope of the line is m. By our definition of slope and the information in the diagram, when the horizontal distance between two points is fixed at one, the slope is m.
- Write the ratio of the corresponding sides. Then, solve for y.
  - $\frac{y}{m} = \frac{x}{1}$
  - $y = mx$
- Therefore, the slope of the graph of $y = mx$ is m.

Point students to their work in Exercise 1 where the graph of $y = 2x$ was a line with a slope of 2.

- We know that the graph of $y = mx$ has slope m, where m is a number. The y in the equation $y = mx$ is equal to the difference in y-coordinates as shown in the diagram below.

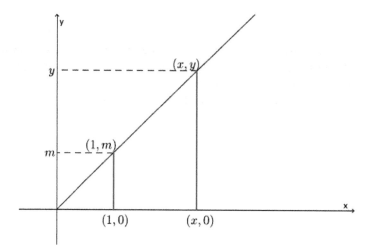

- Consider the diagram below. How does this compare to the graph of $y = mx$ that we just worked on?

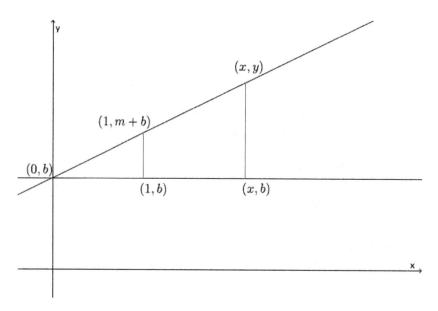

  ▫ The graph is the same except for the fact that the line has been translated b units up the y-axis.

- We want to write the ratio of corresponding sides of the slope triangles. We know the lengths of two sides of the smaller slope triangle; they are 1 and m. We also know one of the lengths of the larger slope triangle, x. What we need to do now is express the length of the larger slope triangle that corresponds to side m. When our line passed through the origin, the length was simply y, but that is not the case here. How can we express the length we need, noted in blue in the diagram below?

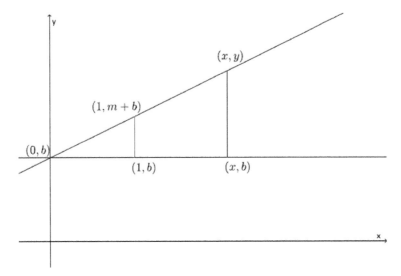

Provide students time to think and share in small groups. If necessary, show students the diagram below, and ask them what the length of the dashed orange segment is. They should recognize that it is the length of b. Then, ask them how they can use that information to determine the length of the blue segment.

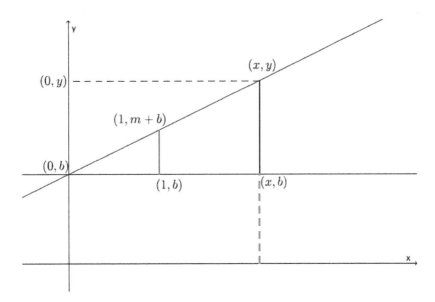

  ▫ The length of the blue segment is $y - b$.

Lesson 17: The Line Joining Two Distinct Points of the Graph $y = mx + b$ Has Slope m

- Now that we know the length of the blue segment, we can write ratios that represent the corresponding sides of the triangles:

$$\frac{m}{y-b} = \frac{1}{x}.$$

Then, we can solve for y.

$$mx = y - b$$
$$mx + b = y - b + b$$
$$mx + b = y$$

Therefore, the slope of the graph of an equation of the form $y = mx + b$ is m.

Point students to their work in Exercises 2 and 3 where the graph of $y = 3x - 1$ and $y = 3x + 1$ was a line with slope 3.

- We can show this algebraically using two distinct points on the graph of the line $y = 3x - 1$. Points $(1, 2)$ and $(2, 5)$ were on the graph of $y = 3x - 1$. Using the slope formula, we see that

$$m = \frac{5-2}{2-1}$$
$$= \frac{3}{1}$$
$$= 3.$$

The next four bullet points generalize the slope between *any* two points of the graph of $y = 3x - 1$ is 3. This is an optional part of the discussion of slope between any two points. The last bullet point of this section should be shared with students because it orients them to the progress toward the ultimate goal, which is to show that the graph of a linear equation is a line.

- In general, let $P(p_1, p_2)$ and $R(r_1, r_2)$ be *any* two distinct points of the graph of $y = 3x - 1$. Recall what is known about points on a graph; it means that they are solutions to the equation. Since P is on the graph of $y = 3x - 1$, then

$$p_2 = 3p_1 - 1.$$

Similarly, since R is on the graph of $y = 3x - 1$, then

$$r_2 = 3r_1 - 1.$$

- By the slope formula, $m = \frac{p_2 - r_2}{p_1 - r_1}$, we can substitute the values of p_2 and r_2 and use our properties of equality to simplify as follows.

$$m = \frac{p_2 - r_2}{p_1 - r_1}$$

$$= \frac{(3p_1 - 1) - (3r_1 - 1)}{p_1 - r_1} \quad \text{By substitution}$$

$$= \frac{3p_1 - 1 - 3r_1 + 1}{p_1 - r_1} \quad \text{By taking the opposite of the terms in the second grouping}$$

$$= \frac{3p_1 - 3r_1}{p_1 - r_1} \quad \text{By simplifying } (-1 + 1)$$

$$= \frac{3(p_1 - r_1)}{p_1 - r_1} \quad \text{By the distributive property "collecting like terms"}$$

$$= 3 \quad \text{By division, the number } p_1 - r_1 \text{ divided by itself}$$

- Thus, we have shown that the line passing through points P and R has a slope of 3.
- To truly generalize our work, we need only replace 3 in the calculations with m and the -1 with b.

Make clear to students that they are still working toward the goal of proving that the graph of a linear equation is a line using the summary of work below.

- Our goal, ultimately, is to prove that the graph of a linear equation is a line. In Lesson 15, we said we have to develop the following tools:

  (1) We must define a number for each non-vertical line that can be used to measure the "steepness" or "slant" of the line. Once defined, this number will be called the **slope** of the line and is often referred to as the *rate of change*.

  (2) We must show that *any* two points on a non-vertical line can be used to find the slope of the line.

  (3) We must show that the line joining two points on a graph of a linear equation of the form $y = mx + b$ has the slope m.

  (4) We must show that there is only one line passing through a given point with a given slope.

- At this point in our work, we just finished (3). (4) is the topic of the next lesson.

Discussion (5 minutes)

- When an equation is in the form $y = mx + b$, the slope m is easily identifiable compared to an equation in the form $ax + by = c$. Note that b in each of the equations is unique. In other words, the number represented by b in the equation $y = mx + b$ is not necessarily the same as the number b in the equation $ax + by = c$. For example, we will solve the equation $8x + 2y = 6$ for y. Our goal is to have y equal to an expression.

Lesson 17: The Line Joining Two Distinct Points of the Graph $y = mx + b$ Has Slope m

- First, we use our properties of equality to remove $8x$ from the left side of the equation.

$$8x + 2y = 6$$
$$8x - 8x + 2y = 6 - 8x$$
$$2y = 6 - 8x$$

Now, we divide both sides by 2.

$$\frac{2y}{2} = \frac{6 - 8x}{2}$$
$$y = \frac{6}{2} - \frac{8x}{2}$$
$$y = 3 - 4x$$
$$y = -4x + 3$$

The slope of the graph of this equation is -4.

- By convention (an agreed-upon way of doing things), we place the term mx before the term b. This is a version of the standard form of a linear equation that is referred to as the *slope-intercept form*. It is called slope-intercept form because it makes clear the number that describes slope (i.e., m) and the y-intercept, which is something that will be discussed later. Also, notice the value of b is different in both forms of the equation.

Exercises 4–11 (11 minutes)

Students work independently or in pairs to identify the slope from an equation and to transform the standard form of an equation into slope-intercept form.

---

4. The graph of the equation $y = 7x - 3$ has what slope?

   The slope is 7.

5. The graph of the equation $y = -\frac{3}{4}x - 3$ has what slope?

   The slope is $-\frac{3}{4}$.

6. You have $20 in savings at the bank. Each week, you add $2 to your savings. Let y represent the total amount of money you have saved at the end of x weeks. Write an equation to represent this situation, and identify the slope of the equation. What does that number represent?

   $y = 2x + 20$

   The slope is 2. It represents how much money is saved each week.

7. A friend is training for a marathon. She can run 4 miles in 28 minutes. Assume she runs at a constant rate. Write an equation to represent the total distance, y, your friend can run in x minutes. Identify the slope of the equation. What does that number represent?

$$\frac{y}{x} = \frac{4}{28}$$
$$y = \frac{4}{28}x$$
$$y = \frac{1}{7}x$$

The slope is $\frac{1}{7}$. It represents the rate at which my friend can run, one mile in seven minutes.

8. Four boxes of pencils cost $5. Write an equation that represents the total cost, y, for x boxes of pencils. What is the slope of the equation? What does that number represent?

$$y = \frac{5}{4}x$$

The slope is $\frac{5}{4}$. It represents the cost of one box of pencils, $1.25.

9. Solve the following equation for y, and then identify the slope of the line: $9x - 3y = 15$.

$$9x - 3y = 15$$
$$9x - 9x - 3y = 15 - 9x$$
$$-3y = 15 - 9x$$
$$\frac{-3}{-3}y = \frac{15 - 9x}{-3}$$
$$y = \frac{15}{-3} - \frac{9x}{-3}$$
$$y = -5 + 3x$$
$$y = 3x - 5$$

The slope of the line is 3.

10. Solve the following equation for y, and then identify the slope of the line: $5x + 9y = 8$.

$$5x + 9y = 8$$
$$5x - 5x + 9y = 8 - 5x$$
$$9y = 8 - 5x$$
$$\frac{9}{9}y = \frac{8 - 5x}{9}$$
$$y = \frac{8}{9} - \frac{5}{9}x$$
$$y = -\frac{5}{9}x + \frac{8}{9}$$

The slope of the line is $-\frac{5}{9}$.

11. Solve the following equation for y, and then identify the slope of the line: $ax + by = c$.

$$ax + by = c$$
$$ax - ax + by = c - ax$$
$$by = c - ax$$
$$\frac{b}{b}y = \frac{c - ax}{b}$$
$$y = \frac{c}{b} - \frac{ax}{b}$$
$$y = \frac{c}{b} - \frac{a}{b}x$$
$$y = -\frac{a}{b}x + \frac{c}{b}$$

The slope of the line is $-\frac{a}{b}$.

Closing (4 minutes)

Summarize, or ask students to summarize, the main points from the lesson:

- We know that the line that joins any two distinct points of the graph $y = mx + b$ has slope m.
- We know how to identify the slope for any equation in the form $y = mx + b$.
- We know how to transform the standard form of a linear equation into another form to more easily identify the slope.

> **Lesson Summary**
>
> The line joining two distinct points of the graph of the linear equation $y = mx + b$ has slope m.
>
> The m of $y = mx + b$ is the number that describes the slope. For example, in the equation $y = -2x + 4$, the slope of the graph of the line is -2.

Exit Ticket (5 minutes)

Name _____ Date _____

Lesson 17: The Line Joining Two Distinct Points of the Graph $y = mx + b$ Has Slope m

Exit Ticket

1. Solve the following equation for y: $35x - 7y = 49$.

2. What is the slope of the equation in Problem 1?

3. Show, using similar triangles, why the graph of an equation of the form $y = mx$ is a line with slope m.

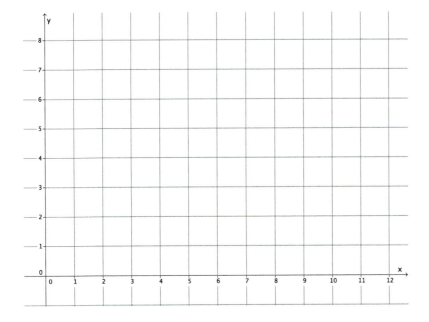

Exit Ticket Sample Solutions

1. **Solve the following equation for y: $35x - 7y = 49$.**

   $$35x - 7y = 49$$
   $$35x - 35x - 7y = 49 - 35x$$
   $$-7y = 49 - 35x$$
   $$\frac{-7}{-7}y = \frac{49}{-7} - \frac{35}{-7}x$$
   $$y = -7 - (-5x)$$
   $$y = 5x - 7$$

2. **What is the slope of the equation in Problem 1?**

   The slope of $y = 5x - 7$ is 5.

3. **Show, using similar triangles, why the graph of an equation of the form $y = mx$ is a line with slope m.**

   Solutions will vary. A sample solution is shown below.

   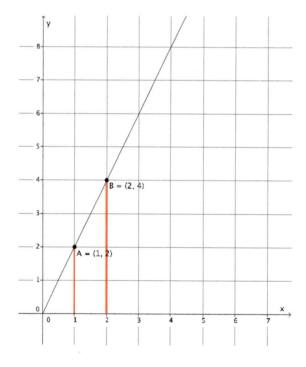

   The line shown has slope 2. When we compare the corresponding side lengths of the similar triangles, we have the ratios $\frac{2}{1} = \frac{4}{2} = 2$. In general, the ratios would be $\frac{x}{1} = \frac{y}{m}$, equivalently $y = mx$, which is a line with slope m.

Problem Set Sample Solutions

Students practice transforming equations from standard form into slope-intercept form and showing that the line joining two distinct points of the graph $y = mx + b$ has slope m. Students graph the equation and informally note the y-intercept.

1. Solve the following equation for y: $-4x + 8y = 24$. Then, answer the questions that follow.

    $$-4x + 8y = 24$$
    $$-4x + 4x + 8y = 24 + 4x$$
    $$8y = 24 + 4x$$
    $$\frac{8}{8}y = \frac{24}{8} + \frac{4}{8}x$$
    $$y = 3 + \frac{1}{2}x$$
    $$y = \frac{1}{2}x + 3$$

    a. Based on your transformed equation, what is the slope of the linear equation $-4x + 8y = 24$?

       The slope is $\frac{1}{2}$.

    b. Complete the table to find solutions to the linear equation.

x	Transformed Linear Equation: $y = \frac{1}{2}x + 3$	y
-2	$y = \frac{1}{2}(-2) + 3$ $= -1 + 3$ $= 2$	2
0	$y = \frac{1}{2}(0) + 3$ $= 0 + 3$ $= 3$	3
2	$y = \frac{1}{2}(2) + 3$ $= 1 + 3$ $= 4$	4
4	$y = \frac{1}{2}(4) + 3$ $= 2 + 3$ $= 5$	5

Lesson 17: The Line Joining Two Distinct Points of the Graph $y = mx + b$ Has Slope m

c. Graph the points on the coordinate plane.

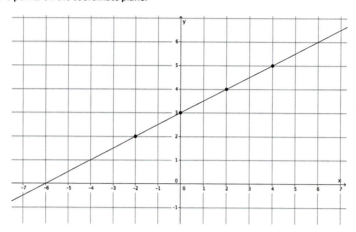

d. Find the slope between any two points.

Using points $(0, 3)$ and $(2, 4)$:

$$m = \frac{4-3}{2-0}$$
$$= \frac{1}{2}$$

e. The slope you found in part (d) should be equal to the slope you noted in part (a). If so, connect the points to make the line that is the graph of an equation of the form $y = mx + b$ that has slope m.

f. Note the location (ordered pair) that describes where the line intersects the y-axis.

$(0, 3)$ is the location where the line intersects the y-axis.

2. Solve the following equation for y: $9x + 3y = 21$. Then, answer the questions that follow.

$$9x + 3y = 21$$
$$9x - 9x + 3y = 21 - 9x$$
$$3y = 21 - 9x$$
$$\frac{3}{3}y = \frac{21}{3} - \frac{9}{3}x$$
$$y = 7 - 3x$$
$$y = -3x + 7$$

a. Based on your transformed equation, what is the slope of the linear equation $9x + 3y = 21$?

The slope is -3.

b. Complete the table to find solutions to the linear equation.

x	Transformed Linear Equation: $y = -3x + 7$	y
-1	$y = -3(-1) + 7$ $= 3 + 7$ $= 10$	10
0	$y = -3(0) + 7$ $= 0 + 7$ $= 7$	7
1	$y = -3(1) + 7$ $= -3 + 7$ $= 4$	4
2	$y = -3(2) + 7$ $= -6 + 7$ $= 1$	1

c. Graph the points on the coordinate plane.

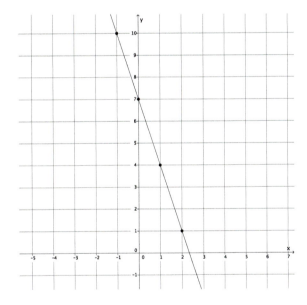

d. Find the slope between any two points.

Using points $(1, 4)$ and $(2, 1)$:

$m = \dfrac{4 - 1}{1 - 2}$

$= \dfrac{3}{-1}$

$= -3$

e. The slope you found in part (d) should be equal to the slope you noted in part (a). If so, connect the points to make the line that is the graph of an equation of the form $y = mx + b$ that has slope m.

Lesson 17: The Line Joining Two Distinct Points of the Graph $y = mx + b$ Has Slope m

f. Note the location (ordered pair) that describes where the line intersects the y-axis.

$(0, 7)$ *is the location where the line intersects the y-axis.*

3. Solve the following equation for y: $2x + 3y = -6$. Then, answer the questions that follow.

$$2x + 3y = -6$$
$$2x - 2x + 3y = -6 - 2x$$
$$3y = -6 - 2x$$
$$\frac{3}{3}y = \frac{-6}{3} - \frac{2}{3}x$$
$$y = -2 - \frac{2}{3}x$$
$$y = -\frac{2}{3}x - 2$$

a. Based on your transformed equation, what is the slope of the linear equation $2x + 3y = -6$?

The slope is $-\frac{2}{3}$.

b. Complete the table to find solutions to the linear equation.

x	Transformed Linear Equation: $y = -\frac{2}{3}x - 2$	y
-6	$y = -\frac{2}{3}(-6) - 2$ $= 4 - 2$ $= 2$	2
-3	$y = -\frac{2}{3}(-3) - 2$ $= 2 - 2$ $= 0$	0
0	$y = -\frac{2}{3}(0) - 2$ $= 0 - 2$ $= -2$	-2
3	$y = -\frac{2}{3}(3) - 2$ $= -2 - 2$ $= -4$	-4

c. Graph the points on the coordinate plane.

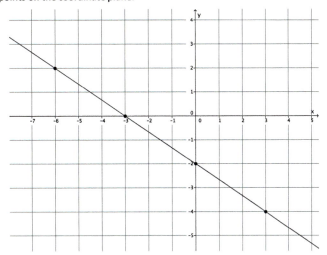

d. Find the slope between any two points.

Using points $(-6, 2)$ and $(3, -4)$:

$$m = \frac{2-(-4)}{-6-3}$$
$$= \frac{6}{-9}$$
$$= -\frac{2}{3}$$

e. The slope you found in part (d) should be equal to the slope you noted in part (a). If so, connect the points to make the line that is the graph of an equation of the form $y = mx + b$ that has slope m.

f. Note the location (ordered pair) that describes where the line intersects the y-axis.

$(0, -2)$ is the location where the line intersects the y-axis.

4. Solve the following equation for y: $5x - y = 4$. Then, answer the questions that follow.

$$5x - y = 4$$
$$5x - 5x - y = 4 - 5x$$
$$-y = 4 - 5x$$
$$y = -4 + 5x$$
$$y = 5x - 4$$

a. Based on your transformed equation, what is the slope of the linear equation $5x - y = 4$?

The slope is 5.

b. Complete the table to find solutions to the linear equation.

x	Transformed Linear Equation: $y = 5x - 4$	y
-1	$y = 5(-1) - 4$ $= -5 - 4$ $= -9$	-9
0	$y = 5(0) - 4$ $= 0 - 4$ $= -4$	-4
1	$y = 5(1) - 4$ $= 5 - 4$ $= 1$	1
2	$y = 5(2) - 4$ $= 10 - 4$ $= 6$	6

c. Graph the points on the coordinate plane.

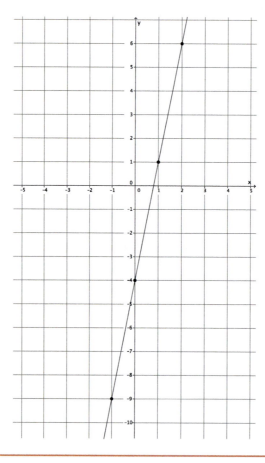

d. Find the slope between any two points.

Using points $(0, -4)$ and $(1, 1)$:

$$m = \frac{-4 - 1}{0 - 1}$$
$$= \frac{-5}{-1}$$
$$= 5$$

e. The slope you found in part (d) should be equal to the slope you noted in part (a). If so, connect the points to make the line that is the graph of an equation of the form $y = mx + b$ that has slope m.

f. Note the location (ordered pair) that describes where the line intersects the y-axis.

$(0, -4)$ is the location where the line intersects the y-axis.

A STORY OF RATIOS   Lesson 18   8•4

Lesson 18: There Is Only One Line Passing Through a Given Point with a Given Slope

Student Outcomes

- Students graph equations in the form of $y = mx + b$ using information about slope and y-intercept point.
- Students know that if they have two straight lines with the same slope and a common point, the lines are the same.

Lesson Notes

The Opening Exercise requires students to examine part (f) from the Problem Set of Lesson 17. Each part of (f) requires students to identify the point where the graph of the line intersects the y-axis. Knowing that this point represents the y-intercept point and that it is the point $(0, b)$ in the equation $y = mx + b$ is integral for the content in the lesson.

To maintain consistency, throughout the lesson students form the slope triangle with points P, Q, and R. Since slope has been defined in terms of lengths of sides of similar triangles, there is notation and verbiage about numbers in terms of distances. While working through the examples, make clear that the lengths of the slope triangle are positive distances, but that does not necessarily mean that the slope must be positive (see Example 2). Remind students that graphs of lines with positive slopes are *left-to-right inclining*, and graphs of lines with negative slopes are *left-to-right declining*.

Coordinate planes are provided for students in the exercises of this lesson, but they need graph paper to complete the Problem Set.

Classwork

Opening Exercise (4 minutes)

Opening Exercise

Examine each of the graphs and their equations. Identify the coordinates of the point where the line intersects the y-axis. Describe the relationship between the point and the equation $y = mx + b$.

a. $y = \frac{1}{2}x + 3$

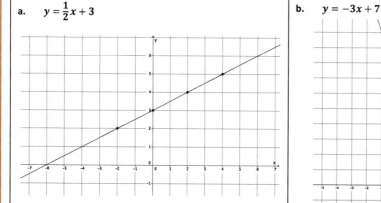

b. $y = -3x + 7$

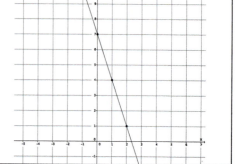

MP.8

A STORY OF RATIOS   Lesson 18   8•4

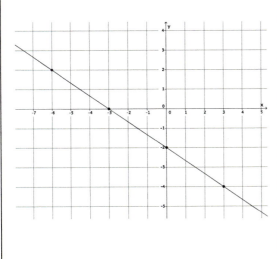
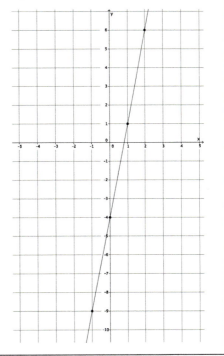

A point is noted in each graph above where the line intersects the y-axis:

a. $y = \frac{1}{2}x + 3$, $(0, 3)$
b. $y = -3x + 7$, $(0, 7)$
c. $y = -\frac{2}{3}x - 2$, $(0, -2)$
d. $y = 5x - 4$, $(0, -4)$

In each equation, the number b was the y-coordinate of the point where the line intersected the y-axis.

Discussion (4 minutes)

- In the last lesson, we transformed the standard form of a linear equation $ax + by = c$ into what was referred to as the *slope-intercept form*, $y = mx + b$. We know that the slope is represented by m, but we did not discuss the meaning of b. In the Opening Exercise, you were asked to note the location (ordered pair) that describes where the line intersected the y-axis.

- What do you notice about the value of b in relation to the point where the graph of the equation intersected the y-axis?

  □ The value of b was the same number as the y-coordinate of each location.

- When a linear equation is in the form $y = mx + b$, it is known as the *slope-intercept form* because this form provides information about the slope, m, and y-intercept point, $(0, b)$, of the graph. The y-intercept point is defined as the location on the graph where a line intersects the y-axis.

- In this lesson, we develop the last tool that we need in order to prove that the graph of a linear equation in two variables $ax + by = c$, where a, b, and c are constants, is a straight line. We will show that if two straight lines have the same slope and pass through the same point, then they are the same line.

Lesson 18:   There Is Only One Line Passing Through a Given Point with a Given Slope

- Since an equation of the form $y = mx + b$ provides information about both the y-intercept point and slope, we will use this equation to graph lines.
- Recall that we began discussing slope graphically, $m = \frac{|QR|}{|PQ|}$.

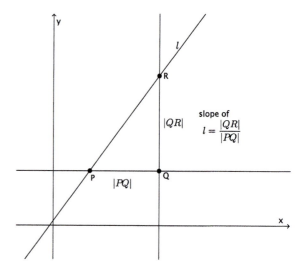

Example 1 (5 minutes)

Graph an equation in the form of $y = mx + b$.

> **Example 1**
>
> Graph the equation $y = \frac{2}{3}x + 1$. Name the slope and y-intercept point.
>
> The slope is $m = \frac{2}{3}$, and the y-intercept point is $(0, 1)$.

- To graph the equation, we must begin with the known point. In this case, that is the y-intercept point. We cannot begin with the slope because the slope describes the rate of change between two points. That means we need a point to begin with. On a graph, we plot the point $(0, 1)$.

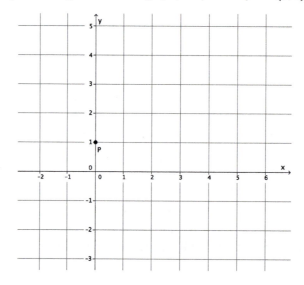

- Next, we use the slope to find the second point. We know that $m = \frac{|QR|}{|PQ|} = \frac{2}{3}$. The slope tells us exactly how many units to go to the right of P to find point Q and then how many vertical units we need to go from Q to find point R. How many units will we go to the right in order to find point Q? How do you know?
    - We need to go 3 units to the right of point P to find Q. We go 3 units because $|PQ| = 3$.

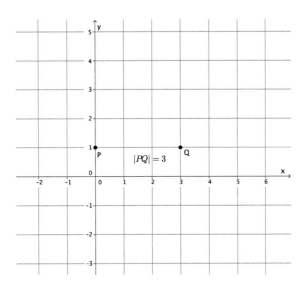

- How many vertical units from point Q must we go to find point R? How do you know?
    - We need to go 2 units from point Q to find R. We go 2 units because $|QR| = 2$.
- Will we go up from point Q or down from point Q to find R? How do you know?
    - We need to go up because the slope is positive. That means that the line will be left-to-right inclining.

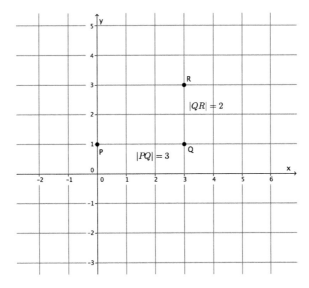

- Since we know that the line joining two distinct points of the form $y = mx + b$ has slope m, and we specifically constructed points P and R with the slope in mind, we can join the points with a line.

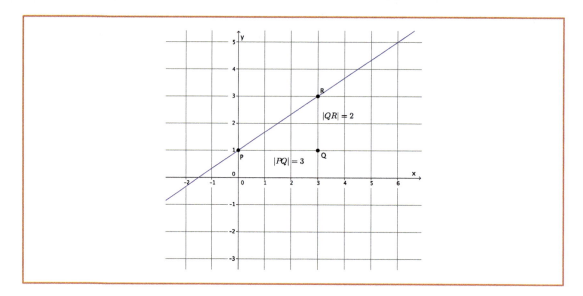

Example 2 (4 minutes)

Graph an equation in the form of $y = mx + b$.

> **Example 2**
>
> Graph the equation $y = -\frac{3}{4}x - 2$. Name the slope and y-intercept point.
>
> The slope is $m = -\frac{3}{4}$, and the y-intercept point is $(0, -2)$.

- How do we begin?
    - We must begin by putting a known point on the graph, $(0, -2)$.

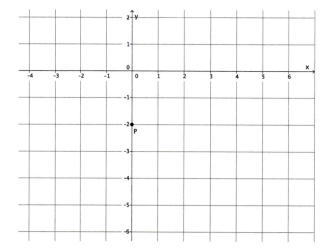

- We know that $m = \frac{|QR|}{|PQ|} = -\frac{3}{4}$. How many units will we go to the right in order to find point Q? How do you know?
  - We need to go 4 units to the right of point P to find Q. We go 4 units because $|PQ| = 4$.

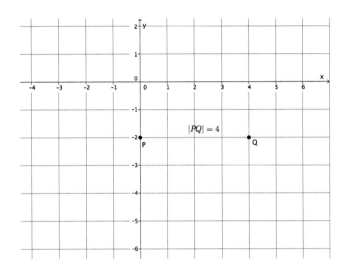

- How many units from point Q must we go to find point R? How do you know?
  - We need to go 3 units from point Q to find R. We go 3 units because $|QR| = 3$.
- Will we go up from point Q or down from point Q to find R? How do you know?
  - We need to go down from point Q to point R because the slope is negative. That means that the line will be left-to-right declining.

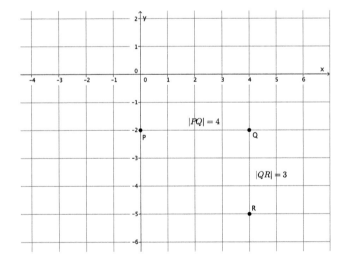

- Now we draw the line through the points P and R.

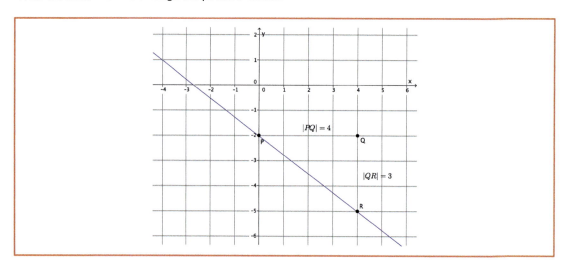

Example 3 (4 minutes)

Graph an equation in the form of $y = mx + b$.

> **Example 3**
>
> Graph the equation $y = 4x - 7$. Name the slope and y-intercept point.
>
> The slope is $m = 4$, and the y-intercept point is $(0, -7)$.

- Graph the equation $y = 4x - 7$. Name the slope and y-intercept point.
    - The slope is $m = 4$, and the y-intercept point is $(0, -7)$.
- How do we begin?
    - We must begin by putting a known point on the graph, $(0, -7)$.

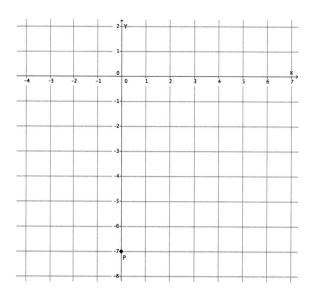

- Notice this time that the slope is the integer 4. In the last two examples, our slopes have been in the form of a fraction so that we can use the information in the numerator and denominator to determine the lengths of $|PQ|$ and $|QR|$. Since $m = \frac{|QR|}{|PQ|} = 4$, what fraction can we use to represent slope to help us graph?
  - The number 4 is equivalent to the fraction $\frac{4}{1}$.
- Using $m = \frac{|QR|}{|PQ|} = \frac{4}{1}$, how many units will we go to the right in order to find point Q? How do you know?
  - We need to go 1 unit to the right of point P to find Q. We go 1 unit because $|PQ| = 1$.

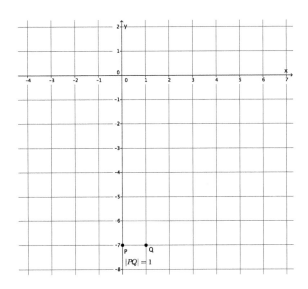

- How many vertical units from point Q must we go to find point R? How do you know?
  - We need to go 4 units from point Q to find R. We go 4 units because $|QR| = 4$.
- Will we go up from point Q or down from point Q to find R? How do you know?
  - We need to go up from point Q to point R because the slope is positive. That means that the line will be left-to-right inclining.

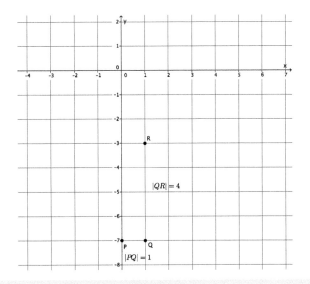

- Now we join the points P and R to make the line.

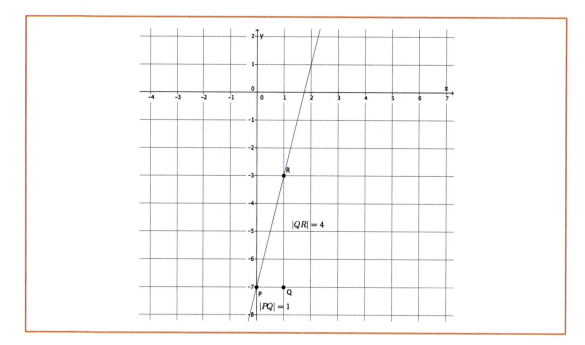

A STORY OF RATIOS   Lesson 18   8•4

Exercises 1–4 (5 minutes)

Students complete Exercises 1–4 individually or in pairs.

Exercises

1. Graph the equation $y = \frac{5}{2}x - 4$.

   a. Name the slope and the y-intercept point.

   The slope is $m = \frac{5}{2}$, and the y-intercept point is $(0, -4)$.

   b. Graph the known point, and then use the slope to find a second point before drawing the line.

   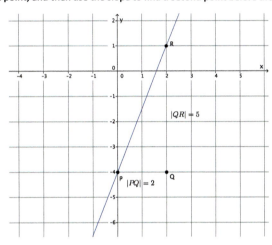

2. Graph the equation $y = -3x + 6$.

   a. Name the slope and the y-intercept point.

   The slope is $m = -3$, and the y-intercept point is $(0, 6)$.

   b. Graph the known point, and then use the slope to find a second point before drawing the line.

   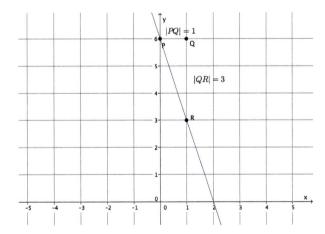

Lesson 18:   There Is Only One Line Passing Through a Given Point with a Given Slope

3. The equation $y = 1x + 0$ can be simplified to $y = x$. Graph the equation $y = x$.

   a. Name the slope and the y-intercept point.

   The slope is $m = 1$, and the y-intercept point is $(0, 0)$.

   b. Graph the known point, and then use the slope to find a second point before drawing the line.

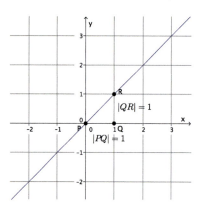

4. Graph the point $(0, 2)$.

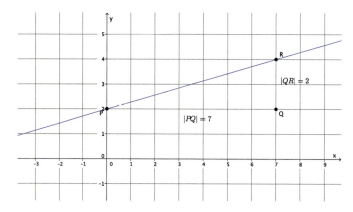

   a. Find another point on the graph using the slope, $m = \frac{2}{7}$.

   b. Connect the points to make the line.

   c. Draw a different line that goes through the point $(0, 2)$ with slope $m = \frac{2}{7}$. What do you notice?

   Only one line can be drawn through the given point with the given slope.

Discussion (5 minutes)

The following proof is optional. Exercises 5 and 6, below the Discussion, can be used as an alternative.

- Now we must show that if two straight lines have the same slope and pass through the same point, then they are the same line, which is what you observed in Exercise 4.

- We let l and l' be two lines with the same slope m passing through the same point P. Since m is a number, m could be positive, negative, or equal to zero. For this proof, we will let $m > 0$. Since $m > 0$, both lines l and l' are left-to-right inclining.

- Since we are trying to show that l and l' are the same line, let's assume that they are not, and prove that our assumption is false.

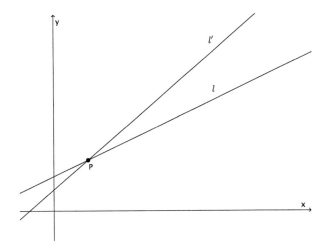

- Using our first understanding of slope, we will pick a point Q one unit to the right of point P. Then, we will draw a line parallel to the y-axis going through point Q, as shown.

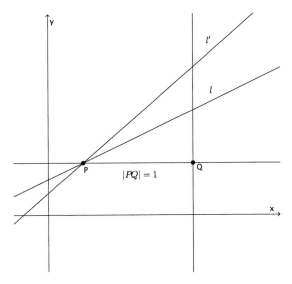

Lesson 18: There Is Only One Line Passing Through a Given Point with a Given Slope

- Now, we label the point of intersection of the line we just drew and line l and l' as R and R', respectively.

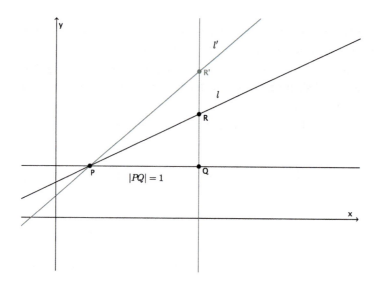

- By definition of slope, the length $|QR|$ is the slope of line l, and the length of $|QR'|$ is the slope of line l'. What do we know about the slopes of lines l and l'?
  - *The slopes are the same.*
- Then what do we know about the lengths $|QR|$ and $|QR'|$?
  - *They must be equal.*
- For that reason, points R and R' must coincide. This means lines l and l' are the same line, not as we originally drew them. Therefore, there is just one line passing through a given point with a given slope.

Exercises 5–6 (6 minutes)

Students complete Exercises 5–6 individually or in pairs.

> 5. A bank put $10 into a savings account when you opened the account. Eight weeks later, you have a total of $24. Assume you saved the same amount every week.
>
>   a. If y is the total amount of money in the savings account and x represents the number of weeks, write an equation in the form $y = mx + b$ that describes the situation.
>
> $24 = m(8) + 10$
> $14 = 8m$
> $\dfrac{14}{8} = m$
> $\dfrac{7}{4} = m$
> $y = \dfrac{7}{4}x + 10$

b. Identify the slope and the y-intercept point. What do these numbers represent?

The slope is $\frac{7}{4}$, and the y-intercept point is $(0, 10)$. The y-intercept point represents the amount of money the bank gave me, in the amount of 10. The slope represents the amount of money I save each week, $\frac{7}{4} = \$1.75$.

c. Graph the equation on a coordinate plane.

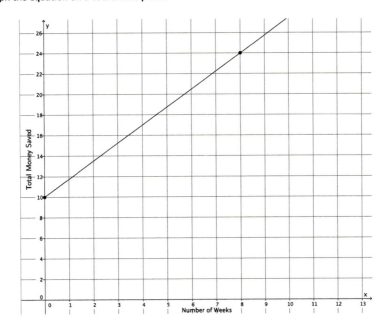

d. Could any other line represent this situation? For example, could a line through point $(0, 10)$ with slope $\frac{7}{5}$ represent the amount of money you save each week? Explain.

No, a line through point $(0, 10)$ with slope $\frac{7}{5}$ cannot represent this situation. That line would show that at the end of the 8 weeks I would have 21.20, but I was told that I would have 24 by the end of the 8 weeks.

6. A group of friends are on a road trip. After 120 miles, they stop to eat lunch. They continue their trip and drive at a constant rate of 50 miles per hour.

   a. Let y represent the total distance traveled, and let x represent the number of hours driven after lunch. Write an equation to represent the total number of miles driven that day.

   $y = 50x + 120$

   b. Identify the slope and the y-intercept point. What do these numbers represent?

   The slope is 50 and represents the rate of driving. The y-intercept point is 120 and represents the number of miles they had already driven before driving at the given constant rate.

c. Graph the equation on a coordinate plane.

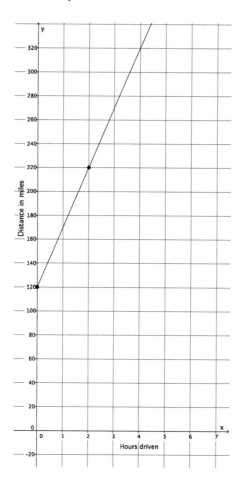

d. Could any other line represent this situation? For example, could a line through point $(0, 120)$ with slope 75 represent the total distance the friends drive? Explain.

No, a line through point $(0, 120)$ with a slope of 75 could not represent this situation. That line would show that after an hour, the friends traveled a total distance of 195 miles. According to the information given, the friends would only have traveled 170 miles after one hour.

Closing (4 minutes)

Summarize, or ask students to summarize, the main points from the lesson:

- We know that in the equation $y = mx + b$, $(0, b)$ is the location where the graph of the line intersects the y-axis.
- We know how to graph a line using a point, namely, the y-intercept point, and the slope.
- We know that there is only one line with a given slope passing through a given point.

Lesson Summary

The equation $y = mx + b$ is in slope-intercept form. The number m represents the slope of the graph, and the point $(0, b)$ is the location where the graph of the line intersects the y-axis.

To graph a line from the slope-intercept form of a linear equation, begin with the known point, $(0, b)$, and then use the slope to find a second point. Connect the points to graph the equation.

There is only one line passing through a given point with a given slope.

Exit Ticket (4 minutes)

Name _____ Date _____

Lesson 18: There Is Only One Line Passing Through a Given Point with a Given Slope

Exit Ticket

Mrs. Hodson said that the graphs of the equations below are incorrect. Find the student's errors, and correctly graph the equations.

1. Student graph of $y = \frac{1}{2}x + 4$:

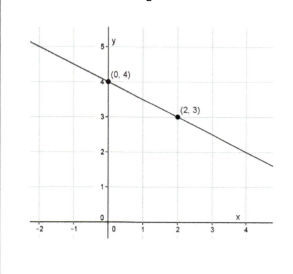

Error:

Correct graph of the equation:

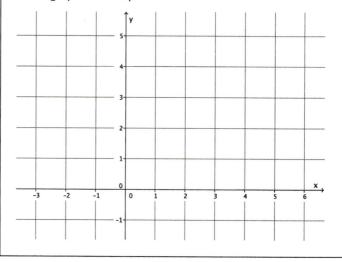

2. Student graph of $y = -\frac{3}{5}x - 1$:

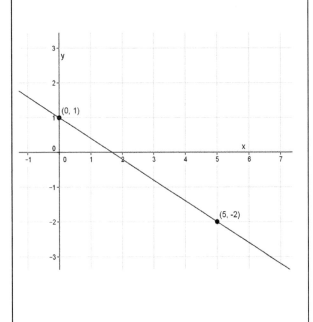

Error:

Correct graph of the equation:

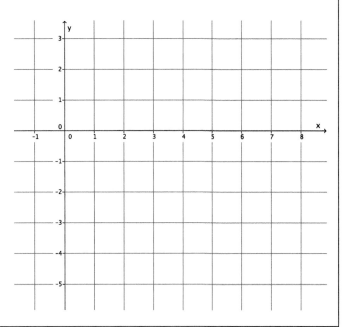

Exit Ticket Sample Solutions

Mrs. Hodson said that the graphs of the equations below are incorrect. Find the student's errors, and correctly graph the equations.

1. Student graph of the equation $y = \frac{1}{2}x + 4$:

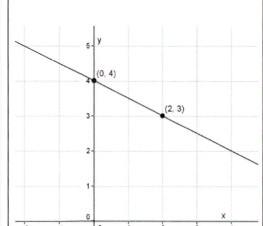

Error: *The student should have gone up 1 unit when finding $|QR|$ since the slope is positive.*

Correct graph of the equation:

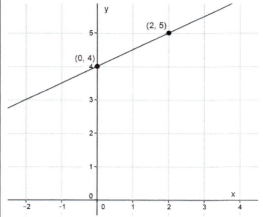

2. Student graph of the equation $y = -\frac{3}{5}x - 1$:

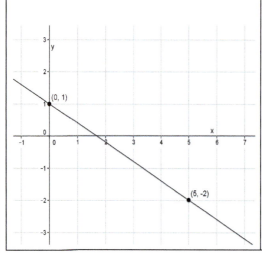

Error: *The student did not find the y-intercept point correctly. It should be the point $(0, -1)$.*

Correct graph of the equation:

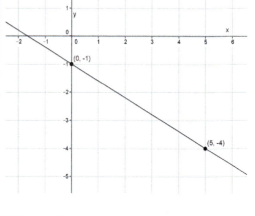

Problem Set Sample Solutions

Students practice graphing equations using y-intercept point and slope. Students need graph paper to complete the Problem Set. Optional Problem 11 has students show that there is only one line passing through a point with a given negative slope.

Graph each equation on a separate pair of x- and y-axes.

1. Graph the equation $y = \frac{4}{5}x - 5$.

   a. Name the slope and the y-intercept point.

      The slope is $m = \frac{4}{5}$, and the y-intercept point is $(0, -5)$.

   b. Graph the known point, and then use the slope to find a second point before drawing the line.

   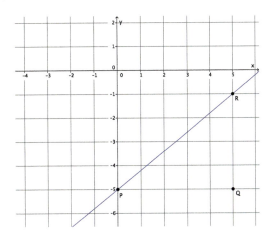

2. Graph the equation $y = x + 3$.

   a. Name the slope and the y-intercept point.

      The slope is $m = 1$, and the y-intercept point is $(0, 3)$.

   b. Graph the known point, and then use the slope to find a second point before drawing the line.

   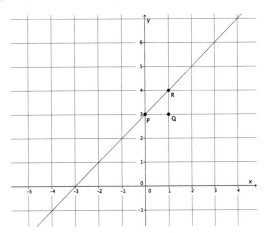

3. Graph the equation $y = -\frac{4}{3}x + 4$.

   a. Name the slope and the y-intercept point.

   The slope is $m = -\frac{4}{3}$, and the y-intercept point is $(0, 4)$.

   b. Graph the known point, and then use the slope to find a second point before drawing the line.

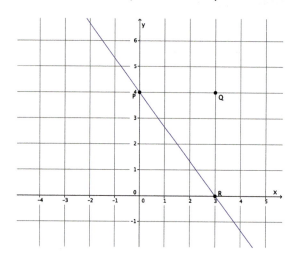

4. Graph the equation $y = \frac{5}{2}x$.

   a. Name the slope and the y-intercept point.

   The slope is $m = \frac{5}{2}$, and the y-intercept point is $(0, 0)$.

   b. Graph the known point, and then use the slope to find a second point before drawing the line.

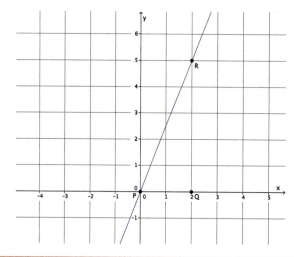

5. Graph the equation $y = 2x - 6$.
   a. Name the slope and the y-intercept point.

   The slope is $m = 2$, and the y-intercept point is $(0, -6)$.

   b. Graph the known point, and then use the slope to find a second point before drawing the line.

   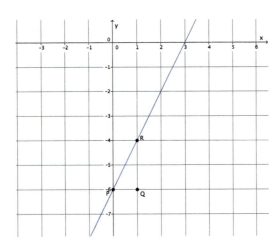

6. Graph the equation $y = -5x + 9$.
   a. Name the slope and the y-intercept point.

   The slope is $m = -5$, and the y-intercept point is $(0, 9)$.

   b. Graph the known point, and then use the slope to find a second point before drawing the line.

   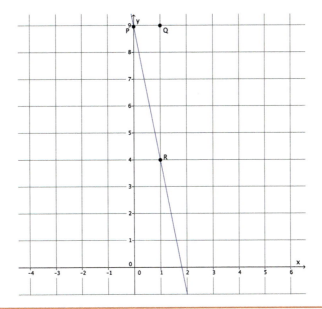

7. Graph the equation $y = \frac{1}{3}x + 1$.

   a. Name the slope and the y-intercept point.

   The slope is $m = \frac{1}{3}$, and the y-intercept point is $(0, 1)$.

   b. Graph the known point, and then use the slope to find a second point before drawing the line.

   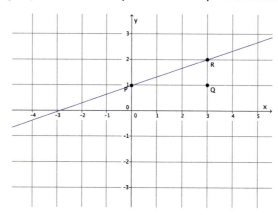

8. Graph the equation $5x + 4y = 8$. (Hint: Transform the equation so that it is of the form $y = mx + b$.)

   a. Name the slope and the y-intercept point.

   $$5x + 4y = 8$$
   $$5x - 5x + 4y = 8 - 5x$$
   $$4y = 8 - 5x$$
   $$\frac{4}{4}y = \frac{8}{4} - \frac{5}{4}x$$
   $$y = 2 - \frac{5}{4}x$$
   $$y = -\frac{5}{4}x + 2$$

   The slope is $m = -\frac{5}{4}$, and the y-intercept point is $(0, 2)$.

   b. Graph the known point, and then use the slope to find a second point before drawing the line.

   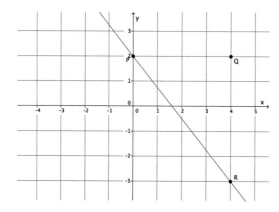

9. Graph the equation $-2x + 5y = 30$.

   a. Name the slope and the y-intercept point.

   $$-2x + 5y = 30$$
   $$2x + 2x + 5y = 30 + 2x$$
   $$5y = 30 + 2x$$
   $$\frac{5}{5}y = \frac{30}{5} + \frac{2}{5}x$$
   $$y = 6 + \frac{2}{5}x$$
   $$y = \frac{2}{5}x + 6$$

   The slope is $m = \frac{2}{5}$, and the y-intercept point is $(0, 6)$.

   b. Graph the known point, and then use the slope to find a second point before drawing the line.

   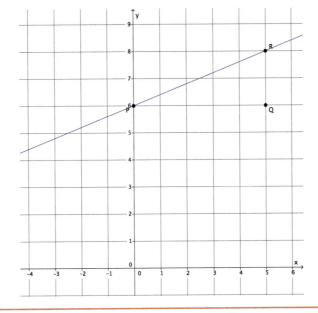

Lesson 18: There Is Only One Line Passing Through a Given Point with a Given Slope

10. Let l and l' be two lines with the same slope m passing through the same point P. Show that there is only one line with a slope m, where $m < 0$, passing through the given point P. Draw a diagram if needed.

First, assume that there are two different lines l and l' with the same negative slope passing through P. From point P, I mark a point Q one unit to the right. Then, I draw a line parallel to the y-axis through point Q. The intersection of this line and line l and l' are noted with points R and R', respectively. By definition of slope, the lengths $|QR|$ and $|QR'|$ represent the slopes of lines l and l', respectively. We are given that the lines have the same slope, which means that lengths $|QR|$ and $|QR'|$ are equal. Since that is true, then points R and R' coincide and so do lines l and l'. Therefore, our assumption that they are different lines is false; l and l' must be the same line. Therefore, there is only one line with slope m passing through the given point P.

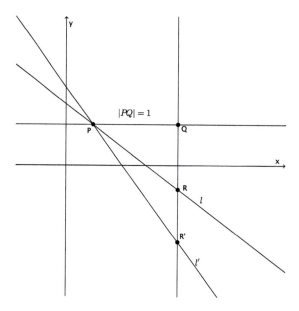

Lesson 19: The Graph of a Linear Equation in Two Variables Is a Line

Student Outcomes

- Students prove that any point on the graph of $y = mx + b$ is on a line l and that any point on a line l is a point on the graph of $y = mx + b$.
- Students graph linear equations on the coordinate plane.

Lesson Notes

Included in this and the next several lessons is an analytical approach to understanding linear equations and their graphs. Though the analytical understandings are not *required* for Grade 8 students, it is of tremendous value for students to understand the proofs of common assertions about lines. For that reason, these lessons are structured to offer extensions (i.e., the proofs) in cases where students are ready and also to provide alternate activities that deepen conceptual understanding. The theorems are presented to students and, based on the readiness of each class, the teacher can choose to discuss the proof of the theorem, or students can explore the theorem experimentally by completing the exercises of the alternate activity.

This lesson is the result of many lessons leading up to this one. In previous lessons, it was stated that there were four things that needed to be proved about lines before it could be stated, definitively, that the graph of a linear equation in two variables is, in fact, a line.

(1) A number must be defined for each non-vertical line that can be used to measure the "steepness" or "slant" of the line. Once defined, this number is called the *slope* of the line.

(2) It must be shown that *any* two points on a non-vertical line can be used to find the slope of the line.

(3) It must be shown that the line joining two points on the graph of a linear equation of the form $y = mx + b$ has slope m.

(4) It must be shown that there is only one line passing through a given point with a given slope.

This lesson is the conclusion of that work, and it is an important result for students to observe. There are two options for the first part of the lesson. The first option is to use the formal proof in the Discussion and Exercise 1. This lesson path skips Exercises 2–5 and resumes the lesson with Example 1. The other option is to skip the Discussion and Exercise 1 on the first five pages, having students examine several equations in the form of $ax + by = c$, create their graphs, and informally conclude that the graphs of equations of this type always result in lines. The alternate activity begins with Exercises 2–8. Following these exercises is Example 1 and Exercises 9–11 where students learn how to graph a linear equation using intercepts. Students need graph paper to complete the Exercises and the Problem Set.

Classwork

Discussion (15 minutes)

Now that all of the necessary tools have been developed, we are ready to show that the graph of a linear equation in two variables in the form $y = mx + b$, where b is always a constant, is a non-vertical line.

- To prove that the graph of a linear equation is a line, we need to show the following:

  (1) Any point on the graph of $y = mx + b$ is a point on a line l, and

  (2) Any point on the line l is a point on the graph of $y = mx + b$.

- Why do we need to show both (1) and (2)? Consider the following: Let P and Q be two distinct points on the graph of $y = mx + b$. Let l be the line passing through points P and Q. Also, let the graph of $y = mx + b$ be the collection of red segments denoted by ⇒, as shown.

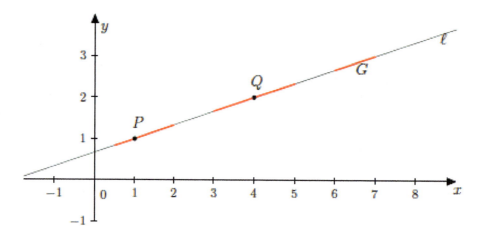

If ⇒ is the collection of red segments, as shown, then (1) is true, but (2) is not. Therefore, if we can show that both (1) and (2) are true, then we know that the graph of $y = mx + b$ is exactly the same as the graph of line l.

- Let's look at a specific case before proving (1) and (2) in general. Show that the graph of $y = 2x + 3$ is a non-vertical line.

- We need two points that we know are on the graph of $y = 2x + 3$. Are points $(0, 3)$ and $(4, 11)$ solutions to the equation $y = 2x + 3$? Explain.

  □ Yes, points $(0, 3)$ and $(4, 11)$ are solutions because $3 = 2(0) + 3$ and $11 = 2(4) + 3$ are true statements.

- Let line l pass through the points $P(0, 3)$ and $Q(1, 5)$. By our work in Lesson 17, we know that an equation in the form of $y = mx + b$ has slope m. Then, the slope of $y = 2x + 3$ is 2; therefore, the slope of line l is 2. We claim that the graph of $y = 2x + 3$ is the line l.

- Now we will show that *any* point on the graph of $y = 2x + 3$ is on line l. Let R be any point on the graph of $y = 2x + 3$ (other than points P and Q because if $R = P$ or if $R = Q$, there would be nothing to prove; we already showed that they are solutions to the equation $y = 2x + 3$). Let l' be the line passing through points P and R, as shown.

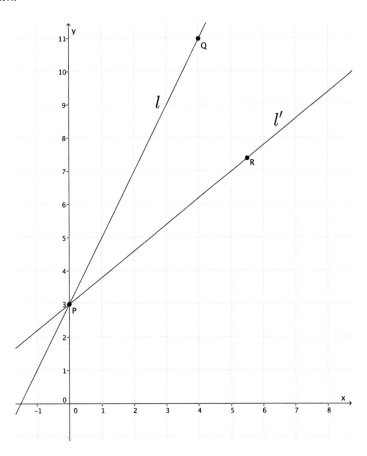

- We need to show that line l' has a slope of 2. We have done this before in Lesson 16. Since P and R are on the graph of $y = 2x + 3$, then the coordinates of P and R are $(p_1, 2p_1 + 3)$ and $(r_1, 2r_1 + 3)$, respectively. We also know the following:

$$\text{slope of } l' = \frac{(2p_1 + 3) - (2r_1 + 3)}{p_1 - r_1}$$
$$= \frac{2p_1 - 2r_1}{p_1 - r_1}$$
$$= \frac{2(p_1 - r_1)}{p_1 - r_1}$$
$$= 2$$

- Now that we know the slope of l' is 2, what can we say about lines l and l'? Hint: What do we know about lines that have the same slope and pass through the same point?
    - *Lines l and l' must be the same line. There can be only one line with a slope of 2 that passes through a given point, in this case P.*

Lesson 19: The Graph of a Linear Equation in Two Variables Is a Line

- Therefore, point R must be on line l, and the proof that any point on the graph of $y = 2x + 3$ is on line l is finished.

- Now we must show that *any* point on l is on the graph of $y = 2x + 3$. Let S be *any* point on l (except for P and Q, for the same reason as before). Using $S(s_1, s_2)$ and $P(0, 3)$ in the slope formula, we get the following:

$$\text{slope of } l = \frac{s_2 - 3}{s_1 - 0}.$$

We know the slope of l is 2.

$$2 = \frac{s_2 - 3}{s_1 - 0}$$
$$2(s_1 - 0) = s_2 - 3$$
$$2s_1 = s_2 - 3$$
$$2s_1 + 3 = s_2 - 3 + 3$$
$$2s_1 + 3 = s_2$$

The equation above shows that point $S(s_1, s_2)$ is a solution to the equation $y = 2x + 3$ and must be on the graph of $y = 2x + 3$.

- We have just shown that any point R on the graph of $y = 2x + 3$ is on line l, and any point S on line l is on the graph of $y = 2x + 3$. Therefore, the proof is complete.

Exercise 1 (10 minutes)

Students work in pairs or small groups to prove the theorem in general. This is an optional exercise and related directly to the optional Discussion. If the teacher chose the alternative path for this lesson, then this exercise should not be assigned.

Exercises 1–11

THEOREM: The graph of a linear equation $y = mx + b$ is a non-vertical line with slope m and passing through $(0, b)$, where b is a constant.

1. Prove the theorem by completing parts (a)–(c). Given two distinct points, P and Q, on the graph of $y = mx + b$, and let l be the line passing through P and Q. You must show the following:

   (1) Any point on the graph of $y = mx + b$ is on line l, and

   (2) Any point on the line l is on the graph of $y = mx + b$.

   a. Proof of (1): Let R be any point on the graph of $y = mx + b$. Show that R is on l. Begin by assuming it is not. Assume the graph looks like the diagram below where R is on l'.

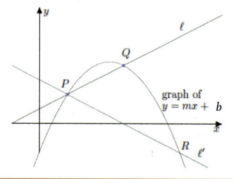

What is the slope of line l?

Since the points P and Q are on the graph of $y = mx + b$, then we know that the slope of the line passing through those points must have slope m. Therefore, line l has slope m.

What is the slope of line l'?

We know that point R is on the graph of $y = mx + b$. Then the coordinates of point R are $(r_1, mr_1 + b)$ because R is a solution to $y = mx + b$, and $r_2 = mr_1 + b$. Similarly, the coordinates of P are $(p_1, mp_1 + b)$.

$$\text{slope of } l' = \frac{(mp_1 + b) - (mr_1 + b)}{p_1 - r_1}$$
$$= \frac{mp_1 + b - mr_1 - b}{p_1 - r_1}$$
$$= \frac{mp_1 - mr_1}{p_1 - r_1}$$
$$= \frac{m(p_1 - r_1)}{p_1 - r_1}$$
$$= m$$

What can you conclude about lines l and l'? Explain.

Lines l and l' are the same line. Both lines go through point P and have slope m. There can be only one line with a given slope going through a given point; therefore, line l is the same as l'.

b. Proof of (2): Let S be any point on line l, as shown.

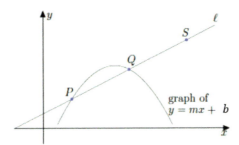

Show that S is a solution to $y = mx + b$. Hint: Use the point $(0, b)$.

Point S is on line l. Let $S = (s_1, s_2)$.

$$\text{slope of } l = \frac{s_2 - b}{s_1 - 0}$$

We know the slope of l is m.

$$m = \frac{s_2 - b}{s_1 - 0}$$
$$m(s_1 - 0) = s_2 - b$$
$$ms_1 = s_2 - b$$
$$ms_1 + b = s_2 - b + b$$
$$ms_1 + b = s_2$$
$$s_2 = ms_1 + b,$$

which shows S is a solution to $y = mx + b$.

Lesson 19: The Graph of a Linear Equation in Two Variables Is a Line

c. Now that you have shown that any point on the graph of $y = mx + b$ is on line l in part (a), and any point on line l is on the graph of $y = mx + b$ in part (b), what can you conclude about the graphs of linear equations?

The graph of a linear equation is a line.

Exercises 2–8 (25 minutes)

These exercises should be completed in place of the Discussion and Exercise 1. Students need graph paper to complete the following exercises. This is the alternate activity that was described in the Lesson Notes. For these exercises, students graph a linear equation and informally verify that any point on the graph of the equation is on the line and that any point on the line is a point on the graph of the equation. Exercises 6–8 ask students to draw these conclusions based on their work in Exercises 2–5. Consider having a class discussion based on Exercises 6–8, where students share their conclusions. End the discussion by making it clear that the graph of a linear equation is a line. Then, proceed with Example 1 and Exercises 9–11.

2. Use $x = 4$ and $x = -4$ to find two solutions to the equation $x + 2y = 6$. Plot the solutions as points on the coordinate plane, and connect the points to make a line.

   The solutions are $(4, 1)$ and $(-4, 5)$.

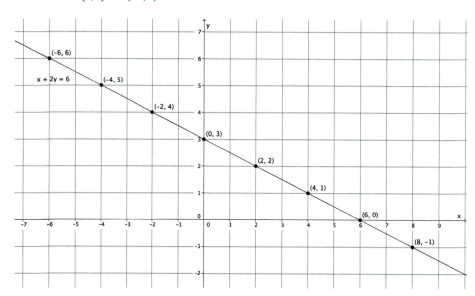

   a. Identify two other points on the line with integer coordinates. Verify that they are solutions to the equation $x + 2y = 6$.

   The choice of points and verifications will vary. Several possibilities are noted in the graph above.

   b. When $x = 1$, what is the value of y? Does this solution appear to be a point on the line?

   $$1 + 2y = 6$$
   $$2y = 5$$
   $$y = \frac{5}{2}$$

   Yes, $\left(1, \frac{5}{2}\right)$ does appear to be a point on the line.

c. When $x = -3$, what is the value of y? Does this solution appear to be a point on the line?

$$-3 + 2y = 6$$
$$2y = 9$$
$$y = \frac{9}{2}$$

Yes, $\left(-3, \frac{9}{2}\right)$ does appear to be a point on the line.

d. Is the point $(3, 2)$ on the line?

No, $(3, 2)$ is not a point on the line.

e. Is the point $(3, 2)$ a solution to the linear equation $x + 2y = 6$?

No, $(3, 2)$ is not a solution to $x + 2y = 6$.
$$3 + 2(2) = 6$$
$$3 + 4 = 6$$
$$7 \neq 6$$

3. Use $x = 4$ and $x = 1$ to find two solutions to the equation $3x - y = 9$. Plot the solutions as points on the coordinate plane, and connect the points to make a line.

The solutions are $(4, 3)$ and $(1, -6)$.

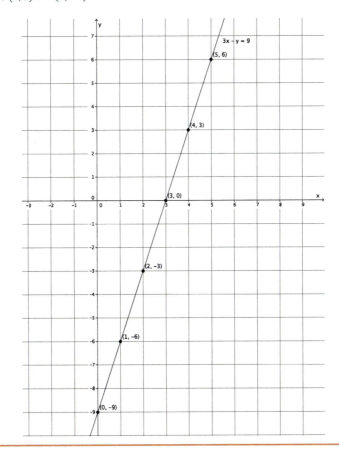

a. Identify two other points on the line with integer coordinates. Verify that they are solutions to the equation $3x - y = 9$.

The choice of points and verifications will vary. Several possibilities are noted in the graph above.

b. When $x = 4.5$, what is the value of y? Does this solution appear to be a point on the line?

$$3(4.5) - y = 9$$
$$13.5 - y = 9$$
$$-y = -4.5$$
$$y = 4.5$$

Yes, $(4.5, 4.5)$ does appear to be a point on the line.

c. When $x = \frac{1}{2}$, what is the value of y? Does this solution appear to be a point on the line?

$$3\left(\frac{1}{2}\right) - y = 9$$
$$\frac{3}{2} - y = 9$$
$$-y = \frac{15}{2}$$
$$y = -\frac{15}{2}$$

Yes, $\left(\frac{1}{2}, -\frac{15}{2}\right)$ does appear to be a point on the line.

d. Is the point $(2, 4)$ on the line?

No, $(2, 4)$ is not a point on the line.

e. Is the point $(2, 4)$ a solution to the linear equation $3x - y = 9$?

No, $(2, 4)$ is not a solution to $3x - y = 9$.

$$3(2) - 4 = 9$$
$$6 - 4 = 9$$
$$2 \neq 9$$

4. Use $x = 3$ and $x = -3$ to find two solutions to the equation $2x + 3y = 12$. Plot the solutions as points on the coordinate plane, and connect the points to make a line.

   The solutions are $(-3, 6)$ and $(3, 2)$.

   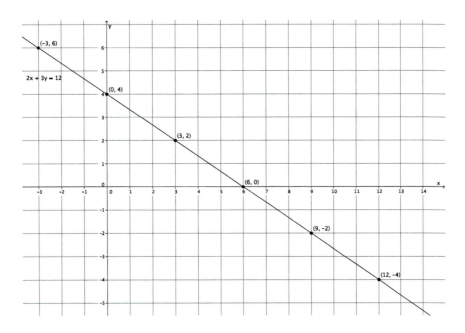

   a. Identify two other points on the line with integer coordinates. Verify that they are solutions to the equation $2x + 3y = 12$.

   The choice of points and verifications will vary. Several possibilities are noted in the graph above.

   b. When $x = 2$, what is the value of y? Does this solution appear to be a point on the line?

   $$2(2) + 3y = 12$$
   $$4 + 3y = 12$$
   $$3y = 8$$
   $$y = \frac{8}{3}$$

   Yes, $\left(2, \frac{8}{3}\right)$ does appear to be a point on the line.

   c. When $x = -2$, what is the value of y? Does this solution appear to be a point on the line?

   $$2(-2) + 3y = 12$$
   $$-4 + 3y = 12$$
   $$3y = 16$$
   $$y = \frac{16}{3}$$

   Yes, $\left(-2, \frac{16}{3}\right)$ does appear to be a point on the line.

   d. Is the point $(8, -3)$ on the line?

   No, $(8, -3)$ is not a point on the line.

e. Is the point $(8, -3)$ a solution to the linear equation $2x + 3y = 12$?

No, $(8, -3)$ is not a solution to $2x + 3y = 12$.

$$2(8) + 3(-3) = 12$$
$$16 - 9 = 12$$
$$7 \neq 12$$

5. Use $x = 4$ and $x = -4$ to find two solutions to the equation $x - 2y = 8$. Plot the solutions as points on the coordinate plane, and connect the points to make a line.

The solutions are $(4, -2)$ and $(-4, -6)$.

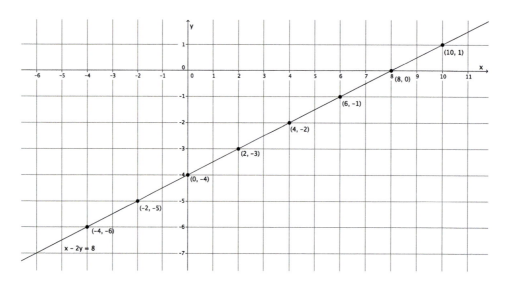

a. Identify two other points on the line with integer coordinates. Verify that they are solutions to the equation $x - 2y = 8$.

The choice of points and verifications will vary. Several possibilities are noted in the graph above.

b. When $x = 7$, what is the value of y? Does this solution appear to be a point on the line?

$$7 - 2y = 8$$
$$-2y = 1$$
$$y = -\frac{1}{2}$$

Yes, $\left(7, -\frac{1}{2}\right)$ does appear to be a point on the line.

c. When $x = -3$, what is the value of y? Does this solution appear to be a point on the line?

$$-3 - 2y = 8$$
$$-2y = 11$$
$$y = -\frac{11}{2}$$

Yes, $\left(-3, -\frac{11}{2}\right)$ does appear to be a point on the line.

d. Is the point $(-2, -3)$ on the line?

No, $(-2, -3)$ is not a point on the line.

e. Is the point $(-2, -3)$ a solution to the linear equation $x - 2y = 8$?

No, $(-2, -3)$ is not a solution to $x - 2y = 8$.

$$-2 - 2(-3) = 8$$
$$-2 + 6 = 8$$
$$4 \neq 8$$

6. Based on your work in Exercises 2–5, what conclusions can you draw about the points on a line and solutions to a linear equation?

   It appears that all points on the line represent a solution to the equation. In other words, any point identified on the line is a solution to the linear equation.

MP.8

7. Based on your work in Exercises 2–5, will a point that is not a solution to a linear equation be a point on the graph of a linear equation? Explain.

   No. Each time we were given a point off the line in part (d), we verified that it was not a solution to the equation in part (e). For that reason, I would expect that all points not on the line would not be a solution to the equation.

8. Based on your work in Exercises 2–5, what conclusions can you draw about the graph of a linear equation?

   The graph of a linear equation is a line.

Example 1 (5 minutes)

- Now that we know that the graph of a linear equation in two variables is a line and that there is only one line that can pass through two points, then we can easily graph equations using intercepts.
- We already know that the y-intercept point is the location on the graph of a line where the line intersects the y-axis. The point of intersection will have coordinates $(0, y)$. Similarly, the x-intercept point is the location on the graph of a line where the line intersects the x-axis. The point of intersection will have coordinates $(x, 0)$.
- To graph using intercepts, simply replace the symbols x and y with zero, one at a time, and solve.
- Graph the equation: $2x + 3y = 9$.

Replace x with zero, and solve for y to determine the y-intercept point.

$$2(0) + 3y = 9$$
$$3y = 9$$
$$y = 3$$

The y-intercept point is at $(0, 3)$.

Replace y with zero, and solve for x to determine the x-intercept point.

$$2x + 3(0) = 9$$
$$2x = 9$$
$$x = \frac{9}{2}$$

The x-intercept point is at $\left(\frac{9}{2}, 0\right)$.

Lesson 19: The Graph of a Linear Equation in Two Variables Is a Line

- Now that we know the intercepts, we can place those two points on the graph and connect them to graph the linear equation $2x + 3y = 9$.

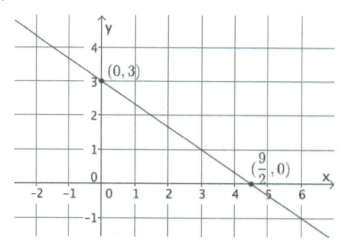

- Graphing using intercepts is an efficient way of graphing linear equations that are in standard form. Graphing using the slope and y-intercept point is the most efficient way of graphing linear equations that are in slope-intercept form. Creating a table and finding solutions is another way that we learned to graph linear equations. All three methods work, but some methods will save time depending on the form of the equation.

Exercises 9–11 (5 minutes)

9. Graph the equation $-3x + 8y = 24$ using intercepts.

$$-3x + 8y = 24$$
$$-3(0) + 8y = 24$$
$$8y = 24$$
$$y = 3$$

The y-intercept point is $(0, 3)$.

$$-3x + 8y = 24$$
$$-3x + 8(0) = 24$$
$$-3x = 24$$
$$x = -8$$

The x-intercept point is $(-8, 0)$.

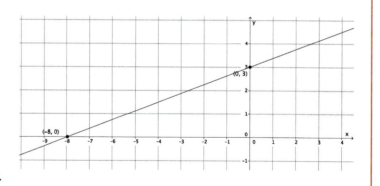

10. Graph the equation $x - 6y = 15$ using intercepts.

$$x - 6y = 15$$
$$0 - 6y = 15$$
$$-6y = 15$$
$$y = -\frac{15}{6}$$
$$y = -\frac{5}{2}$$

The y-intercept point is $\left(0, -\frac{5}{2}\right)$.

$$x - 6y = 15$$
$$x - 6(0) = 15$$
$$x = 15$$

The x-intercept point is $(15, 0)$.

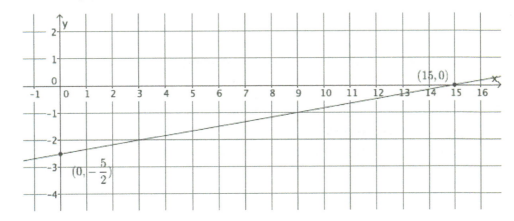

11. Graph the equation $4x + 3y = 21$ using intercepts.

$$4x + 3y = 21$$
$$4(0) + 3y = 21$$
$$3y = 21$$
$$y = 7$$

The y-intercept point is $(0, 7)$.

$$4x + 3y = 21$$
$$4x + 3(0) = 21$$
$$4x = 21$$
$$x = \frac{21}{4}$$

The x-intercept point is $\left(\frac{21}{4}, 0\right)$.

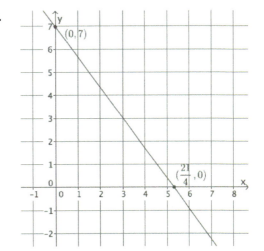

Closing (5 minutes)

Summarize, or ask students to summarize, the main points from the lesson:

- We know that to prove the graph of a linear equation is a line, we have to show that any point on the graph of the equation is on the line, and any point on the line is on the graph of the equation.
- We can use what we know about slope and the fact that there is only one line with a given slope that goes through a given point to prove that the graph of a linear equation is a line.
- We have another method for graphing linear equations: using intercepts.

Lesson Summary

The graph of a linear equation is a line. A linear equation can be graphed using two points: the x-intercept point and the y-intercept point.

Example:

Graph the equation: $2x + 3y = 9$.

Replace x with zero, and solve for y to determine the y-intercept point.

$$2(0) + 3y = 9$$
$$3y = 9$$
$$y = 3$$

The y-intercept point is at $(0, 3)$.

Replace y with zero, and solve for x to determine the x-intercept point.

$$2x + 3(0) = 9$$
$$2x = 9$$
$$x = \frac{9}{2}$$

The x-intercept point is at $\left(\frac{9}{2}, 0\right)$.

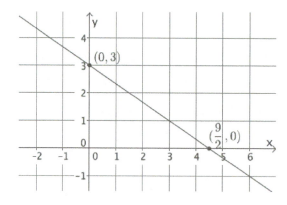

Exit Ticket (5 minutes)

Name _____ Date _____

Lesson 19: The Graph of a Linear Equation in Two Variables Is a Line

Exit Ticket

1. Graph the equation $y = \frac{5}{4}x - 10$ using the y-intercept point and slope.

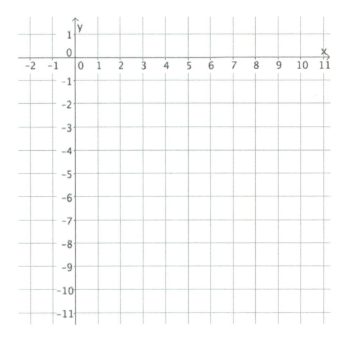

2. Graph the equation $5x - 4y = 40$ using intercepts.

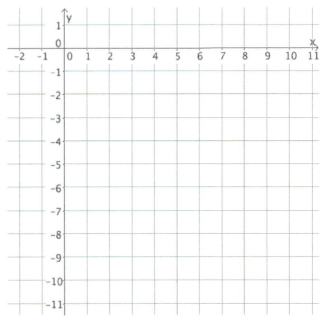

3. What can you conclude about the equations $y = \frac{5}{4}x - 10$ and $5x - 4y = 40$?

Exit Ticket Sample Solutions

1. Graph the equation $y = \frac{5}{4}x - 10$ using the y-intercept point and slope.

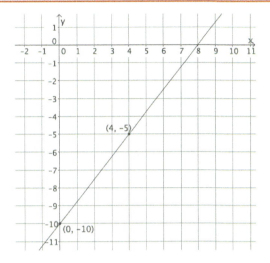

2. Graph the equation $5x - 4y = 40$ using intercepts.

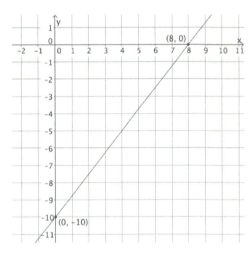

3. What can you conclude about the equations $y = \frac{5}{4}x - 10$ and $5x - 4y = 40$?

   Since the points $(0, -10)$, $(4, -5)$, and $(8, 0)$ are common to both graphs, then the lines must be the same. There is only one line that can pass through two points. If you transform the equation $y = \frac{5}{4}x - 10$ so that it is in standard form, it is the equation $5x - 4y = 40$.

Lesson 19: The Graph of a Linear Equation in Two Variables Is a Line

Problem Set Sample Solutions

Now that students know the graph of a linear equation is a line, students practice graphing linear equations in two variables using an appropriate method. Students need graph paper to complete the Problem Set.

1. Graph the equation: $y = -6x + 12$.

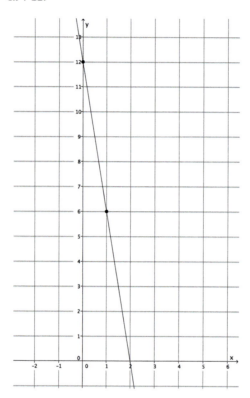

2. Graph the equation: $9x + 3y = 18$.

$$9(0) + 3y = 18$$
$$3y = 18$$
$$y = 6$$

The y-intercept point is $(0, 6)$.

$$9x + 3(0) = 18$$
$$9x = 18$$
$$x = 2$$

The x-intercept point is $(2, 0)$.

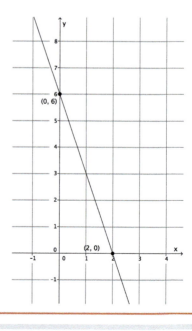

3. Graph the equation: $y = 4x + 2$.

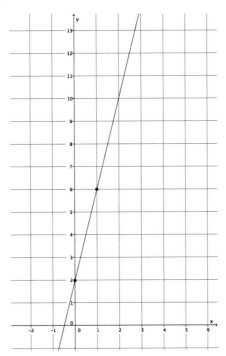

4. Graph the equation: $y = -\frac{5}{7}x + 4$.

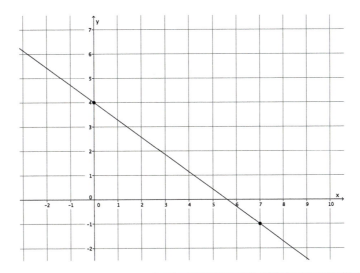

5. Graph the equation: $\frac{3}{4}x + y = 8$.

$$\frac{3}{4}(0) + y = 8$$
$$y = 8$$

The y-intercept point is $(0, 8)$.

$$\frac{3}{4}x + 0 = 8$$
$$\frac{3}{4}x = 8$$
$$x = \frac{32}{3}$$

The x-intercept point is $\left(\frac{32}{3}, 0\right)$.

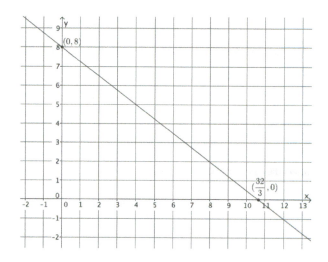

6. Graph the equation: $2x - 4y = 12$.

$$2(0) - 4y = 12$$
$$-4y = 12$$
$$y = -3$$

The y-intercept point is $(0, -3)$.

$$2x - 4(0) = 12$$
$$2x = 12$$
$$x = 6$$

The x-intercept point is $(6, 0)$.

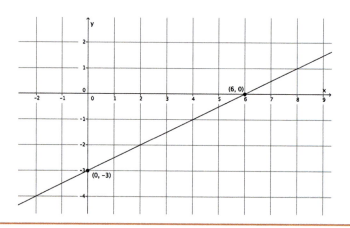

7. Graph the equation: $y = 3$. What is the slope of the graph of this line?

   The slope of this line is zero.

   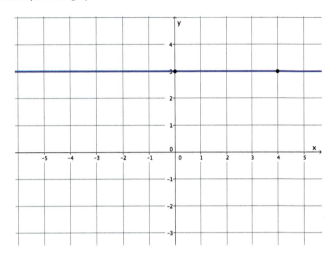

8. Graph the equation: $x = -4$. What is the slope of the graph of this line?

   The slope of this line is undefined.

   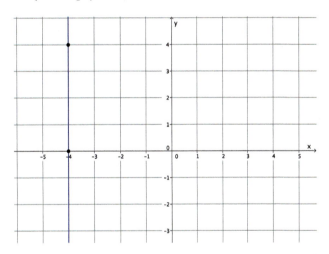

9. Is the graph of $4x + 5y = \frac{3}{7}$ a line? Explain.

   Yes, the graph of $4x + 5y = \frac{3}{7}$ is a line because it is a linear equation comprising linear expressions on both sides of the equal sign.

10. Is the graph of $6x^2 - 2y = 7$ a line? Explain.

    Maybe. The equation $6x^2 - 2y = 7$ is not a linear equation because the expression on the left side of the equal sign is not a linear expression. If this were a linear equation, then I would be sure that it graphs as a line, but because it is not, I am not sure what the graph of this equation would look like.

A STORY OF RATIOS   Lesson 20   8•4

Lesson 20: Every Line Is a Graph of a Linear Equation

Student Outcomes

- Students know that any non-vertical line is the graph of a linear equation in the form of $y = mx + b$, where b is a constant.
- Students write the linear equation whose graph is a given line.

Lesson Notes

The proof that every line is the graph of a linear equation in the Discussion is optional. If using the Discussion, skip the Opening Exercise, and resume the lesson with Example 1. Complete all other examples and exercises that follow. As an alternative to the Discussion, complete the Opening Exercise by showing a graph of a line on the coordinate plane and having students attempt to name the equation of the line. Two graphs are provided beginning on page 317. Have students write their equations and strategies for determining the equation of the line; then, lead the discussion described on page 317. Once students complete the Opening Exercise, work through Example 1 and the remaining examples and exercises in the lesson. Revisit the equations and strategies students developed by having them critique their reasoning in comparison to the work in the example; then, continue with the remainder of the lesson.

Classwork

Discussion (10 minutes)

- Now that we are confident that the graph of every linear equation is a line, can we say that every line is the graph of a linear equation? We can say yes with respect to vertical and horizontal lines; recall $x = c$ and $y = c$. But what about other non-vertical lines?
- We must prove that any non-vertical (and non-horizontal) line is a graph of a linear equation in the form of $y = mx + b$, where $m \neq 0$ and b are constants.
- Let l be any non-vertical (and non-horizontal) line. Suppose the slope of the line is m and that the line intersects the y-axis at point $Q(0, b)$.
- First, we show that any point on the line l is a point on the graph of the linear equation $y = mx + b$.
- Let $P(x, y)$ be any point on line l. We need to show that P is a solution to $y = mx + b$. Think about how we did this in the last lesson. What should we do?

    □ Use the points P and Q in the slope formula.

MP.3

$$m = \frac{y - b}{x - 0}$$
$$mx = y - b$$
$$mx + b = y - b + b$$
$$mx + b = y$$

- That shows that point P is a point on the graph of $y = mx + b$. Point Q is also on the graph of $y = mx + b$ because $b = m \cdot 0 + b$. Therefore, any point on the line l is a point on the graph of the linear equation $y = mx + b$.

- Now we want to show that any point on the graph of $y = mx + b$ is on l.
- Let R be any point on the graph of the linear equation $y = mx + b$. We know that the graph of $y = mx + b$ is a line with slope m. Let's call this line l'. We know that Q is on l' because $b = m \cdot 0 + b$. Therefore, l' is a line with slope m that passes through point Q. However, l is a line with slope m that passes through point Q. What does that mean about lines l and l'?
  - *The lines l and l' are the same line because there is only one line with a given slope that can go through a given point.*
- Therefore, R is a point on l.
- Now we can be certain that every line is a graph of a linear equation.

Opening Exercise (10 minutes)

Show students Figure 1 below, and challenge them to write the equation for the line. Provide students time to work independently and then in pairs. Lead a discussion where students share their strategies for developing the equation of the line. Ask students how they knew their equations were correct; that is, did they verify that the points with integer coordinates were solutions to the equations they wrote? Ask students what kind of equation they wrote: linear or nonlinear. Ask students if they were given another line, could they write an equation for it using their strategy. Show them Figure 2 and, again, ask them to write the equation of the line. Verify that they wrote the correct equation, and conclude the discussion by stating that every line is the graph of a linear equation.

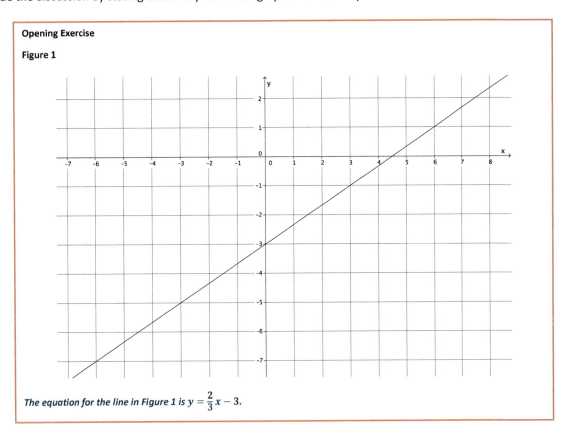

Opening Exercise

Figure 1

The equation for the line in Figure 1 is $y = \frac{2}{3}x - 3$.

Lesson 20: Every Line Is a Graph of a Linear Equation

A STORY OF RATIOS   Lesson 20   8•4

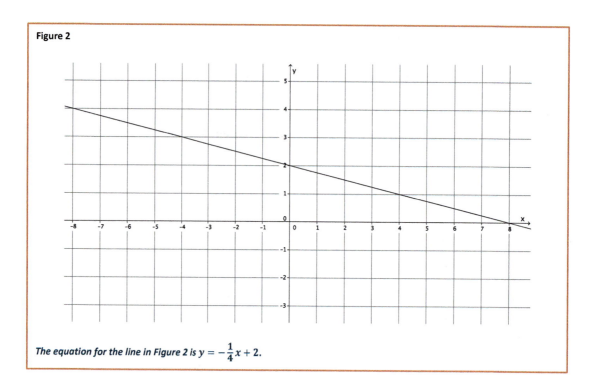

Figure 2

The equation for the line in Figure 2 is $y = -\frac{1}{4}x + 2$.

Example 1 (5 minutes)

- Given a line, we want to be able to write the equation that represents it.
- Which form of a linear equation do you think will be most valuable for this task: the standard form $ax + by = c$, or the slope-intercept form $y = mx + b$?
    - The slope-intercept form because we can easily identify the slope and y-intercept point from both the equation and the graph.
- Write the equation that represents the line shown below.

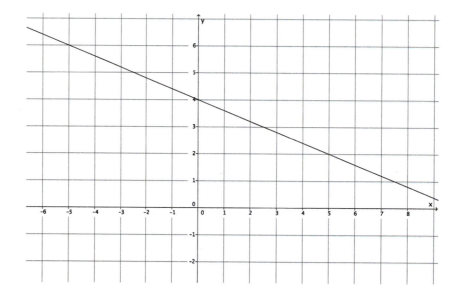

Lesson 20: Every Line Is a Graph of a Linear Equation

- First, identify the y-intercept point.
  - *The line intersects the y-axis at $(0, 4)$.*
- Now we must use what we know about slope to determine the slope of the line. Recall the following:

$$m = \frac{|QR|}{|PQ|}.$$

The point P represents our y-intercept point. Let's locate a point R on the line with integer coordinates.
  - *We can use the point $(5, 2)$ or $(-5, 6)$.*
- We can use either point. For this example, let's use $(5, 2)$.

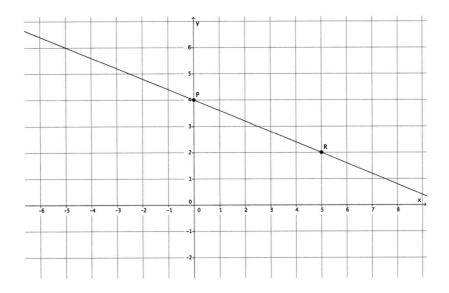

- Now we can locate point Q. It must be to the right of point P and be on a line parallel to the y-axis that goes through point R. What is the location of Q?
  - *Point Q must be $(5, 4)$.*

- What fraction represents the slope of this line?

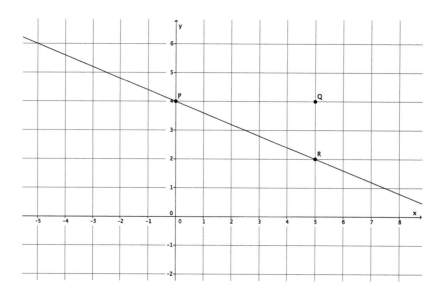

  □ The slope of the line is $m = -\frac{2}{5}$.

- The slope of the line is $m = -\frac{2}{5}$, and the y-intercept point is $(0, 4)$. What must the equation of the line be?

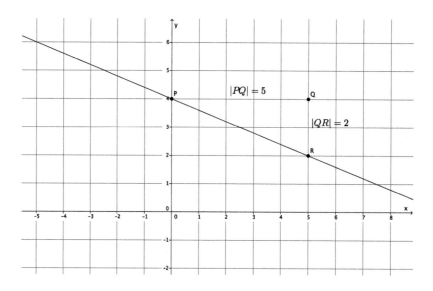

  □ The line is the graph of $y = -\frac{2}{5}x + 4$.

Example 2 (5 minutes)

- What is the y-intercept point of the line?

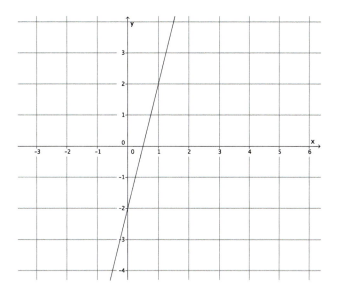

  - The y-intercept point is $(0, -2)$.

- Select another point, R, on the line with integer coordinates.

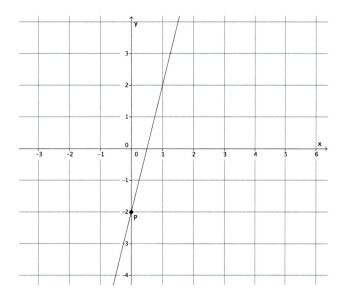

  - Let R be the point located at $(1, 2)$.

- Now, place point Q, and find the slope of the line.

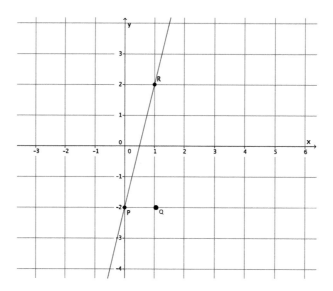

  ▫ The slope of the line is $m = 4$.
- Write the equation for the line.

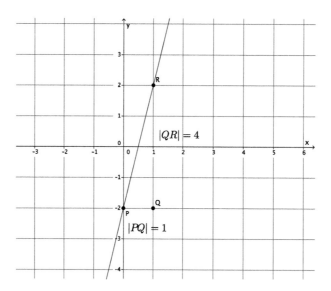

  ▫ The line is the graph of $y = 4x - 2$.

Example 3 (5 minutes)

- What is the y-intercept point of the line? Notice the units on the coordinate plane have increased.

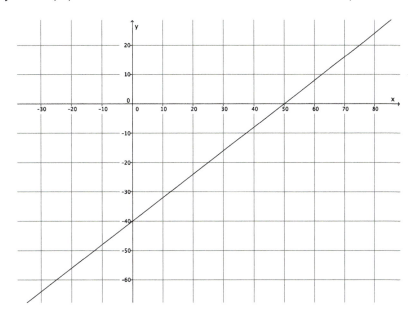

  □ The y-intercept point is $(0, -40)$.

- Select another point, R, on the line with integer coordinates.

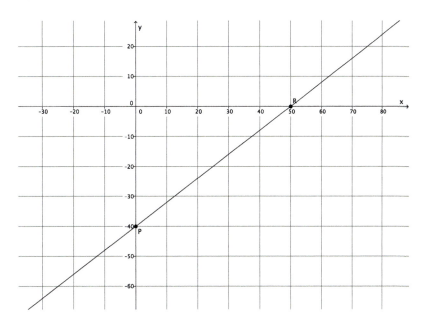

  □ Let R be $(50, 0)$.

- Now, place point Q, and find the slope of the line.

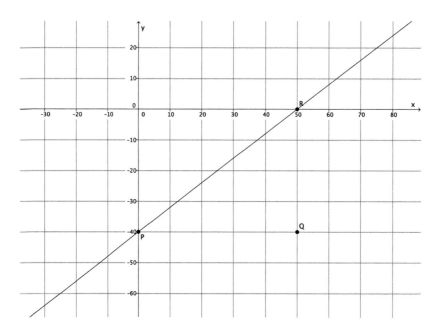

  □ The slope of the line is $m = \dfrac{40}{50} = \dfrac{4}{5}$.

- Write the equation for the line.

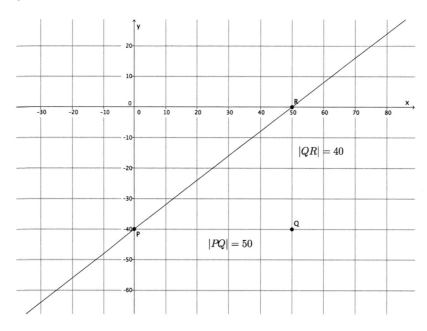

  □ The line is the graph of $y = \dfrac{4}{5}x - 40$.

- The last thing we will do to this linear equation is rewrite it in standard form $ax + by = c$, where a, b, and c are integers, and a is not negative. That means we must multiply the entire equation by a number that will turn $\frac{4}{5}$ into an integer. What number should we multiply by?
  - $\frac{4}{5}(5) = 4$
- We multiply the entire equation by 5.

$$\left(y = \frac{4}{5}x - 40\right)5$$
$$5y = 4x - 200$$
$$-4x + 5y = 4x - 4x - 200$$
$$-4x + 5y = -200$$
$$-1(-4x + 5y = -200)$$
$$4x - 5y = 200$$

The standard form of the linear equation is $4x - 5y = 200$.

Exercises (10 minutes)

Students complete Exercises 1–6 independently.

Exercises

1. Write the equation that represents the line shown.

   $y = 3x + 2$

   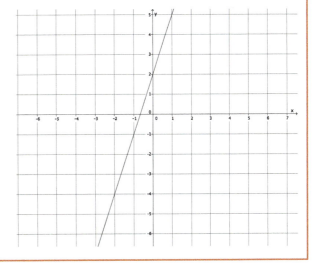

   Use the properties of equality to change the equation from slope-intercept form, $y = mx + b$, to standard form, $ax + by = c$, where a, b, and c are integers, and a is not negative.

   $y = 3x + 2$
   $-3x + y = 3x - 3x + 2$
   $-3x + y = 2$
   $-1(-3x + y = 2)$
   $3x - y = -2$

Lesson 20: Every Line Is a Graph of a Linear Equation

2. Write the equation that represents the line shown.

$$y = -\frac{2}{3}x - 1$$

Use the properties of equality to change the equation from slope-intercept form, $y = mx + b$, to standard form, $ax + by = c$, where a, b, and c are integers, and a is not negative.

$$y = -\frac{2}{3}x - 1$$
$$\left(y = -\frac{2}{3}x - 1\right)3$$
$$3y = -2x - 3$$
$$2x + 3y = -2x + 2x - 3$$
$$2x + 3y = -3$$

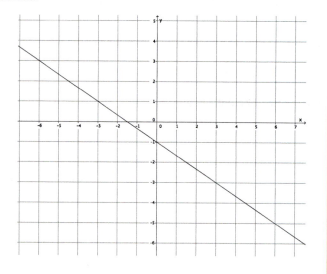

3. Write the equation that represents the line shown.

$$y = -\frac{1}{5}x - 4$$

Use the properties of equality to change the equation from slope-intercept form, $y = mx + b$, to standard form, $ax + by = c$, where a, b, and c are integers, and a is not negative.

$$y = -\frac{1}{5}x - 4$$
$$\left(y = -\frac{1}{5}x - 4\right)5$$
$$5y = -x - 20$$
$$x + 5y = -x + x - 20$$
$$x + 5y = -20$$
$$x + 5$$

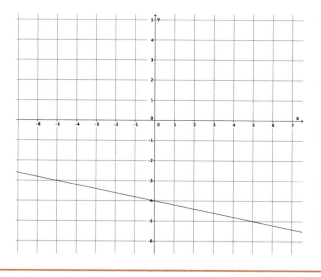

4. Write the equation that represents the line shown.

$$y = x$$

Use the properties of equality to change the equation from slope-intercept form, $y = mx + b$, to standard form, $ax + by = c$, where a, b, and c are integers, and a is not negative.

$$y = x$$
$$-x + y = x - x$$
$$-x + y = 0$$
$$-1(-x + y = 0)$$
$$x - y = 0$$

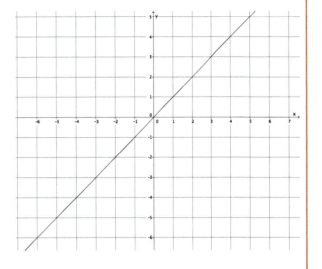

5. Write the equation that represents the line shown.

$$y = \frac{1}{4}x + 5$$

Use the properties of equality to change the equation from slope-intercept form, $y = mx + b$, to standard form, $ax + by = c$, where a, b, and c are integers, and a is not negative.

$$y = \frac{1}{4}x + 5$$
$$\left(y = \frac{1}{4}x + 5\right)4$$
$$4y = x + 20$$
$$-x + 4y = x - x + 20$$
$$-x + 4y = 20$$
$$-1(-x + 4y = 20)$$
$$x - 4y = -20$$

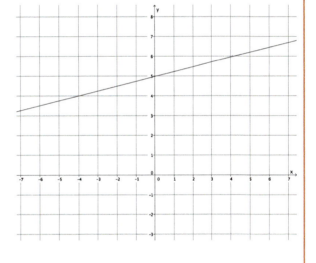

Lesson 20: Every Line Is a Graph of a Linear Equation

6. Write the equation that represents the line shown.

$$y = -\frac{8}{5}x - 7$$

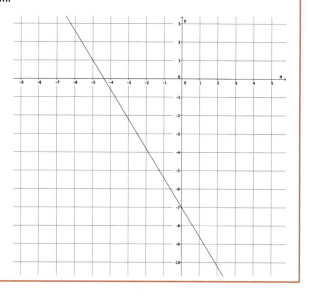

Use the properties of equality to change the equation from slope-intercept form, $y = mx + b$, to standard form, $ax + by = c$, where a, b, and c are integers, and a is not negative.

$$y = -\frac{8}{5}x - 7$$
$$\left(y = -\frac{8}{5}x - 7\right)5$$
$$5y = -8x - 35$$
$$8x + 5y = -8x + 8x - 35$$
$$8x + 5y = -35$$

Closing (5 minutes)

Summarize, or ask students to summarize, the main points from the lesson:

- We know that every line is a graph of a linear equation.
- We know how to use the y-intercept point and the slope of a line to write the equation of a line.

Lesson Summary

Write the equation of a line by determining the y-intercept point, $(0, b)$, and the slope, m, and replacing the numbers b and m into the equation $y = mx + b$.

Example:

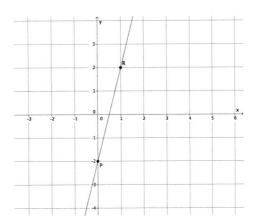

The y-intercept point of this graph is $(0, -2)$.

The slope of this graph is $m = \frac{4}{1} = 4$.

The equation that represents the graph of this line is $y = 4x - 2$.

Use the properties of equality to change the equation from slope-intercept form, $y = mx + b$, to standard form, $ax + by = c$, where a, b, and c are integers, and a is not negative.

Exit Ticket (5 minutes)

Lesson 20: Every Line Is a Graph of a Linear Equation

Exit Ticket

1. Write an equation in slope-intercept form that represents the line shown.

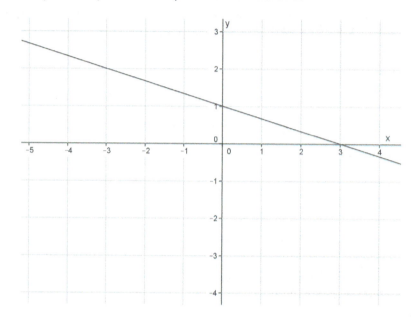

2. Use the properties of equality to change the equation you wrote for Problem 1 from slope-intercept form, $y = mx + b$, to standard form, $ax + by = c$, where a, b, and c are integers, and a is not negative.

3. Write an equation in slope-intercept form that represents the line shown.

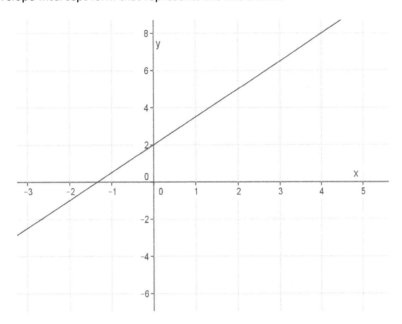

4. Use the properties of equality to change the equation you wrote for Problem 3 from slope-intercept form, $y = mx + b$, to standard form, $ax + by = c$, where a, b, and c are integers, and a is not negative.

Exit Ticket Sample Solutions

1. Write an equation in slope-intercept form that represents the line shown.

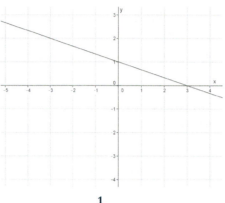

$$y = -\frac{1}{3}x + 1$$

2. Use the properties of equality to change the equation you wrote for Problem 1 from slope-intercept form, $y = mx + b$, to standard form, $ax + by = c$, where a, b, and c are integers, and a is not negative.

$$y = -\frac{1}{3}x + 1$$
$$\left(y = -\frac{1}{3}x + 1\right)3$$
$$3y = -x + 3$$
$$x + 3y = -x + x + 3$$
$$x + 3y = 3$$

3. Write an equation in slope-intercept form that represents the line shown.

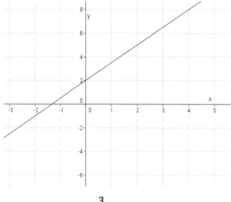

$$y = \frac{3}{2}x + 2$$

4. Use the properties of equality to change the equation you wrote for Problem 3 from slope-intercept form, $y = mx + b$, to standard form, $ax + by = c$, where a, b, and c are integers, and a is not negative.

$$y = \frac{3}{2}x + 2$$
$$\left(y = \frac{3}{2}x + 2\right)2$$
$$2y = 3x + 4$$
$$-3x + 2y = 3x - 3x + 4$$
$$-3x + 2y = 4$$
$$-1(-3x + 2y = 4)$$
$$3x - 2y = -4$$

Problem Set Sample Solutions

Students practice writing equations for lines.

1. Write the equation that represents the line shown.

$$y = -\frac{2}{3}x - 4$$

Use the properties of equality to change the equation from slope-intercept form, $y = mx + b$, to standard form, $ax + by = c$, where a, b, and c are integers, and a is not negative.

$$y = -\frac{2}{3}x - 4$$
$$\left(y = -\frac{2}{3}x - 4\right)3$$
$$3y = -2x - 12$$
$$2x + 3y = -2x + 2x - 12$$
$$2x + 3y = -12$$

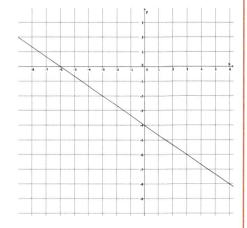

2. Write the equation that represents the line shown.

$$y = 8x + 1$$

Use the properties of equality to change the equation from slope-intercept form, $y = mx + b$, to standard form, $ax + by = c$, where a, b, and c are integers, and a is not negative.

$$y = 8x + 1$$
$$-8x + y = 8x - 8x + 1$$
$$-8x + y = 1$$
$$-1(-8x + y = 1)$$
$$8x - y = -1$$

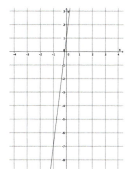

Lesson 20: Every Line Is a Graph of a Linear Equation

3. Write the equation that represents the line shown.

$$y = \frac{1}{2}x - 4$$

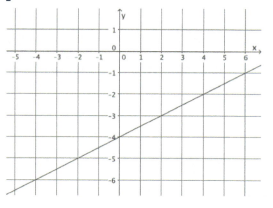

Use the properties of equality to change the equation from slope-intercept form, $y = mx + b$, to standard form, $ax + by = c$, where a, b, and c are integers, and a is not negative.

$$y = \frac{1}{2}x - 4$$
$$\left(y = \frac{1}{2}x - 4\right)2$$
$$2y = x - 8$$
$$-x + 2y = x - x - 8$$
$$-x + 2y = -8$$
$$-1(-x + 2y = -8)$$
$$x - 2y = 8$$

4. Write the equation that represents the line shown.

$$y = -9x - 8$$

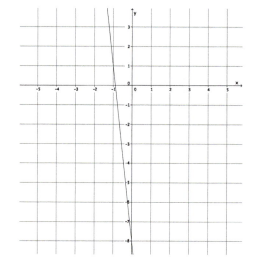

Use the properties of equality to change the equation from slope-intercept form, $y = mx + b$, to standard form, $ax + by = c$, where a, b, and c are integers, and a is not negative.

$$y = -9x - 8$$
$$9x + y = -9x + 9x - 8$$
$$9x + y = -8$$

5. Write the equation that represents the line shown.

$$y = 2x - 14$$

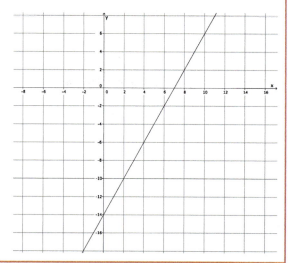

Use the properties of equality to change the equation from slope-intercept form, $y = mx + b$, to standard form, $ax + by = c$, where a, b, and c are integers, and a is not negative.

$$y = 2x - 14$$
$$-2x + y = 2x - 2x - 14$$
$$-2x + y = -14$$
$$-1(-2x + y = -14)$$
$$2x - y = 14$$

Lesson 20: Every Line Is a Graph of a Linear Equation

6. Write the equation that represents the line shown.

$$y = -5x + 45$$

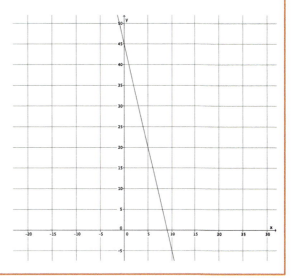

Use the properties of equality to change the equation from slope-intercept form, $y = mx + b$, to standard form, $ax + by = c$, where a, b, and c are integers, and a is not negative.

$$y = -5x + 45$$
$$5x + y = -5x + 5x + 45$$
$$5x + y = 45$$

A STORY OF RATIOS   Lesson 21   8•4

 # Lesson 21: Some Facts About Graphs of Linear Equations in Two Variables

Student Outcomes

- Students write the equation of a line given two points or the slope and a point on the line.
- Students know the traditional forms of the slope formula and slope-intercept equation.

Classwork

Example 1 (10 minutes)

Students determine the equation of a line from a graph by using information about slope and a point.

- Let a line l be given in the coordinate plane. Our goal is to find the equation that represents the line l. Can we use information about the slope and intercept to write the equation of the line like we did in the last lesson?

Provide students time to attempt to write the equation of the line. Ask students to share their equations and explanations. Consider having the class vote on whose explanation/equation they think is correct.

Example 1

Let a line l be given in the coordinate plane. What linear equation is the graph of line l?

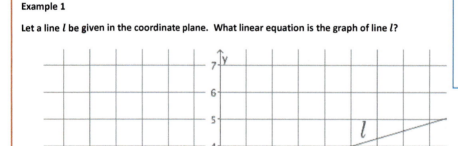

Scaffolding:

If necessary, include another point, as demonstrated in Lesson 15, to help students determine the slope of the line.

- We can pick two points to determine the slope, but the precise location of the y-intercept point cannot be determined from the graph.

336   Lesson 21:   Some Facts About Graphs of Linear Equations in Two Variables

- Calculate the slope of the line.
    - Using points $(-2, 2)$ and $(5, 4)$, the slope of the line is
    $$m = \frac{2-4}{-2-5}$$
    $$= \frac{-2}{-7}$$
    $$= \frac{2}{7}.$$

- Now we need to determine the y-intercept point of the line. We know that it is a point with coordinates $(0, b)$, and we know that the line goes through points $(-2, 2)$ and $(5, 4)$ and has slope $m = \frac{2}{7}$. Using this information, we can determine the coordinates of the y-intercept point and the value of b that we need in order to write the equation of the line.

- Recall what it means for a point to be on a line; the point is a solution to the equation. In the equation $y = mx + b$, (x, y) is a solution, and m is the slope. Can we find the value of b? Explain.
    - Yes. We can substitute one of the points and the slope into the equation and solve for b.

- Do you think it matters which point we choose to substitute into the equation? That is, will we get a different equation if we use the point $(-2, 2)$ compared to $(5, 4)$?
    - No, because there can be only one line with a given slope that goes through a point.

- Verify this claim by using $m = \frac{2}{7}$ and $(-2, 2)$ to find the equation of the line and then by using $m = \frac{2}{7}$ and $(5, 4)$ to see if the result is the same equation.
    - Sample student work:

    $$2 = \frac{2}{7}(-2) + b$$
    $$2 = -\frac{4}{7} + b$$
    $$2 + \frac{4}{7} = -\frac{4}{7} + \frac{4}{7} + b$$
    $$\frac{18}{7} = b$$

    $$4 = \frac{2}{7}(5) + b$$
    $$4 = \frac{10}{7} + b$$
    $$4 - \frac{10}{7} = \frac{10}{7} - \frac{10}{7} + b$$
    $$\frac{18}{7} = b$$

    The y-intercept point is at $\left(0, \frac{18}{7}\right)$, and the equation of the line is $y = \frac{2}{7}x + \frac{18}{7}$.

- The equation of the line is
$$y = \frac{2}{7}x + \frac{18}{7}.$$
- Write it in standard form.
  - *Sample student work:*

$$\left(y = \frac{2}{7}x + \frac{18}{7}\right)7$$
$$7y = 2x + 18$$
$$-2x + 7y = 2x - 2x + 18$$
$$-2x + 7y = 18$$
$$-1(-2x + 7y = 18)$$
$$2x - 7y = -18$$

Example 2 (5 minutes)

Students determine the equation of a line from a graph by using information about slope and a point.

- Let a line l be given in the coordinate plane. What information do we need to write the equation of the line?
  - *We need to know the slope, so we must identify two points we can use to calculate the slope. Then we can use the slope and a point to determine the equation of the line.*

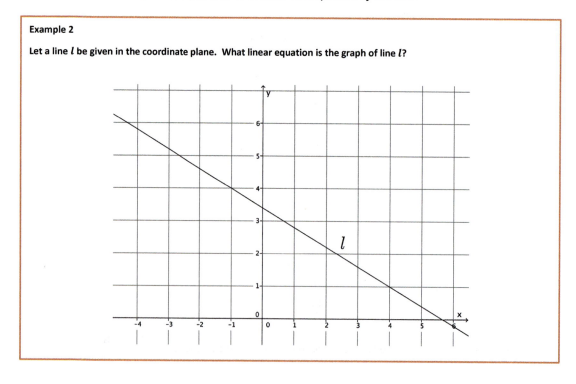

Example 2

Let a line l be given in the coordinate plane. What linear equation is the graph of line l?

Lesson 21: Some Facts About Graphs of Linear Equations in Two Variables

- Determine the slope of the line.
  - Using points $(-1, 4)$ and $(4, 1)$, the slope of the line is
  $$m = \frac{4-1}{-1-4}$$
  $$= \frac{3}{-5}$$
  $$= -\frac{3}{5}.$$

- Determine the y-intercept point of the line.
  - Sample student work:
  $$4 = \left(-\frac{3}{5}\right)(-1) + b$$
  $$4 = \frac{3}{5} + b$$
  $$4 - \frac{3}{5} = \frac{3}{5} - \frac{3}{5} + b$$
  $$\frac{17}{5} = b$$

  The y-intercept point is at $\left(0, \frac{17}{5}\right)$.

- Now that we know the slope, $m = -\frac{3}{5}$, and the y-intercept point, $\left(0, \frac{17}{5}\right)$, write the equation of the line l in slope-intercept form.
  - $y = -\frac{3}{5}x + \frac{17}{5}$

- Transform the equation so that it is written in standard form.
  - Sample student work:
  $$y = -\frac{3}{5}x + \frac{17}{5}$$
  $$\left(y = -\frac{3}{5}x + \frac{17}{5}\right)5$$
  $$5y = -3x + 17$$
  $$3x + 5y = -3x + 3x + 17$$
  $$3x + 5y = 17$$

Example 3 (5 minutes)

Students determine the equation of a line from a graph by using information about slope and a point.

- Let a line l be given in the coordinate plane. Assume the y-axis intervals are units of one (like the visible x-axis). What information do we need to write the equation of the line?
    - *We need to know the slope, so we must identify two points we can use to calculate the slope. Then we can use the slope and a point to determine the equation of the line.*

> **Example 3**
>
> Let a line l be given in the coordinate plane. What linear equation is the graph of line l?
>
>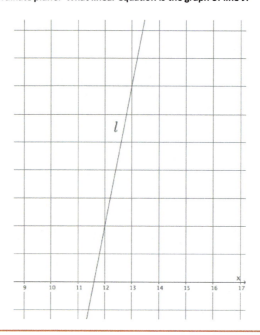

- *Using points $(12, 2)$ and $(13, 7)$, the slope of the line is*

$$m = \frac{2-7}{12-13}$$
$$= \frac{-5}{-1}$$
$$= 5.$$

- Now, determine the y-intercept point of the line, and write the equation of the line in slope-intercept form.
    - *Sample student work:*

$$2 = 5(12) + b$$
$$2 = 60 + b$$
$$b = -58$$

The y-intercept point is at $(0, -58)$, and the equation of the line is $y = 5x - 58$.

- Now that we know the slope, $m = 5$, and the y-intercept point, $(0, -58)$, write the equation of the line l in standard form.
  - Sample student work:
  $$y = 5x - 58$$
  $$-5x + y = 5x - 5x - 58$$
  $$-5x + y = -58$$
  $$-1(-5x + y = -58)$$
  $$5x - y = 58$$

Example 4 (3 minutes)

Students determine the equation of a line from a graph by using information about slope and a point.

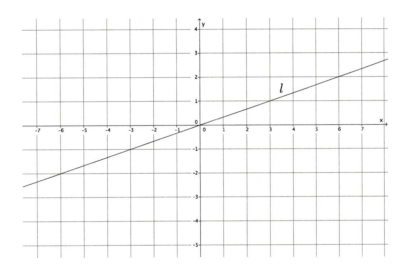

Example 4

Let a line l be given in the coordinate plane. What linear equation is the graph of line l?

Using points $(3, 1)$ and $(-3, -1)$, the slope of the line is
$$m = \frac{-1 - 1}{-3 - 3}$$
$$= \frac{-2}{-6}$$
$$= \frac{1}{3}.$$

The y-intercept point is at $(0, 0)$, and the equation of the line is $y = \frac{1}{3}x$.

- The y-intercept point is the origin of the graph. What value does b have when this occurs?
  - When the line goes through the origin, the value of b is zero.

Lesson 21: Some Facts About Graphs of Linear Equations in Two Variables

- All linear equations that go through the origin have the form $y = mx + 0$ or simply $y = mx$. We have done a lot of work with equations in this form. Which do you remember?
  - *All problems that describe constant rate proportional relationships have equations of this form.*

Concept Development (5 minutes)

- The following are some facts about graphs of linear equations in two variables:
  - Let (x_1, y_1) and (x_2, y_2) be the coordinates of two distinct points on the graph of a line l. We find the slope of the line by

  $$m = \frac{y_2 - y_1}{x_2 - x_1}.$$

  This version of the slope formula, using coordinates of x and y instead of p and r, is a commonly accepted version.

  - As soon as you multiply the slope by the denominator of the fraction above, you get the following equation:

  $$m(x_2 - x_1) = y_2 - y_1.$$

  This form of an equation is referred to as the *point-slope form* of a linear equation. As you can see, it does not convey any more information than the slope formula. It is just another way to look at it.

  - Given a known (x, y), then the equation is written as

  $$m(x - x_1) = (y - y_1).$$

  - The following is the slope-intercept form of a line:

  $$y = mx + b.$$

  In this equation, m is slope, and $(0, b)$ is the y-intercept point.

- What information must you have in order to write the equation of a line?
  - *We need two points or one point and slope.*
- The names and symbols used are not nearly as important as your understanding of the concepts. Basically, if you can remember a few simple facts about lines, namely, the slope formula and the fact that slope is the same between any two points on a line, you can derive the equation of any line.

Exercises (7 minutes)

Students complete Exercises 1–5 independently.

Exercises

1. Write the equation for the line l shown in the figure.

   Using the points $(-1, -3)$ and $(2, -2)$, the slope of the line is

   $$m = \frac{-3 - (-2)}{-1 - 2}$$
   $$= \frac{-1}{-3}$$
   $$= \frac{1}{3}.$$

   $$-2 = \frac{1}{3}(2) + b$$
   $$-2 = \frac{2}{3} + b$$
   $$-2 - \frac{2}{3} = \frac{2}{3} - \frac{2}{3} + b$$
   $$-\frac{8}{3} = b$$

   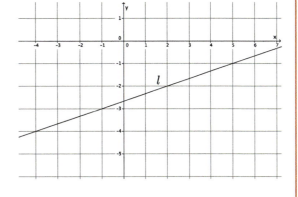

   The equation of the line is $y = \frac{1}{3}x - \frac{8}{3}$.

2. Write the equation for the line l shown in the figure.

   Using the points $(-3, 7)$ and $(2, 8)$, the slope of the line is

   $$m = \frac{7 - 8}{-3 - 2}$$
   $$= \frac{-1}{-5}$$
   $$= \frac{1}{5}.$$

   $$8 = \frac{1}{5}(2) + b$$
   $$8 = \frac{2}{5} + b$$
   $$8 - \frac{2}{5} = \frac{2}{5} - \frac{2}{5} + b$$
   $$\frac{38}{5} = b$$

   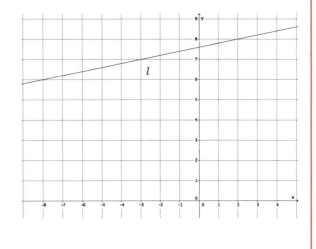

   The equation of the line is $y = \frac{1}{5}x + \frac{38}{5}$.

Lesson 21: Some Facts About Graphs of Linear Equations in Two Variables

3. Determine the equation of the line that goes through points $(-4, 5)$ and $(2, 3)$.

The slope of the line is
$$m = \frac{5-3}{-4-2}$$
$$= \frac{2}{-6}$$
$$= -\frac{1}{3}.$$

The y-intercept point of the line is
$$3 = -\frac{1}{3}(2) + b$$
$$3 = -\frac{2}{3} + b$$
$$\frac{11}{3} = b.$$

The equation of the line is $y = -\frac{1}{3}x + \frac{11}{3}$.

4. Write the equation for the line l shown in the figure.

Using the points $(-7, 2)$ and $(-6, -2)$, the slope of the line is
$$m = \frac{2 - (-2)}{-7 - (-6)}$$
$$= \frac{4}{-1}$$
$$= -4.$$

$$-2 = -4(-6) + b$$
$$-2 = 24 + b$$
$$-26 = b$$

The equation of the line is $y = -4x - 26$.

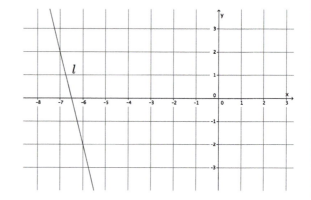

5. A line goes through the point $(8, 3)$ and has slope $m = 4$. Write the equation that represents the line.
$$3 = 4(8) + b$$
$$3 = 32 + b$$
$$-29 = b$$

The equation of the line is $y = 4x - 29$.

Closing (5 minutes)

Summarize, or ask students to summarize, the main points from the lesson:

- We know how to write an equation for a line from a graph, even if the line does not intersect the y-axis at integer coordinates.
- We know how to write the equation for a line given two points or one point and the slope of the line.
- We know other versions of the formulas and equations that we have been using related to linear equations.

Lesson Summary

Let (x_1, y_1) and (x_2, y_2) be the coordinates of two distinct points on a non-vertical line in a coordinate plane. We find the slope of the line by

$$m = \frac{y_2 - y_1}{x_2 - x_1}.$$

This version of the slope formula, using coordinates of x and y instead of p and r, is a commonly accepted version.

As soon as you multiply the slope by the denominator of the fraction above, you get the following equation:

$$m(x_2 - x_1) = y_2 - y_1.$$

This form of an equation is referred to as the *point-slope form* of a linear equation.

Given a known (x, y), then the equation is written as

$$m(x - x_1) = (y - y_1).$$

The following is the slope-intercept form of a line:

$$y = mx + b.$$

In this equation, m is slope, and $(0, b)$ is the y-intercept point.

To write the equation of a line, you must have two points, one point and slope, or a graph of the line.

Exit Ticket (5 minutes)

Name _____   Date _____

Lesson 21: Some Facts About Graphs of Linear Equations in Two Variables

Exit Ticket

1. Write the equation for the line l shown in the figure below.

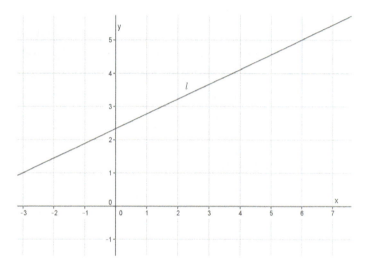

2. A line goes through the point $(5, -7)$ and has slope $m = -3$. Write the equation that represents the line.

Exit Ticket Sample Solutions

Note that some students may write equations in standard form.

1. Write the equation for the line l shown in the figure below.

   Using the points $(-3, 1)$ and $(6, 5)$, the slope of the line is

   $$m = \frac{5-1}{6-(-3)}$$
   $$m = \frac{4}{9}.$$

   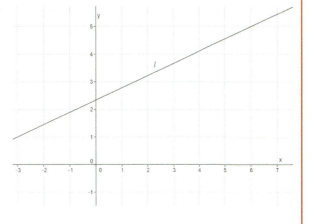

   $$5 = \frac{4}{9}(6) + b$$
   $$5 = \frac{8}{3} + b$$
   $$5 - \frac{8}{3} = \frac{8}{3} - \frac{8}{3} + b$$
   $$\frac{7}{3} = b$$

   The equation of the line is $y = \frac{4}{9}x + \frac{7}{3}$.

2. A line goes through the point $(5, -7)$ and has slope $m = -3$. Write the equation that represents the line.

   $$-7 = -3(5) + b$$
   $$-7 = -15 + b$$
   $$8 = b$$

   The equation of the line is $y = -3x + 8$.

Lesson 21: Some Facts About Graphs of Linear Equations in Two Variables

Problem Set Sample Solutions

Students practice writing equations from graphs of lines. Students write the equation of a line given only the slope and a point.

1. Write the equation for the line l shown in the figure.

   Using the points $(-3, 2)$ and $(2, -2)$, the slope of the line is

   $$m = \frac{2-(-2)}{-3-2}$$
   $$= \frac{4}{-5}$$
   $$= -\frac{4}{5}.$$

   $$2 = \left(-\frac{4}{5}\right)(-3) + b$$
   $$2 = \frac{12}{5} + b$$
   $$2 - \frac{12}{5} = \frac{12}{5} - \frac{12}{5} + b$$
   $$-\frac{2}{5} = b$$

   The equation of the line is $y = -\frac{4}{5}x - \frac{2}{5}$.

   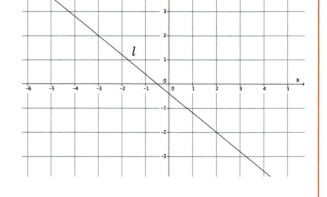

2. Write the equation for the line l shown in the figure.

   Using the points $(-6, 2)$ and $(-5, 5)$, the slope of the line is

   $$m = \frac{2-5}{-6-(-5)}$$
   $$= \frac{-3}{-1}$$
   $$= 3.$$

   $$5 = 3(-5) + b$$
   $$5 = -15 + b$$
   $$20 = b$$

   The equation of the line is $y = 3x + 20$.

   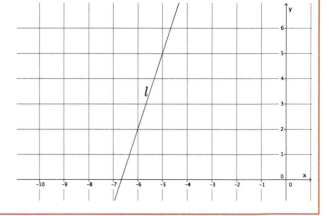

3. Write the equation for the line l shown in the figure.

Using the points $(-3, 1)$ and $(2, 2)$, the slope of the line is
$$m = \frac{1-2}{-3-2}$$
$$= \frac{-1}{-5}$$
$$= \frac{1}{5}.$$

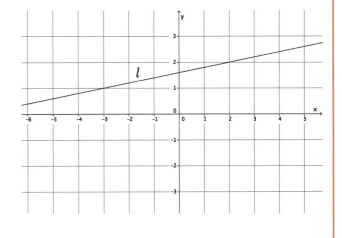

$$2 = \frac{1}{5}(2) + b$$
$$2 = \frac{2}{5} + b$$
$$2 - \frac{2}{5} = \frac{2}{5} - \frac{2}{5} + b$$
$$\frac{8}{5} = b$$

The equation of the line is $y = \frac{1}{5}x + \frac{8}{5}$.

4. Triangle ABC is made up of line segments formed from the intersection of lines L_{AB}, L_{BC}, and L_{AC}. Write the equations that represent the lines that make up the triangle.

$A(-3, -3)$, $B(3, 2)$, $C(5, -2)$

The slope of L_{AB}:
$$m = \frac{-3-2}{-3-3}$$
$$= \frac{-5}{-6}$$
$$= \frac{5}{6}$$

$$2 = \frac{5}{6}(3) + b$$
$$2 = \frac{5}{2} + b$$
$$2 - \frac{5}{2} = \frac{5}{2} - \frac{5}{2} + b$$
$$-\frac{1}{2} = b$$

The equation of L_{AB} is $y = \frac{5}{6}x - \frac{1}{2}$.

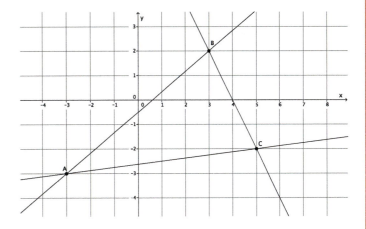

The slope of L_{BC}:
$$m = \frac{2-(-2)}{3-5}$$
$$= \frac{4}{-2}$$
$$= -2$$

$$2 = -2(3) + b$$
$$2 = -6 + b$$
$$8 = b$$

The equation of L_{BC} is $y = -2x + 8$.

The slope of L_{AC}:
$$m = \frac{-3-(-2)}{-3-5}$$
$$= \frac{-1}{-8}$$
$$= \frac{1}{8}$$

$$-2 = \frac{1}{8}(5) + b$$
$$-2 = \frac{5}{8} + b$$
$$-2 - \frac{5}{8} = \frac{5}{8} - \frac{5}{8} + b$$
$$-\frac{21}{8} = b$$

The equation of L_{AC} is $y = \frac{1}{8}x - \frac{21}{8}$.

Lesson 21: Some Facts About Graphs of Linear Equations in Two Variables

5. Write the equation for the line that goes through point $(-10, 8)$ with slope $m = 6$.

$$8 = 6(-10) + b$$
$$8 = -60 + b$$
$$68 = b$$

The equation of the line is $y = 6x + 68$.

6. Write the equation for the line that goes through point $(12, 15)$ with slope $m = -2$.

$$15 = -2(12) + b$$
$$15 = -24 + b$$
$$39 = b$$

The equation of the line is $y = -2x + 39$.

7. Write the equation for the line that goes through point $(1, 1)$ with slope $m = -9$.

$$1 = -9(1) + b$$
$$1 = -9 + b$$
$$10 = b$$

The equation of the line is $y = -9x + 10$.

8. Determine the equation of the line that goes through points $(1, 1)$ and $(3, 7)$.

The slope of the line is

$$m = \frac{1 - 7}{1 - 3}$$
$$= \frac{-6}{-2}$$
$$= 3.$$

The y-intercept point of the line is

$$7 = 3(3) + b$$
$$7 = 9 + b$$
$$-2 = b.$$

The equation of the line is $y = 3x - 2$.

A STORY OF RATIOS   Lesson 22   8•4

Lesson 22: Constant Rates Revisited

Student Outcomes

- Students know that any constant rate problem can be described by a linear equation in two variables where the slope of the graph is the constant rate.
- Students compare two different proportional relationships represented by graphs, equations, and tables to determine which has a greater rate of change.

Classwork

Example 1 (8 minutes)

- Recall our definition of *constant rate*: If the average speed of motion over any time interval is equal to the same constant, then we say the motion has constant speed.
- Erika set her stopwatch to zero and switched it on at the beginning of her walk. She walks at a constant speed of 3 miles per hour. Work in pairs to express this situation as an equation, a table of values, and a graph.

Scaffolding:
It may be necessary to review the definition of *average speed* as well. Average speed is the distance (or area, pages typed, etc.) divided by the time interval spent moving that distance (or painting the given area or typing a specific number of pages, etc.).

Provide students time to analyze the situation and represent it in the requested forms. If necessary, provide support to students with the next few bullet points. Once students have finished working, resume the work on the example (bullets are below the graph).

- Suppose Erika walked a distance of y miles in x hours. Then her average speed in the time interval from 0 to x hours is $\frac{y}{x}$. Since Erika walks at a constant speed of 3 miles per hour, we can express her motion as $\frac{y}{x} = 3$.
- Solve the equation for y.
    - $\frac{y}{x} = 3$
    - $y = 3x$
- We can use the equation to develop a table of values where x is the number of hours Erika walks and y is the distance she walks in miles. Complete the table for the given x-values.

x	y
0	0
1	3
2	6
3	9

- In terms of x and y, $y = 3x$ is a linear equation in two variables. Its graph is a line.

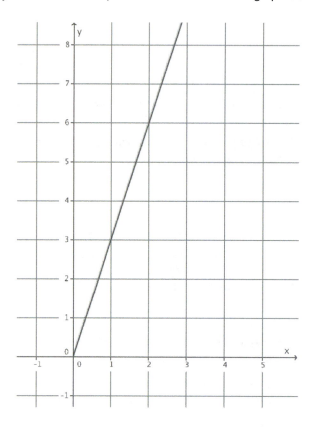

- What is the meaning of the 3 in the equation $y = 3x$?
  - The 3 is the slope of the line. It represents the rate at which Erika walks.

- We can prove that her motion is constant, i.e., she walks at a constant speed, by showing that the slope is 3 over any time interval that she walks:

Assume Erika walks y_1 miles in the time interval from 0 to x_1, and she walks y_2 miles in the time interval from 0 to x_2.

Her average distance between time interval x_1 and x_2 is $\frac{y_2-y_1}{x_2-x_1}$.

We know that $y = 3x$; therefore, $y_2 = 3x_2$ and $y_1 = 3x_1$. Then, by substitution, we have

$$\frac{y_2 - y_1}{x_2 - x_1} = \frac{3x_2 - 3x_1}{x_2 - x_1}$$
$$= \frac{3(x_2 - x_1)}{x_2 - x_1}$$
$$= 3.$$

Scaffolding:
Explain that students are looking at the distance traveled between times x_1 and x_2; therefore, they must subtract the distances traveled to represent the total distance traveled in $x_2 - x_1$. Concrete numbers and a diagram like the following may be useful as well.

Average speed of walking over interval from 1 hour to 4 hours $= \frac{\text{distance traveled in time interval}}{\text{time of interval}}$
$= \frac{12-3}{4-1}$
$= \frac{9}{3}$
$= 3$

Lesson 22: Constant Rates Revisited

- Therefore, Erika's average speed in the time interval between x_1 and x_2 is 3 mph. Since x_1 and x_2 can be any two times, then the average speed she walks over any time interval is 3 mph, and we have confirmed that she walks at a constant speed of 3 mph. We can also show that her motion is constant by looking at the slope between different points on the graph.

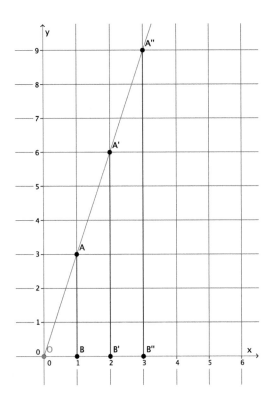

- Are $\triangle OB$, $\triangle 'OB'$, and $\triangle "OB"$ similar?
    - Yes, the triangles are similar because each triangle has a common angle, $\angle AOB$, and a right angle. By the AA criterion, the triangles are similar.
- We used similar triangles before to show that the slope of a line would be the same between any two points. Since we have similar triangles, we can conclude that the rate at which Erika walks is constant because the ratio of the corresponding sides will be equal to a constant (i.e., the scale factor of dilation). For example, when we compare the values of the ratios of corresponding sides for $\triangle "OB"$ and $\triangle 'OB'$ we see that

$$\frac{9}{6} = \frac{3}{2}.$$

Make clear to students that the ratio of corresponding sides is not equal to the slope of the line; rather, the ratios are equal to one another and the scale factor of dilation.

- Or we could compare the values of the ratios of corresponding side lengths for $\triangle 'OB'$ and $\triangle OB$:

$$\frac{6}{3} = \frac{2}{1}.$$

- Again, because the ratio of corresponding sides is equal, we know that the rate at which Erika walks is constant.

A STORY OF RATIOS   Lesson 22   8•4

Example 2 (5 minutes)

- A faucet leaks at a constant rate of 7 gallons per hour. Suppose y gallons leak in x hours from 6:00 a.m. Express the situation as a linear equation in two variables.

- If we say the leak began at 6:00 a.m., and the total number of gallons leaked over the time interval from 6:00 a.m. to x hours is y, then we know the average rate of the leak is $\frac{y}{x}$.

- Since we know that the faucet leaks at a constant rate, then we can write the linear equation in two variables as $\frac{y}{x} = 7$, which can then be transformed to $y = 7x$.

- Again, take note of the fact that the number that represents slope in the equation is the same as the constant rate of water leaking.

- Another faucet leaks at a constant rate, and the table below shows the number of gallons, y, that leak in x hours for four selected hours.

Hours (x)	Gallons (y)
2	13
4	26
7	45.5
10	65

- How can we determine the rate at which this faucet leaks?
  - *If these were points on a graph, we could find the slope between them.*
- Using what you know about slope, determine the rate the faucet leaks.
  - *Sample student work:*

$$m = \frac{26 - 13}{4 - 2}$$
$$= \frac{13}{2}$$
$$= 6.5$$

- The number of gallons that leak from each faucet is dependent on the amount of time the faucet leaks. The information provided for each faucet's rate of change (i.e., slope) allows us to answer a question, such as the following: Which faucet has the worse leak? That is, which faucet leaks more water over a given time interval?
  - *The first faucet has the worse leak because the rate is greater: 7 compared to 6.5.*

354   Lesson 22:   Constant Rates Revisited

Example 3 (4 minutes)

- The graph below represents the constant rate at which Train A travels.

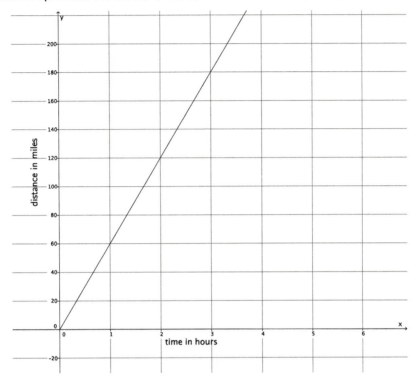

- What is the constant rate of travel for Train A?
  - We know that the constant rate of travel is the same as the slope. On the graph, we can see that the train is traveling 60 miles every hour. That means that the slope of the line is 60.

- Train B travels at a constant rate. The train travels at an average rate of 95 miles every one and a half hours. We want to know which train is traveling at a greater speed. Let's begin by writing a linear equation that represents the constant rate of Train B. If y represents the total distance traveled over any time period x, then

$$\frac{95}{1.5} = \frac{y}{x}$$

$$y = \frac{95}{1.5}x$$

$$y = 63\frac{1}{3}x.$$

- Which train is traveling at a greater speed? Explain.

Provide students time to talk to their partners about how to determine which train is traveling at a greater speed.

  - Train B is traveling at a greater speed. The graph provided the information about the constant rate of Train A; Train A travels at a constant rate of 60 miles per hour. The given rate for Train B, the slope of the graph for Train B, is equal to $63\frac{1}{3}$. Since $63\frac{1}{3} > 60$, Train B is traveling at a greater speed.

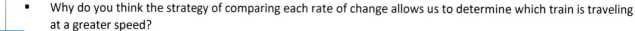

- Why do you think the strategy of comparing each rate of change allows us to determine which train is traveling at a greater speed?

  MP.4
  - *The distance that each train travels is proportional to the amount of time the train travels. Therefore, if we can describe the rate of change as a number (i.e., slope) and compare the numbers, we can determine which train travels at a greater speed.*

Example 4 (5 minutes)

- The graph below represents the constant rate at which Kristina can paint.

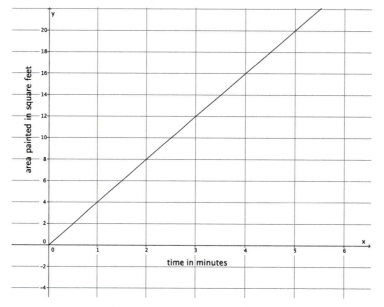

- Her sister Tracee paints at an average rate of 45 square feet in 12 minutes. Assuming Tracee paints at a constant rate, determine which sister paints faster.

Provide students time to determine the linear equation that represents Tracee's constant rate and then discuss with their partners who is the faster painter.

- *If we let y represent the total area Tracee paints in x minutes, then her constant rate is*

  $$\frac{45}{12} = \frac{y}{x}$$

  $$\frac{45}{12}x = y.$$

  We need to compare the slope of the line for Kristina with the slope of the equation for Tracee. The slope of the line is 4, and the slope in the equation that represents Tracee's rate is $\frac{45}{12}$. Since $4 > \frac{45}{12}$, Kristina paints at a faster rate.

- How does the slope provide the information we need to answer the question about which sister paints faster?
  - *The slope describes the rate of change for proportional relationships. If we know which rate of change is greater, then we can determine which sister paints faster.*

A STORY OF RATIOS — Lesson 22 — 8•4

Example 5 (5 minutes)

- The graph below represents the constant rate of watts of energy produced from a single solar panel produced by Company A.

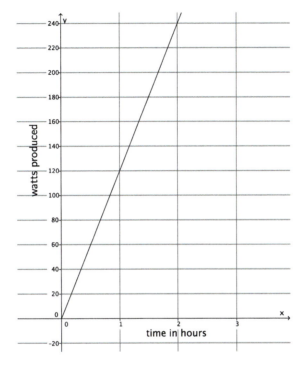

- Company B offers a solar panel that produces energy at an average rate of 325 watts in 2.6 hours. Assuming solar panels produce energy at a constant rate, determine which company produces more efficient solar panels (solar panels that produce more energy per hour).

Provide students time to work with their partners to answer the question.

- If we let y represent the energy produced by a solar panel made by Company B in x minutes, then the constant rate is

$$\frac{325}{2.6} = \frac{y}{x}$$
$$\frac{325}{2.6}x = y$$
$$125x = y.$$

We need to compare the slope of the line for Company A with the slope in the equation that represents the rate for Company B. The slope of the line representing Company A is 120, and the slope of the line representing Company B is 125. Since $125 > 120$, Company B produces the more efficient solar panel.

Lesson 22: Constant Rates Revisited

A STORY OF RATIOS — Lesson 22 — 8•4

Exercises (10 minutes)

Students work in pairs or small groups to complete Exercises 1–5. Discuss Exercise 5 to generalize a strategy for comparing proportional relationships.

Exercises

1. Peter paints a wall at a constant rate of 2 square feet per minute. Assume he paints an area y, in square feet, after x minutes.

   a. Express this situation as a linear equation in two variables.

   $$\frac{y}{x} = \frac{2}{1}$$
   $$y = 2x$$

   b. Sketch the graph of the linear equation.

   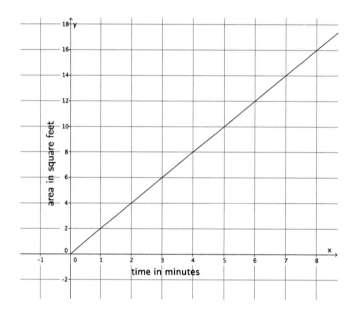

   c. Using the graph or the equation, determine the total area he paints after 8 minutes, $1\frac{1}{2}$ hours, and 2 hours. Note that the units are in minutes and hours.

   In 8 minutes, he paints 16 square feet.

   $$y = 2(90)$$
   $$= 180$$

   In $1\frac{1}{2}$ hours, he paints 180 square feet.

   $$y = 2(120)$$
   $$= 240$$

   In 2 hours, he paints 240 square feet.

Lesson 22: Constant Rates Revisited

2. The figure below represents Nathan's constant rate of walking.

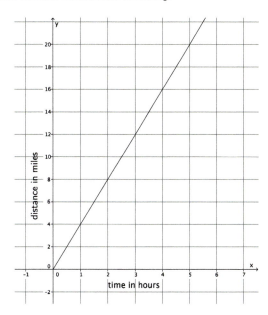

a. Nicole just finished a 5-mile walkathon. It took her 1.4 hours. Assume she walks at a constant rate. Let y represent the distance Nicole walks in x hours. Describe Nicole's walking at a constant rate as a linear equation in two variables.

$$\frac{y}{x} = \frac{5}{1.4}$$
$$y = \frac{25}{7}x$$

b. Who walks at a greater speed? Explain.

Nathan walks at a greater speed. The slope of the graph for Nathan is 4, and the slope or rate for Nicole is $\frac{25}{7}$. When you compare the slopes, you see that $4 > \frac{25}{7}$.

3.
a. Susan can type 4 pages of text in 10 minutes. Assuming she types at a constant rate, write the linear equation that represents the situation.

Let y represent the total number of pages Susan can type in x minutes. We can write $\frac{y}{x} = \frac{4}{10}$ and $y = \frac{2}{5}x$.

Lesson 22: Constant Rates Revisited

b. The table of values below represents the number of pages that Anne can type, y, in a few selected x minutes. Assume she types at a constant rate.

Minutes (x)	Pages Typed (y)
3	2
5	$\frac{10}{3}$
8	$\frac{16}{3}$
10	$\frac{20}{3}$

Who types faster? Explain.

Anne types faster. Using the table, we can determine that the slope that represents Anne's constant rate of typing is $\frac{2}{3}$. The slope or rate for Nicole is $\frac{2}{5}$. When you compare the slopes, you see that $\frac{2}{3} > \frac{2}{5}$.

4.
a. Phil can build 3 birdhouses in 5 days. Assuming he builds birdhouses at a constant rate, write the linear equation that represents the situation.

Let y represent the total number of birdhouses Phil can build in x days. We can write $\frac{y}{x} = \frac{3}{5}$ and $y = \frac{3}{5}x$.

b. The figure represents Karl's constant rate of building the same kind of birdhouses.
Who builds birdhouses faster? Explain.

Karl can build birdhouses faster. The slope of the graph for Karl is $\frac{3}{4}$, and the slope or rate of change for Phil is $\frac{3}{5}$. When you compare the slopes, $\frac{3}{4} > \frac{3}{5}$.

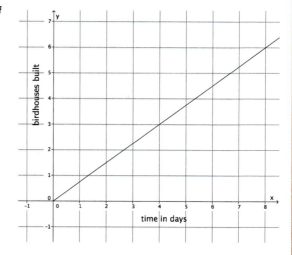

5. Explain your general strategy for comparing proportional relationships.

When comparing proportional relationships, we look specifically at the rate of change for each situation. The relationship with the greater rate of change will end up producing more, painting a greater area, or walking faster when compared to the same amount of time with the other proportional relationship.

Closing (4 minutes)

Summarize, or ask students to summarize, the main points from the lesson:

- We know how to write a constant rate problem as a linear equation in two variables.
- We know that in order to determine who has a greater speed (or is faster at completing a task), we need to compare the rates of changes, which corresponds to the slopes of the graphs of the proportional relationships. Whichever has the greater slope is the proportional relationship with the greater rate of change. We can determine rate of change from an equation, a graph, or a table.

> **Lesson Summary**
>
> Problems involving constant rate can be expressed as linear equations in two variables.
>
> When given information about two proportional relationships, their rates of change can be compared by comparing the slopes of the graphs of the two proportional relationships.

Exit Ticket (4 minutes)

Lesson 22: Constant Rates Revisited

Exit Ticket

1. Water flows out of Pipe A at a constant rate. Pipe A can fill 3 buckets of the same size in 14 minutes. Write a linear equation that represents the situation.

2. The figure below represents the rate at which Pipe B can fill the same-sized buckets.

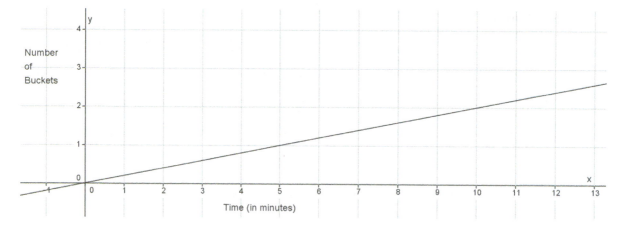

Which pipe fills buckets faster? Explain.

A STORY OF RATIOS Lesson 22 8•4

Exit Ticket Sample Solutions

1. Water flows out of Pipe A at a constant rate. Pipe A can fill 3 buckets of the same size in 14 minutes. Write a linear equation that represents the situation.

   Let y represent the total number of buckets that Pipe A can fill in x minutes. We can write $\frac{y}{x} = \frac{3}{14}$ and $y = \frac{3}{14}x$.

2. The figure below represents the rate at which Pipe B can fill the same-sized buckets.

   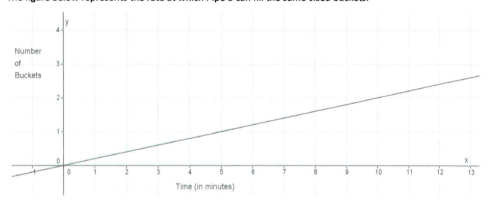

   Which pipe fills buckets faster? Explain.

   Pipe A fills the same-sized buckets faster than Pipe B. The slope of the graph for Pipe B is $\frac{1}{5}$, and the slope or rate for Pipe A is $\frac{3}{14}$. When you compare the slopes, you see that $\frac{3}{14} > \frac{1}{5}$.

Problem Set Sample Solutions

Students practice writing constant rate problems as linear equations in two variables. Students determine which of two proportional relationships is greater.

1.
   a. Train A can travel a distance of 500 miles in 8 hours. Assuming the train travels at a constant rate, write the linear equation that represents the situation.

   Let y represent the total number of miles Train A travels in x minutes. We can write $\frac{y}{x} = \frac{500}{8}$ and $y = \frac{125}{2}x$.

Lesson 22: Constant Rates Revisited

b. The figure represents the constant rate of travel for Train B.

Which train is faster? Explain.

Train B is faster than Train A. The slope or rate for Train A is $\frac{125}{2}$, and the slope of the line for Train B is $\frac{200}{3}$. When you compare the slopes, you see that $\frac{200}{3} > \frac{125}{2}$.

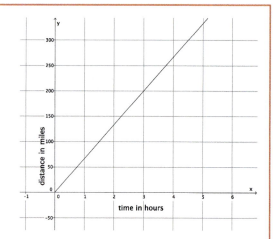

2.
a. Natalie can paint 40 square feet in 9 minutes. Assuming she paints at a constant rate, write the linear equation that represents the situation.

Let y represent the total square feet Natalie can paint in x minutes. We can write $\frac{y}{x} = \frac{40}{9}$, and $y = \frac{40}{9}x$.

b. The table of values below represents the area painted by Steven for a few selected time intervals. Assume Steven is painting at a constant rate.

Minutes (x)	Area Painted (y)
3	10
5	$\frac{50}{3}$
6	20
8	$\frac{80}{3}$

Who paints faster? Explain.

Natalie paints faster. Using the table of values, I can find the slope that represents Steven's constant rate of painting: $\frac{10}{3}$. The slope or rate for Natalie is $\frac{40}{9}$. When you compare the slopes, you see that $\frac{40}{9} > \frac{10}{3}$.

3.
a. Bianca can run 5 miles in 41 minutes. Assuming she runs at a constant rate, write the linear equation that represents the situation.

Let y represent the total number of miles Bianca can run in x minutes. We can write $\frac{y}{x} = \frac{5}{41}$, and $y = \frac{5}{41}x$.

b. The figure below represents Cynthia's constant rate of running.

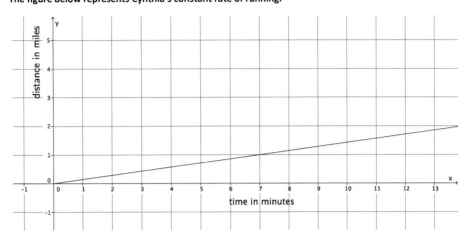

Who runs faster? Explain.

Cynthia runs faster. The slope of the graph for Cynthia is $\frac{1}{7}$, and the slope or rate for Nicole is $\frac{5}{41}$. When you compare the slopes, you see that $\frac{1}{7} > \frac{5}{41}$.

4.
a. Geoff can mow an entire lawn of 450 square feet in 30 minutes. Assuming he mows at a constant rate, write the linear equation that represents the situation.

Let y represent the total number of square feet Geoff can mow in x minutes. We can write $\frac{y}{x} = \frac{450}{30}$, and $y = 15x$.

b. The figure represents Mark's constant rate of mowing a lawn.

Who mows faster? Explain.

Geoff mows faster. The slope of the graph for Mark is $\frac{14}{2} = 7$, and the slope or rate for Geoff is $\frac{450}{30} = 15$. When you compare the slopes, you see that $15 > 7$.

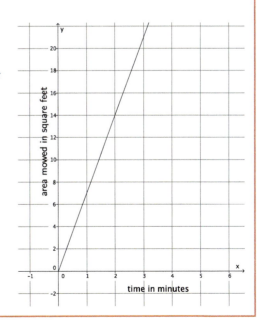

Lesson 22: Constant Rates Revisited

5.

a. Juan can walk to school, a distance of 0.75 mile, in 8 minutes. Assuming he walks at a constant rate, write the linear equation that represents the situation.

Let y represent the total distance in miles that Juan can walk in x minutes. We can write $\frac{y}{x} = \frac{0.75}{8}$, and $y = \frac{3}{32}x$.

b. The figure below represents Lena's constant rate of walking.

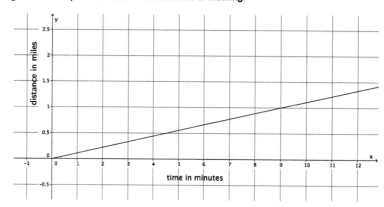

Who walks faster? Explain.

Lena walks faster. The slope of the graph for Lena is $\frac{1}{9}$, and the slope of the equation for Juan is $\frac{0.75}{8}$, or $\frac{3}{32}$. When you compare the slopes, you see that $\frac{1}{9} > \frac{3}{32}$.

A STORY OF RATIOS   Lesson 23   8•4

Lesson 23: The Defining Equation of a Line

Student Outcomes

- Students know that two equations in the form of $ax + by = c$ and $a'x + b'y = c'$ graph as the same line when $\dfrac{a'}{a} = \dfrac{b'}{b} = \dfrac{c'}{c}$ and at least one of a or b is nonzero.
- Students know that the graph of a linear equation $ax + by = c$, where a, b, and c are constants and at least one of a or b is nonzero, is the line defined by the equation $ax + by = c$.

Lesson Notes

Following the Exploratory Challenge is a Discussion that presents a theorem about the defining equation of a line and then a proof of the theorem. The proof of the theorem is optional. The Discussion can end with the theorem, and in place of the proof, students can complete Exercises 4–8. Whether the teacher chooses to discuss the proof or have students complete Exercises 4–8, it is important that students understand that two equations that are written differently can be the same, and their graph is the same line. This reasoning becomes important when considering systems of linear equations. In order to make sense of "infinitely many solutions" to a system of linear equations, students must know that equations that might appear to be different can have the same graph and represent the same line. Further, students should be able to recognize when two equations define the same line without having to graph each equation, which is the goal of this lesson. Students need graph paper to complete the Exploratory Challenge.

Classwork

Exploratory Challenge/Exercises 1–3 (20 minutes)

Students need graph paper to complete the exercises in the Exploratory Challenge. Students complete Exercises 1–3 in pairs or small groups.

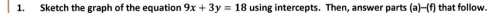

Exploratory Challenge/Exercises 1–3

1. Sketch the graph of the equation $9x + 3y = 18$ using intercepts. Then, answer parts (a)–(f) that follow.

    $9(0) + 3y = 18$
    $3y = 18$
    $y = 6$

    The y-intercept point is $(0, 6)$.

    $9x + 3(0) = 18$
    $9x = 18$
    $x = 2$

    The x-intercept point is $(2, 0)$.

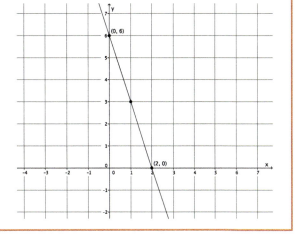

a. Sketch the graph of the equation $y = -3x + 6$ on the same coordinate plane.

b. What do you notice about the graphs of $9x + 3y = 18$ and $y = -3x + 6$? Why do you think this is so?

The graphs of the equations produce the same line. Both equations go through the same two points, so they are the same line.

c. Rewrite $y = -3x + 6$ in standard form.

$y = -3x + 6$
$3x + y = 6$

d. Identify the constants a, b, and c of the equation in standard form from part (c).

$a = 3$, $b = 1$, and $c = 6$

e. Identify the constants of the equation $9x + 3y = 18$. Note them as a', b', and c'.

$a' = 9$, $b' = 3$, and $c' = 18$

f. What do you notice about $\dfrac{a'}{a}$, $\dfrac{b'}{b}$, and $\dfrac{c'}{c}$?

$\dfrac{a'}{a} = \dfrac{9}{3} = 3$, $\dfrac{b'}{b} = \dfrac{3}{1} = 3$, and $\dfrac{c'}{c} = \dfrac{18}{6} = 3$

Each fraction is equal to the number 3.

2. Sketch the graph of the equation $y = \dfrac{1}{2}x + 3$ using the y-intercept point and the slope. Then, answer parts (a)–(f) that follow.

a. Sketch the graph of the equation $4x - 8y = -24$ using intercepts on the same coordinate plane.

$4(0) - 8y = -24$
$-8y = -24$
$y = 3$

The y-intercept point is $(0, 3)$.

$4x - 8(0) = -24$
$4x = -24$
$x = -6$

The x-intercept point is $(-6, 0)$.

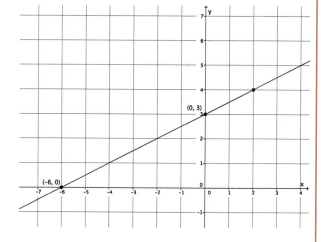

b. What do you notice about the graphs of $y = \dfrac{1}{2}x + 3$ and $4x - 8y = -24$? Why do you think this is so?

The graphs of the equations produce the same line. Both equations go through the same two points, so they are the same line.

c. Rewrite $y = \frac{1}{2}x + 3$ in standard form.

$$y = \frac{1}{2}x + 3$$
$$\left(y = \frac{1}{2}x + 3\right)2$$
$$2y = x + 6$$
$$-x + 2y = 6$$
$$-1(-x + 2y = 6)$$
$$x - 2y = -6$$

d. Identify the constants a, b, and c of the equation in standard form from part (c).

$a = 1, b = -2,$ and $c = -6$

e. Identify the constants of the equation $4x - 8y = -24$. Note them as a', b', and c'.

$a' = 4, b' = -8,$ and $c' = -24$

f. What do you notice about $\frac{a'}{a}, \frac{b'}{b},$ and $\frac{c'}{c}$?

$\frac{a'}{a} = \frac{4}{1} = 4, \frac{b'}{b} = \frac{-8}{-2} = 4,$ and $\frac{c'}{c} = \frac{-24}{-6} = 4$

Each fraction is equal to the number 4.

3. The graphs of the equations $y = \frac{2}{3}x - 4$ and $6x - 9y = 36$ are the same line.

a. Rewrite $y = \frac{2}{3}x - 4$ in standard form.

$$y = \frac{2}{3}x - 4$$
$$\left(y = \frac{2}{3}x - 4\right)3$$
$$3y = 2x - 12$$
$$-2x + 3y = -12$$
$$-1(-2x + 3y = -12)$$
$$2x - 3y = 12$$

b. Identify the constants a, b, and c of the equation in standard form from part (a).

$a = 2, b = -3,$ and $c = 12$

c. Identify the constants of the equation $6x - 9y = 36$. Note them as a', b', and c'.

$a' = 6, b' = -9,$ and $c' = 36$

d. What do you notice about $\frac{a'}{a}, \frac{b'}{b},$ and $\frac{c'}{c}$?

$\frac{a'}{a} = \frac{6}{2} = 3, \frac{b'}{b} = \frac{-9}{-3} = 3,$ and $\frac{c'}{c} = \frac{36}{12} = 3$

Each fraction is equal to the number 3.

Lesson 23: The Defining Equation of a Line

> e. You should have noticed that each fraction was equal to the same constant. Multiply that constant by the standard form of the equation from part (a). What do you notice?
>
> $$2x - 3y = 12$$
> $$3(2x - 3y = 12)$$
> $$6x - 9y = 36$$
>
> After multiplying the equation from part (a) by 3, I noticed that it is the exact same equation that was given.

Discussion (15 minutes)

Following the statement of the theorem is an optional proof of the theorem. Below the proof are Exercises 4–8 that can be completed instead of the proof.

- What did you notice about the equations you graphed in each of Exercises 1–3?
  - *In each case, the graphs of the equations are the same line.*
- What you observed in Exercises 1–3 can be summarized in the following theorem:

THEOREM: Suppose a, b, c, a', b', and c' are constants, where at least one of a or b is nonzero, and one of a' or b' is nonzero.

(1) If there is a nonzero number s so that $a' = sa$, $b' = sb$, and $c' = sc$, then the graphs of the equations $ax + by = c$ and $a'x + b'y = c'$ are the same line.

(2) If the graphs of the equations $ax + by = c$ and $a'x + b'y = c'$ are the same line, then there exists a nonzero number s so that $a' = sa$, $b' = sb$, and $c' = sc$.

The optional part of the Discussion begins here.

- We want to show that (1) is true. We need to show that the graphs of the equations $ax + by = c$ and $a'x + b'y = c'$ are the same. What information are we given in (1) that will be useful in showing that the two equations are the same?
  - *We are given that $a' = sa$, $b' = sb$, and $c' = sc$. We can use substitution in the equation $a'x + b'y = c'$ since we know what a', b', and c' are equal to.*

- Then, by substitution, we have

$$a'x + b'y = c'$$
$$sax + sby = sc.$$

By the distributive property,

$$s(ax + by) = sc.$$

Divide both sides of the equation by s:

$$ax + by = c,$$

which means the graph of $a'x + b'y = c'$ is equal to the graph of $ax + by = c$. That is, they represent the same line. Therefore, we have proved (1).

Lesson 23: The Defining Equation of a Line

- To prove (2), we will assume that $a, b \neq 0$; that is, we are not dealing with horizontal or vertical lines. Proving (2) will require us to rewrite the given equations $ax + by = c$ and $a'x + b'y = c'$ in slope-intercept form. Rewrite the equations.
  - $ax + by = c$
  $$by = -ax + c$$
  $$\frac{b}{b}y = -\frac{a}{b}x + \frac{c}{b}$$
  $$y = -\frac{a}{b}x + \frac{c}{b}$$
  - $a'x + b'y = c'$
  $$b'y = -a'x + c'$$
  $$\frac{b'}{b'}y = -\frac{a'}{b'}x + \frac{c'}{b'}$$
  $$y = -\frac{a'}{b'}x + \frac{c'}{b'}$$

- We will refer to $y = -\frac{a}{b}x + \frac{c}{b}$ as (A) and $y = -\frac{a'}{b'}x + \frac{c'}{b'}$ as (B).
- What are the slopes of (A) and (B)?
  - The slope of (A) is $-\frac{a}{b}$, and the slope of (B) is $-\frac{a'}{b'}$.
- Since we know that the two lines are the same, we know the slopes must be the same.
$$-\frac{a}{b} = -\frac{a'}{b'}$$
When we multiply both sides of the equation by -1, we have
$$\frac{a}{b} = \frac{a'}{b'}.$$
By the multiplication property of equality, we can rewrite $\frac{a}{b} = \frac{a'}{b'}$ as
$$\frac{a'}{a} = \frac{b'}{b}.$$
Notice that this proportion is equivalent to the original form.

Scaffolding:
Students may need to see the intermediate steps in the rewriting of $\frac{a}{b} = \frac{a'}{b'}$ as $\frac{a'}{a} = \frac{b'}{b}$.

- What are the y-intercept points of $y = -\frac{a}{b}x + \frac{c}{b}$ (A) and $y = -\frac{a'}{b'}x + \frac{c'}{b'}$ (B)?
  - The y-intercept point of (A) is $\frac{c}{b}$, and the y-intercept point of (B) is $\frac{c'}{b'}$.

Lesson 23: The Defining Equation of a Line

- Because we know that the two lines are the same, the y-intercept points will be the same.

$$\frac{c}{b} = \frac{c'}{b'}$$

We can rewrite the proportion.

$$\frac{c'}{c} = \frac{b'}{b}$$

Using the transitive property and the fact that $\frac{a'}{a} = \frac{b'}{b}$ and $\frac{c'}{c} = \frac{b'}{b}$, we can make the following statement:

$$\frac{a'}{a} = \frac{b'}{b} = \frac{c'}{c}.$$

Therefore, all three ratios are equal to the same number. Let this number be s.

$$\frac{a'}{a} = s, \quad \frac{b'}{b} = s, \quad \frac{c'}{c} = s$$

We can rewrite this to state that $a' = sa$, $b' = sb$, and $c' = cs$. Therefore, (2) is proved.

- When two equations are the same (i.e., their graphs are the same line), we say that any one of those equations is the *defining equation* of the line.

Exercises 4–8 (15 minutes)

Students complete Exercises 4–8 independently or in pairs. Consider having students share the equations they write for each exercise while others in the class verify which equations have the same line as their graphs.

> **Exercises 4–8**
>
> 4. Write three equations whose graphs are the same line as the equation $3x + 2y = 7$.
>
>   *Answers will vary. Verify that students have multiplied a, b, and c by the same constant when they write the new equation.*
>
> 5. Write three equations whose graphs are the same line as the equation $x - 9y = \frac{3}{4}$.
>
>   *Answers will vary. Verify that students have multiplied a, b, and c by the same constant when they write the new equation.*
>
> 6. Write three equations whose graphs are the same line as the equation $-9x + 5y = -4$.
>
>   *Answers will vary. Verify that students have multiplied a, b, and c by the same constant when they write the new equation.*

7. Write at least two equations in the form $ax + by = c$ whose graphs are the line shown below.

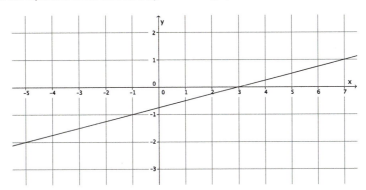

Answers will vary. Verify that students have the equation $-x + 4y = -3$ in some form.

8. Write at least two equations in the form $ax + by = c$ whose graphs are the line shown below.

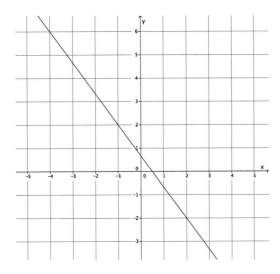

Answers will vary. Verify that students have the equation $4x + 3y = 2$ in some form.

Closing (5 minutes)

Summarize, or ask students to summarize, the main points from the lesson:

- We know that when the graphs of two equations are the same line, it is because they are the same equation in different forms.
- We know that even if the equations with the same line as their graph look different (i.e., different constants or different forms) that any one of those equations can be referred to as the *defining equation* of the line.

Lesson Summary

Two equations define the same line if the graphs of those two equations are the same given line. Two equations that define the same line are the same equation, just in different forms. The equations may look different (different constants, different coefficients, or different forms).

When two equations are written in standard form, $ax + by = c$ and $a'x + b'y = c'$, they define the same line when $\frac{a'}{a} = \frac{b'}{b} = \frac{c'}{c}$ is true.

Exit Ticket (5 minutes)

A STORY OF RATIOS — Lesson 23 8•4

Name _____ Date _____

Lesson 23: The Defining Equation of a Line

Exit Ticket

1. Do the graphs of the equations $-16x + 12y = 33$ and $-4x + 3y = 8$ graph as the same line? Why or why not?

2. Given the equation $3x - y = 11$, write another equation that will have the same graph. Explain why.

Exit Ticket Sample Solutions

1. Do the graphs of the equations $-16x + 12y = 33$ and $-4x + 3y = 8$ graph as the same line? Why or why not?

   No. In the first equation, $a = -16$, $b = 12$, and $c = 33$, and in the second equation, $a' = -4$, $b' = 3$, and $c' = 8$. Then,

   $$\frac{a'}{a} = \frac{-4}{-16} = \frac{1}{4}, \frac{b'}{b} = \frac{3}{12} = \frac{1}{4}, \text{ but } \frac{c'}{c} = \frac{8}{33} = \frac{8}{33}.$$

   Since each fraction does not equal the same number, then they do not have the same graph.

2. Given the equation $3x - y = 11$, write another equation that will have the same graph. Explain why.

   Answers will vary. Verify that students have written an equation that defines the same line by showing that the fractions $\frac{a'}{a} = \frac{b'}{b} = \frac{c'}{c} = s$, where s is some constant.

Problem Set Sample Solutions

Students practice identifying pairs of equations as the defining equation of a line or two distinct lines.

1. Do the equations $x + y = -2$ and $3x + 3y = -6$ define the same line? Explain.

   Yes, these equations define the same line. When you compare the constants from each equation, you get

   $$\frac{a'}{a} = \frac{3}{1} = 3, \frac{b'}{b} = \frac{3}{1} = 3, \text{ and } \frac{c'}{c} = \frac{-6}{-2} = 3.$$

   When I multiply the first equation by 3, I get the second equation.

   $$(x + y = -2)3$$
   $$3x + 3y = -6$$

   Therefore, these equations define the same line.

2. Do the equations $y = -\frac{5}{4}x + 2$ and $10x + 8y = 16$ define the same line? Explain.

   Yes, these equations define the same line. When you rewrite the first equation in standard form, you get

   $$y = -\frac{5}{4}x + 2$$
   $$\left(y = -\frac{5}{4}x + 2\right)4$$
   $$4y = -5x + 8$$
   $$5x + 4y = 8.$$

   When you compare the constants from each equation, you get

   $$\frac{a'}{a} = \frac{10}{5} = 2, \frac{b'}{b} = \frac{8}{4} = 2, \text{ and } \frac{c'}{c} = \frac{16}{8} = 2.$$

   When I multiply the first equation by 2, I get the second equation.

   $$(5x + 4y = 8)2$$
   $$10x + 8y = 16$$

   Therefore, these equations define the same line.

3. **Write an equation that would define the same line as $7x - 2y = 5$.**

   Answers will vary. Verify that students have written an equation that defines the same line by showing that the fractions $\frac{a'}{a} = \frac{b'}{b} = \frac{c'}{c} = s$, where s is some constant.

4. **Challenge: Show that if the two lines given by $ax + by = c$ and $a'x + b'y = c'$ are the same when $b = 0$ (vertical lines), then there exists a nonzero number s so that $a' = sa$, $b' = sb$, and $c' = sc$.**

   When $b = 0$, then the equations are $ax = c$ and $a'x = c'$. We can rewrite the equations as $x = \frac{c}{a}$ and $x = \frac{c'}{a'}$. Because the equations graph as the same line, then we know that

   $$\frac{c}{a} = \frac{c'}{a'}$$

   and we can rewrite those fractions as

   $$\frac{a'}{a} = \frac{c'}{c}.$$

   These fractions are equal to the same number. Let that number be s. Then $\frac{a'}{a} = s$ and $\frac{c'}{c} = s$. Therefore, $a' = sa$ and $c' = sc$.

5. **Challenge: Show that if the two lines given by $ax + by = c$ and $a'x + b'y = c'$ are the same when $a = 0$ (horizontal lines), then there exists a nonzero number s so that $a' = sa$, $b' = sb$, and $c' = sc$.**

   When $a = 0$, then the equations are $by = c$ and $b'y = c'$. We can rewrite the equations as $y = \frac{c}{b}$ and $y = \frac{c'}{b'}$. Because the equations graph as the same line, then we know that their slopes are the same.

   $$\frac{c}{b} = \frac{c'}{b'}$$

   We can rewrite the proportion.

   $$\frac{b'}{b} = \frac{c'}{c}$$

   These fractions are equal to the same number. Let that number be s. Then $\frac{b'}{b} = s$ and $\frac{c'}{c} = s$. Therefore, $b' = sb$ and $c' = sc$.

Lesson 23: The Defining Equation of a Line

A STORY OF RATIOS

Mathematics Curriculum

GRADE 8 • MODULE 4

Topic D

Systems of Linear Equations and Their Solutions

8.EE.B.5, 8.EE.C.8

Focus Standards:	8.EE.B.5	Graph proportional relationships, interpreting the unit rate as the slope of the graph. Compare two different proportional relationships represented in different ways. *For example, compare a distance-time graph to a distance-time equation to determine which of two moving objects has greater speed.*
	8.EE.C.8	Analyze and solve pairs of simultaneous linear equations.
	a.	Understand that solutions to a system of two linear equations in two variables correspond to points of intersection of their graphs, because points of intersection satisfy both equations simultaneously.
	b.	Solve systems of two linear equations in two variables algebraically, and estimate solutions by graphing the equations. Solve simple cases by inspection. *For example, $3x + 2y = 5$ and $3x + 2y = 6$ have no solution because $3x + 2y$ cannot simultaneously be 5 and 6.*
	c.	Solve real-world and mathematical problems leading to two linear equations in two variables. *For example, given coordinates for two pairs of points, determine whether the line through the first pair of points intersects the line through the second pair.*
Instructional Days:	7	
Lesson 24:	Introduction to Simultaneous Equations (P)[1]	
Lesson 25:	Geometric Interpretation of the Solutions of a Linear System (E)	
Lesson 26:	Characterization of Parallel Lines (S)	
Lesson 27:	Nature of Solutions of a System of Linear Equations (P)	
Lesson 28:	Another Computational Method of Solving a Linear System (P)	
Lesson 29:	Word Problems (P)	
Lesson 30:	Conversion Between Celsius and Fahrenheit (M)	

[1]Lesson Structure Key: **P**-Problem Set Lesson, **M**-Modeling Cycle Lesson, **E**-Exploration Lesson, **S**-Socratic Lesson

Topic D: Systems of Linear Equations and Their Solutions

Topic D introduces students to systems of linear equations by comparing distance-time graphs to determine which of two objects has greater speed (**8.EE.B.5**, **8.EE.C.8c**). Lessons 25–27 expose students to the possibilities for solutions of a system of linear equations. In Lesson 25, students graph two linear equations on a coordinate plane and identify the point of intersection of the two lines as the solution to the system (**8.EE.C.8a**). Next, students look at systems of equations that graph as parallel lines (**8.EE.C.8b**). In Lesson 26, students learn that a system can have no solutions because parallel lines do not have a point of intersection (**8.EE.C.8b**). Lesson 27 continues this thinking with respect to systems that have infinitely many solutions (**8.EE.C.8b**). In Lesson 28, students learn how to solve a system of equations using computational methods, such as elimination and substitution (**8.EE.C.8b**). In Lesson 29, students must use all of the skills of the module to transcribe written statements into a system of linear equations, find the solution(s) if it exists, and then verify that it is correct. Lesson 30 is an application of what students have learned about linear equations. Students develop a linear equation that represents the conversion between temperatures in Celsius to temperatures in Fahrenheit.

Topic D: Systems of Linear Equations and Their Solutions

Lesson 24: Introduction to Simultaneous Equations

Student Outcomes

- Students know that a system of linear equations, also known as *simultaneous equations,* is when two or more equations are involved in the same problem and work must be completed on them simultaneously. Students also learn the notation for simultaneous equations.
- Students compare the graphs that comprise a system of linear equations in the context of constant rates to answer questions about time and distance.

Lesson Notes

Students complete Exercises 1–3 as an introduction to simultaneous linear equations in a familiar context. Example 1 demonstrates what happens to the graph of a line when there is change in the circumstances involving time with constant rate problems. It is in preparation for the examples that follow. It is not necessary that Example 1 be shown to students, but it is provided as a scaffold. If Example 1 is used, consider skipping Example 3.

Classwork

Exercises 1–3 (5 minutes)

Students complete Exercises 1–3 in pairs. Once students are finished, continue with the Discussion about systems of linear equations.

Exercises

1. Derek scored 30 points in the basketball game he played, and not once did he go to the free throw line. That means that Derek scored two-point shots and three-point shots. List as many combinations of two- and three-pointers as you can that would total 30 points.

Number of Two-Pointers	Number of Three-Pointers
15	0
0	10
12	2
9	4
6	6
3	8

Write an equation to describe the data.

Let x represent the number of 2-pointers and y represent the number of 3-pointers.

$30 = 2x + 3y$

A STORY OF RATIOS   Lesson 24   8•4

2. Derek tells you that the number of two-point shots that he made is five more than the number of three-point shots. How many combinations can you come up with that fit this scenario? (Don't worry about the total number of points.)

Number of Two-Pointers	Number of Three-Pointers
6	1
7	2
8	3
9	4
10	5
11	6

Write an equation to describe the data.

Let x represent the number of two-pointers and y represent the number of three-pointers.

$x = 5 + y$

3. Which pair of numbers from your table in Exercise 2 would show Derek's actual score of 30 points?

The pair 9 and 4 would show Derek's actual score of 30 points.

Discussion (5 minutes)

- There are situations where we need to work with two linear equations simultaneously. Hence, the phrase *simultaneous linear equations*. Sometimes a pair of linear equations is referred to as a *system of linear equations*.
- The situation with Derek can be represented as a system of linear equations.

  Let x represent the number of two-pointers and y represent the number of three-pointers; then

  $$\begin{cases} 2x + 3y = 30 \\ x = 5 + y. \end{cases}$$

- The notation for simultaneous linear equations lets us know that we are looking for the ordered pair (x, y) that makes both equations true. That point is called the *solution* to the system.
- Just like equations in one variable, some systems of equations have exactly one solution, no solution, or infinitely many solutions. This is a topic for later lessons.
- Ultimately, our goal is to determine the exact location on the coordinate plane where the graphs of the two linear equations intersect, giving us the ordered pair (x, y) that is the solution to the system of equations. This, too, is a topic for a later lesson.

Lesson 24:   Introduction to Simultaneous Equations

- We can graph both equations on the same coordinate plane.

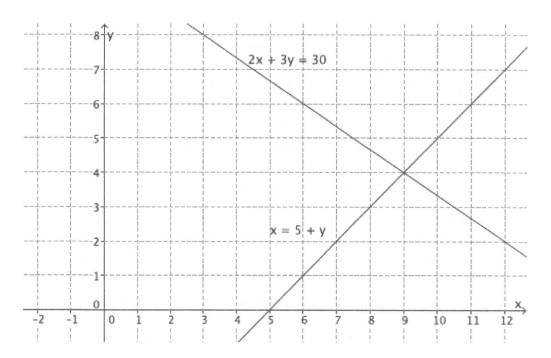

- Because we are graphing two distinct lines on the same graph, we identify the lines in some manner. In this case, we identify them by the equations.
- Note the point of intersection. Does it satisfy both equations in the system?
  - *The point of intersection of the two lines is $(9, 4)$.*

$$2(9) + 3(4) = 30$$
$$18 + 12 = 30$$
$$30 = 30$$
$$9 = 4 + 5$$
$$9 = 9$$

  Yes, $x = 9$ and $y = 4$ satisfies both equations of the system.
- Therefore, Derek made 9 two-point shots and 4 three-point shots.

Example 1 (6 minutes)

- Pia types at a constant rate of 3 pages every 15 minutes. Pia's rate is $\frac{1}{5}$ pages per minute. If she types y pages in x minutes at that constant rate, then $y = \frac{1}{5}x$. Pia's rate represents the proportional relationship $y = \frac{1}{5}x$.
- The following table displays the number of pages Pia typed at the end of certain time intervals.

Number of Minutes (x)	Pages Typed (y)
0	0
5	1
10	2
15	3
20	4
25	5

- The following is the graph of $y = \frac{1}{5}x$.

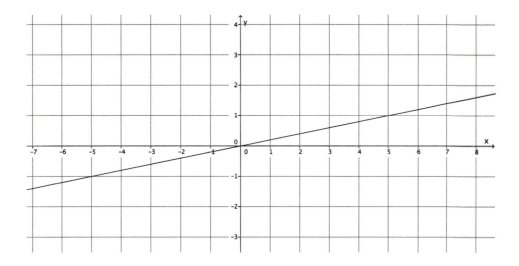

- Pia typically begins work at 8:00 a.m. every day. On our graph, her start time is reflected as the origin of the graph $(0, 0)$, that is, zero minutes worked and zero pages typed. For some reason, she started working 5 minutes earlier today. How can we reflect the extra 5 minutes she worked on our graph?
    - *The x-axis represents the time worked, so we need to do something on the x-axis to reflect the additional 5 minutes of work.*

Lesson 24: Introduction to Simultaneous Equations

- If we translate the graph of $y = \frac{1}{5}x$ to the right 5 units to reflect the additional 5 minutes of work, then we have the following graph:

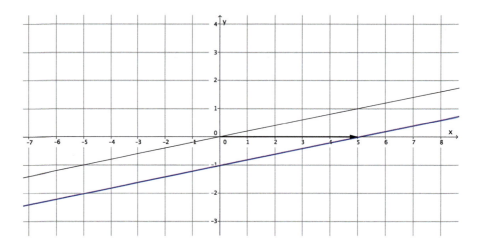

- Does a translation of 5 units to the right reflect her working an additional 5 minutes?
  - No. It makes it look like she got nothing done the first 5 minutes she was at work.
- Let's see what happens when we translate 5 units to the left.

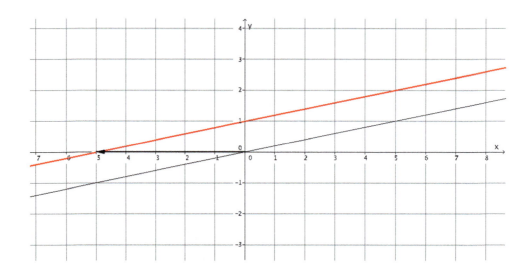

- Does a translation of 5 units to the left reflect her working an additional 5 minutes?
  - Yes. It shows that she was able to type 1 page by the time she normally begins work.
- What is the equation that represents the graph of the translated line?
  - The equation that represents the graph of the red line is $y = \frac{1}{5}x + 1$.
- Note again that even though the graph has been translated, the slope of the line is still equal to the constant rate of typing.

- When we factor out $\frac{1}{5}$ from $\frac{1}{5}x + 1$ to give us $y = \frac{1}{5}(x + 5)$, we can better see the additional 5 minutes of work time. Pia typed for an additional 5 minutes, so it makes sense that we are adding 5 to the number of minutes, x, that she types. However, on the graph, we translated 5 units to the left of zero. How can we make sense of that?

    ▫ *Since her normal start time was 8:00 a.m., then 5 minutes before 8:00 a.m. is 5 minutes less than 8:00 a.m., which means we would need to go to the left of 8:00 a.m. (in this case, the origin of the graph) to mark her start time.*

- If Pia started work 20 minutes early, what equation would represent the number of pages she could type in x minutes?

    ▫ *The equation that represents the graph of the red line is $y = \frac{1}{5}(x + 20)$.*

Example 2 (8 minutes)

- Now we will look at an example of a situation that requires simultaneous linear equations.

    Sandy and Charlie run at constant speeds. Sandy runs from their school to the train station in 15 minutes, and Charlie runs the same distance in 10 minutes. Charlie starts 4 minutes after Sandy left the school. Can Charlie catch up to Sandy? The distance between the school and the station is 2 miles.

- What is Sandy's average speed in 15 minutes? Explain.

    ▫ *Sandy's average speed in 15 minutes is $\frac{2}{15}$ miles per minute because she runs 2 miles in 15 minutes.*

- Since we know Sandy runs at a constant speed, then her constant speed is $\frac{2}{15}$ miles per minute.

- What is Charlie's average speed in 10 minutes? Explain.

    ▫ *Charlie's average speed in 10 minutes is $\frac{2}{10}$ miles per minute, which is the same as $\frac{1}{5}$ miles per minute because he walks runs 1 mile in 5 minutes.*

- Since we know Charlie runs at a constant speed, then his constant speed is $\frac{1}{5}$ miles per minute.

- Suppose Charlie ran y miles in x minutes at that constant speed; then $y = \frac{1}{5}x$. Charlie's speed represents the proportional relationship $y = \frac{1}{5}x$.

- Let's put some information about Charlie's run in a table:

Number of Minutes (x)	Miles Run (y)
0	0
5	1
10	2
15	3
20	4
25	5

Lesson 24: Introduction to Simultaneous Equations

- At x minutes, Sandy has run 4 minutes longer than Charlie. Then the distance that Sandy ran in $x + 4$ minutes is y miles. Then the linear equation that represents Sandy's motion is

$$\frac{y}{x+4} = \frac{2}{15}$$
$$y = \frac{2}{15}(x+4)$$
$$y = \frac{2}{15}x + \frac{8}{15}.$$

- How many miles had Sandy run by the time Charlie left the school?
  - At zero minutes in Charlie's run, Sandy will have run $\frac{8}{15}$ miles.

Let's put some information about Sandy's run in a table:

Number of Minutes (x)	Miles Run (y)
5	$\frac{18}{15} = 1\frac{3}{15} = 1\frac{1}{5}$
10	$\frac{28}{15} = 1\frac{13}{15}$
15	$\frac{38}{15} = 2\frac{8}{15}$
20	$\frac{48}{15} = 3\frac{3}{15} = 3\frac{1}{5}$
25	$\frac{58}{15} = 3\frac{13}{15}$

- Now let's sketch the graphs of each linear equation on a coordinate plane.

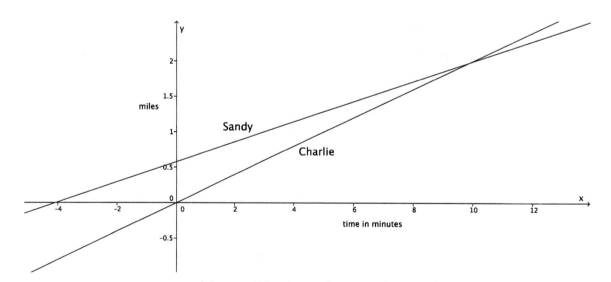

- A couple of comments about our graph:
  - The y-intercept point of the graph of Sandy's run shows the exact distance she has run at the moment that Charlie starts walking. Notice that the x-intercept point of the graph of Sandy's run shows that she starts running 4 minutes before Charlie.
  - Since the y-axis represents the distance traveled, the point of intersection of the graphs of the two lines represents the moment they have both traveled the same distance.
- Recall the original question that was asked: Can Charlie catch up to Sandy? Keep in mind that the train station is 2 miles from the school.
  - It looks like the lines intersect at a point between 1.5 and 2 miles; therefore, the answer is yes, Charlie can catch up to Sandy.
- At approximately what point do the graphs of the lines intersect?
  - The lines intersect at approximately $(10, 1.8)$.
- A couple of comments about our equations $y = \frac{1}{5}x$ and $y = \frac{2}{15}x + \frac{8}{15}$:
  - Notice that x (the representation of time) is the same in both equations.
  - Notice that we are interested in finding out when $y = y$ because this is when the distance traveled by both Sandy and Charlie is the same (i.e., when Charlie catches up to Sandy).
  - We write the pair of simultaneous linear equations as

$$\begin{cases} y = \frac{2}{15}x + \frac{8}{15} \\ y = \frac{1}{5}x \end{cases}.$$

Example 3 (5 minutes)

- Randi and Craig ride their bikes at constant speeds. It takes Randi 25 minutes to bike 4 miles. Craig can bike 4 miles in 32 minutes. If Randi gives Craig a 20-minute head start, about how long will it take Randi to catch up to Craig?
- We want to express the information about Randi and Craig in terms of a system of linear equations. Write the linear equations that represent their constant speeds.
  - Randi's rate is $\frac{4}{25}$ miles per minute. If she bikes y miles in x minutes at that constant rate, then $y = \frac{4}{25}x$. Randi's rate represents the proportional relationship $y = \frac{4}{25}x$.
  - Craig's rate is $\frac{4}{32}$ miles per minute, which is equivalent to $\frac{1}{8}$ miles per minute. If he bikes y miles in x minutes at that constant rate, then $y = \frac{1}{8}x$. Craig's rate represents the proportional relationship $y = \frac{1}{8}x$.

- We want to account for the head start that Craig is getting. Since Craig gets a head start of 20 minutes, we need to add that time to his total number of minutes traveled:

$$y = \frac{1}{8}(x + 20)$$
$$y = \frac{1}{8}x + \frac{20}{8}$$
$$y = \frac{1}{8}x + \frac{5}{2}$$

- The system of linear equations that represents this situation is

$$\begin{cases} y = \frac{1}{8}x + \frac{5}{2} \\ y = \frac{4}{25}x \end{cases}.$$

- Now we can sketch the graphs of the system of equations on a coordinate plane.

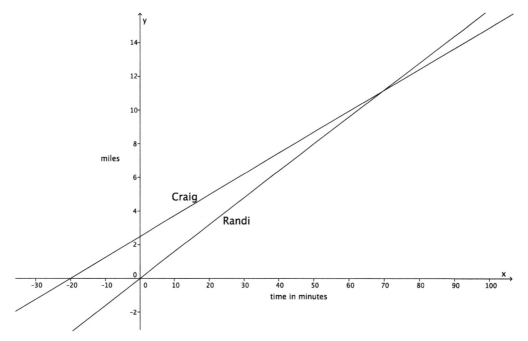

- Notice again that the y-intercept point of Craig's graph shows the distance that Craig was able to travel at the moment Randi began biking. Also notice that the x-intercept point of Craig's graph shows us that he started biking 20 minutes before Randi.
- Now, answer the question: About how long will it take Randi to catch up to Craig? We can give two answers: one in terms of time and the other in terms of distance. What are those answers?
    □ *It will take Randi about 70 minutes or about 11 miles to catch up to Craig.*
- At approximately what point do the graphs of the lines intersect?
    □ *The lines intersect at approximately (70, 11).*

Exercises 4–5 (7 minutes)

Students complete Exercises 4–5 individually or in pairs.

4. Efrain and Fernie are on a road trip. Each of them drives at a constant speed. Efrain is a safe driver and travels 45 miles per hour for the entire trip. Fernie is not such a safe driver. He drives 70 miles per hour throughout the trip. Fernie and Efrain left from the same location, but Efrain left at 8:00 a.m., and Fernie left at 11:00 a.m. Assuming they take the same route, will Fernie ever catch up to Efrain? If so, approximately when?

   a. Write the linear equation that represents Efrain's constant speed. Make sure to include in your equation the extra time that Efrain was able to travel.

   Efrain's rate is $\frac{45}{1}$ miles per hour, which is the same as 45 miles per hour. If he drives y miles in x hours at that constant rate, then $y = 45x$. To account for his additional 3 hours of driving time that Efrain gets, we write the equation $y = 45(x + 3)$.

   $$y = 45x + 135$$

   b. Write the linear equation that represents Fernie's constant speed.

   Fernie's rate is $\frac{70}{1}$ miles per hour, which is the same as 70 miles per hour. If he drives y miles in x hours at that constant rate, then $y = 70x$.

   c. Write the system of linear equations that represents this situation.

   $$\begin{cases} y = 45x + 135 \\ y = 70x \end{cases}$$

   d. Sketch the graphs of the two linear equations.

   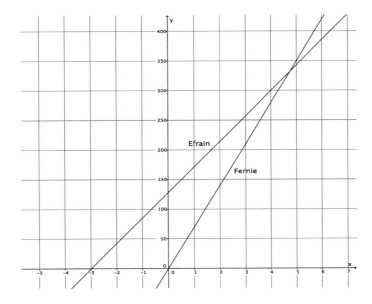

   e. Will Fernie ever catch up to Efrain? If so, approximately when?

   Yes, Fernie will catch up to Efrain after about $4\frac{1}{2}$ hours of driving or after traveling about 325 miles.

f. At approximately what point do the graphs of the lines intersect?

The lines intersect at approximately $(4.5, 325)$.

5. Jessica and Karl run at constant speeds. Jessica can run 3 miles in 24 minutes. Karl can run 2 miles in 14 minutes. They decide to race each other. As soon as the race begins, Karl trips and takes 2 minutes to recover.

   a. Write the linear equation that represents Jessica's constant speed. Make sure to include in your equation the extra time that Jessica was able to run.

   Jessica's rate is $\frac{3}{24}$ miles per minute, which is equivalent to $\frac{1}{8}$ miles per minute. If Jessica runs y miles x minutes at that constant speed, then $y = \frac{1}{8}x$. To account for her additional 2 minute of running that Jessica gets, we write the equation

   $$y = \frac{1}{8}(x+2)$$
   $$y = \frac{1}{8}x + \frac{1}{4}$$

   b. Write the linear equation that represents Karl's constant speed.

   Karl's rate is $\frac{2}{14}$ miles per minute, which is the same as $\frac{1}{7}$ miles per minute. If Karl runs y miles in x minutes at that constant speed, then $y = \frac{1}{7}x$.

   c. Write the system of linear equations that represents this situation.

   $$\begin{cases} y = \frac{1}{8}x + \frac{1}{8} \\ y = \frac{1}{7}x \end{cases}$$

   d. Sketch the graphs of the two linear equations.

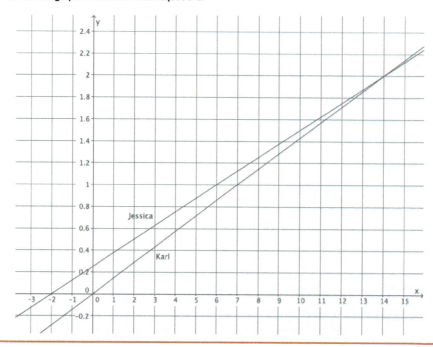

e. Use the graph to answer the questions below.

i. If Jessica and Karl raced for 3 miles, who would win? Explain.

If the race were 3 miles, then Karl would win. It only takes Karl 21 minutes to run 3 miles, but it takes Jessica 24 minutes to run the distance of 3 miles.

ii. At approximately what point would Jessica and Karl be tied? Explain.

Jessica and Karl would be tied after about 4 minutes or a distance of 1 mile. That is where the graphs of the lines intersect.

Closing (4 minutes)

Summarize, or ask students to summarize, the main points from the lesson:

- We know that some situations require two linear equations. In those cases, we have what is called a *system of linear equations* or *simultaneous linear equations*.
- The solution to a system of linear equations, similar to a linear equation, is all of the points that make the equations true.
- We can recognize a system of equations by the notation used, for example: $\begin{cases} y = \frac{1}{8}x + \frac{5}{2} \\ y = \frac{4}{25}x \end{cases}$.

Lesson Summary

A *system of linear equations* is a set of two or more linear equations. When graphing a pair of linear equations in two variables, both equations in the system are graphed on the same coordinate plane.

A *solution to a system of two linear equations in two variables* is an ordered pair of numbers that is a solution to both equations. For example, the solution to the system of linear equations $\begin{cases} x + y = 6 \\ x - y = 4 \end{cases}$ is the ordered pair $(5, 1)$ because substituting 5 in for x and 1 in for y results in two true equations: $5 + 1 = 6$ and $5 - 1 = 4$.

Systems of linear equations are notated using brackets to group the equations, for example: $\begin{cases} y = \frac{1}{8}x + \frac{5}{2} \\ y = \frac{4}{25}x \end{cases}$.

Exit Ticket (5 minutes)

A STORY OF RATIOS   Lesson 24 8•4

Name _____ Date _____

Lesson 24: Introduction to Simultaneous Equations

Exit Ticket

Darnell and Hector ride their bikes at constant speeds. Darnell leaves Hector's house to bike home. He can bike the 8 miles in 32 minutes. Five minutes after Darnell leaves, Hector realizes that Darnell left his phone. Hector rides to catch up. He can ride to Darnell's house in 24 minutes. Assuming they bike the same path, will Hector catch up to Darnell before he gets home?

    a.   Write the linear equation that represents Darnell's constant speed.

    b.   Write the linear equation that represents Hector's constant speed. Make sure to take into account that Hector left after Darnell.

    c.   Write the system of linear equations that represents this situation.

d. Sketch the graphs of the two equations.

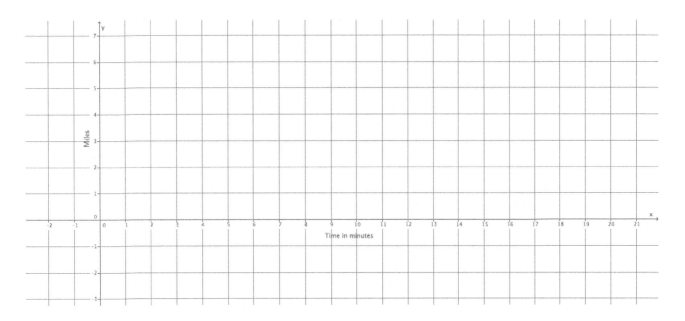

e. Will Hector catch up to Darnell before he gets home? If so, approximately when?

f. At approximately what point do the graphs of the lines intersect?

Lesson 24: Introduction to Simultaneous Equations

Exit Ticket Sample Solutions

Darnell and Hector ride their bikes at constant speeds. Darnell leaves Hector's house to bike home. He can bike the 8 miles in 32 minutes. Five minutes after Darnell leaves, Hector realizes that Darnell left his phone. Hector rides to catch up. He can ride to Darnell's house in 24 minutes. Assuming they bike the same path, will Hector catch up to Darnell before he gets home?

a. Write the linear equation that represents Darnell's constant speed.

Darnell's rate is $\frac{1}{4}$ miles per minute. If he bikes y miles in x minutes at that constant speed, then $y = \frac{1}{4}x$.

b. Write the linear equation that represents Hector's constant speed. Make sure to take into account that Hector left after Darnell.

Hector's rate is $\frac{1}{3}$ miles per minute. If he bikes y miles in x minutes, then $y = \frac{1}{3}x$. To account for the extra time Darnell has to bike, we write the equation

$$y = \frac{1}{3}(x - 5)$$
$$y = \frac{1}{3}x - \frac{5}{3}.$$

c. Write the system of linear equations that represents this situation.

$$\begin{cases} y = \frac{1}{4}x \\ y = \frac{1}{3}x - \frac{5}{3} \end{cases}$$

d. Sketch the graphs of the two equations.

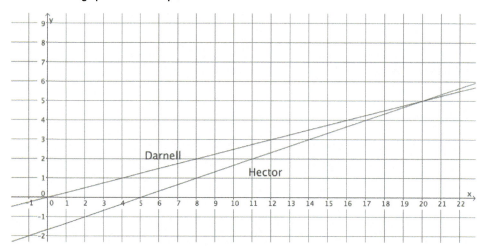

e. Will Hector catch up to Darnell before he gets home? If so, approximately when?

Hector will catch up 20 minutes after Darnell left his house (or 15 minutes of biking by Hector) or approximately 5 miles.

f. At approximately what point do the graphs of the lines intersect?

The lines intersect at approximately $(20, 5)$.

Problem Set Sample Solutions

1. Jeremy and Gerardo run at constant speeds. Jeremy can run 1 mile in 8 minutes, and Gerardo can run 3 miles in 33 minutes. Jeremy started running 10 minutes after Gerardo. Assuming they run the same path, when will Jeremy catch up to Gerardo?

    a. Write the linear equation that represents Jeremy's constant speed.

    Jeremy's rate is $\frac{1}{8}$ miles per minute. If he runs y miles in x minutes, then $y = \frac{1}{8}x$.

    b. Write the linear equation that represents Gerardo's constant speed. Make sure to include in your equation the extra time that Gerardo was able to run.

    Gerardo's rate is $\frac{3}{33}$ miles per minute, which is the same as $\frac{1}{11}$ miles per minute. If he runs y miles in x minutes, then $y = \frac{1}{11}x$. To account for the extra time that Gerardo gets to run, we write the equation

    $$y = \frac{1}{11}(x + 10)$$
    $$y = \frac{1}{11}x + \frac{10}{11}$$

    c. Write the system of linear equations that represents this situation.

    $$\begin{cases} y = \frac{1}{8}x \\ y = \frac{1}{11}x + \frac{10}{11} \end{cases}$$

    d. Sketch the graphs of the two equations.

    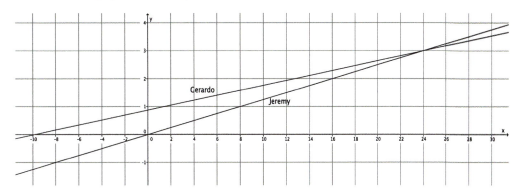

    e. Will Jeremy ever catch up to Gerardo? If so, approximately when?

    Yes, Jeremy will catch up to Gerardo after about 24 minutes or about 3 miles.

    f. At approximately what point do the graphs of the lines intersect?

    The lines intersect at approximately $(24, 3)$.

2. Two cars drive from town A to town B at constant speeds. The blue car travels 25 miles per hour, and the red car travels 60 miles per hour. The blue car leaves at 9:30 a.m., and the red car leaves at noon. The distance between the two towns is 150 miles.

   a. Who will get there first? Write and graph the system of linear equations that represents this situation.

   The linear equation that represents the distance traveled by the blue car is $y = 25(x + 2.5)$, which is the same as $y = 25x + 62.5$. The linear equation that represents the distance traveled by the red car is $y = 60x$. The system of linear equations that represents this situation is

   $$\begin{cases} y = 25x + 62.5 \\ y = 60x \end{cases}.$$

   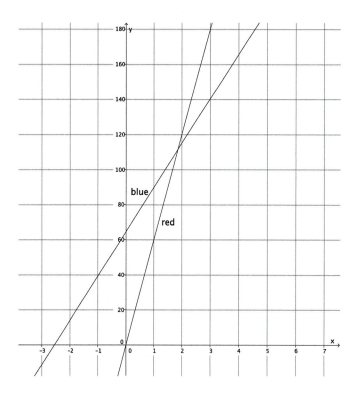

   The red car will get to town B first.

   b. At approximately what point do the graphs of the lines intersect?

   The lines intersect at approximately $(1.8, 110)$.

A STORY OF RATIOS   Lesson 25   8•4

Lesson 25: Geometric Interpretation of the Solutions of a Linear System

Student Outcomes

- Students sketch the graphs of two linear equations and find the point of intersection.
- Students identify the point of intersection of the two lines as the solution to the system.
- Students verify by computation that the point of intersection is a solution to each of the equations in the system.

Lesson Notes

In the last lesson, students were introduced to the concept of simultaneous equations. Students compared the graphs of two different equations and found the point of intersection. Students estimated the coordinates of the point of intersection of the lines in order to answer questions about the context related to time and distance. In this lesson, students first graph systems of linear equations and find the precise point of intersection. Then, students verify that the ordered pair that is the intersection of the two lines is a solution to *each* equation in the system. Finally, students informally verify that there can be only one point of intersection of two distinct lines in the system by checking to see if another point that is a solution to one equation is a solution to both equations.

MP.6

Students need graph paper to complete the Problem Set.

Classwork

Exploratory Challenge/Exercises 1–5 (25 minutes)

Students work independently on Exercises 1–5.

Exploratory Challenge/Exercises 1–5

1. Sketch the graphs of the linear system on a coordinate plane: $\begin{cases} 2y + x = 12 \\ y = \frac{5}{6}x - 2 \end{cases}$.

   For the equation $2y + x = 12$:
   $$2y + 0 = 12$$
   $$2y = 12$$
   $$y = 6$$
   The y-intercept point is $(0, 6)$.
   $$2(0) + x = 12$$
   $$x = 12$$
   The x-intercept point is $(12, 0)$.

   For the equation $y = \frac{5}{6}x - 2$:
   The slope is $\frac{5}{6}$, and the y-intercept point is $(0, -2)$.

Lesson 25:   Geometric Interpretation of the Solutions of a Linear System   397

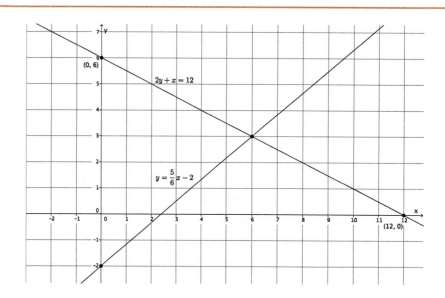

a. Name the ordered pair where the graphs of the two linear equations intersect.

$(6, 3)$

b. Verify that the ordered pair named in part (a) is a solution to $2y + x = 12$.

$$2(3) + 6 = 12$$
$$6 + 6 = 12$$
$$12 = 12$$

The left and right sides of the equation are equal.

c. Verify that the ordered pair named in part (a) is a solution to $y = \frac{5}{6}x - 2$.

$$3 = \frac{5}{6}(6) - 2$$
$$3 = 5 - 2$$
$$3 = 3$$

The left and right sides of the equation are equal.

d. Could the point $(4, 4)$ be a solution to the system of linear equations? That is, would $(4, 4)$ make both equations true? Why or why not?

No. The graphs of the equations represent all of the possible solutions to the given equations. The point $(4, 4)$ is a solution to the equation $2y + x = 12$ because it is on the graph of that equation. However, the point $(4, 4)$ is not on the graph of the equation $y = \frac{5}{6}x - 2$. Therefore, $(4, 4)$ cannot be a solution to the system of equations.

2. Sketch the graphs of the linear system on a coordinate plane: $\begin{cases} x + y = -2 \\ y = 4x + 3 \end{cases}$.

For the equation $x + y = -2$:

$$0 + y = -2$$
$$y = -2$$

The y-intercept point is $(0, -2)$.

$$x + 0 = -2$$
$$x = -2$$

The x-intercept point is $(-2, 0)$.

For the equation $y = 4x + 3$:

The slope is $\frac{4}{1}$, and the y-intercept point is $(0, 3)$.

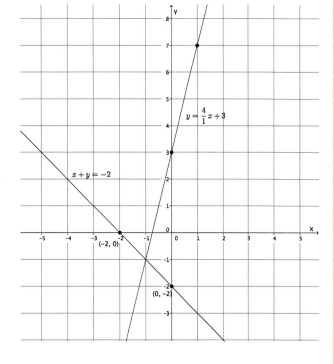

a. Name the ordered pair where the graphs of the two linear equations intersect.

$(-1, -1)$

b. Verify that the ordered pair named in part (a) is a solution to $x + y = -2$.

$$-1 + (-1) = -2$$
$$-2 = -2$$

The left and right sides of the equation are equal.

c. Verify that the ordered pair named in part (a) is a solution to $y = 4x + 3$.

$$-1 = 4(-1) + 3$$
$$-1 = -4 + 3$$
$$-1 = -1$$

The left and right sides of the equation are equal.

d. Could the point $(-4, 2)$ be a solution to the system of linear equations? That is, would $(-4, 2)$ make both equations true? Why or why not?

No. The graphs of the equations represent all of the possible solutions to the given equations. The point $(-4, 2)$ is a solution to the equation $x + y = -2$ because it is on the graph of that equation. However, the point $(-4, 2)$ is not on the graph of the equation $y = 4x + 3$. Therefore, $(-4, 2)$ cannot be a solution to the system of equations.

Lesson 25: Geometric Interpretation of the Solutions of a Linear System

3. Sketch the graphs of the linear system on a coordinate plane: $\begin{cases} 3x + y = -3 \\ -2x + y = 2 \end{cases}$.

For the equation $3x + y = -3$:

$$3(0) + y = -3$$
$$y = -3$$

The y-intercept point is $(0, -3)$.

$$3x + 0 = -3$$
$$3x = -3$$
$$x = -1$$

The x-intercept point is $(-1, 0)$.

For the equation $-2x + y = 2$:

$$-2(0) + y = 2$$
$$y = 2$$

The y-intercept point is $(0, 2)$.

$$-2x + 0 = 2$$
$$-2x = 2$$
$$x = -1$$

The x-intercept point is $(-1, 0)$.

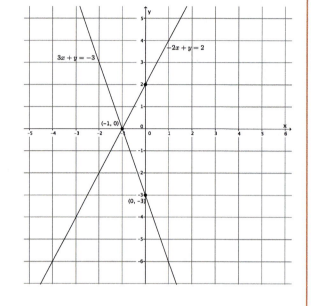

a. Name the ordered pair where the graphs of the two linear equations intersect.

$(-1, 0)$

b. Verify that the ordered pair named in part (a) is a solution to $3x + y = -3$.

$$3(-1) + 0 = -3$$
$$-3 = -3$$

The left and right sides of the equation are equal.

c. Verify that the ordered pair named in part (a) is a solution to $-2x + y = 2$.

$$-2(-1) + 0 = 2$$
$$2 = 2$$

The left and right sides of the equation are equal.

d. Could the point $(1, 4)$ be a solution to the system of linear equations? That is, would $(1, 4)$ make both equations true? Why or why not?

No. The graphs of the equations represent all of the possible solutions to the given equations. The point $(1, 4)$ is a solution to the equation $-2x + y = 2$ because it is on the graph of that equation. However, the point $(1, 4)$ is not on the graph of the equation $3x + y = -3$. Therefore, $(1, 4)$ cannot be a solution to the system of equations.

4. Sketch the graphs of the linear system on a coordinate plane: $\begin{cases} 2x - 3y = 18 \\ 2x + y = 2 \end{cases}$.

For the equation $2x - 3y = 18$:

$$2(0) - 3y = 18$$
$$-3y = 18$$
$$y = -6$$

The y-intercept point is $(0, -6)$.

$$2x - 3(0) = 18$$
$$2x = 18$$
$$x = 9$$

The x-intercept point is $(9, 0)$.

For the equation $2x + y = 2$:

$$2(0) + y = 2$$
$$y = 2$$

The y-intercept point is $(0, 2)$.

$$2x + 0 = 2$$
$$2x = 2$$
$$x = 1$$

The x-intercept point is $(1, 0)$.

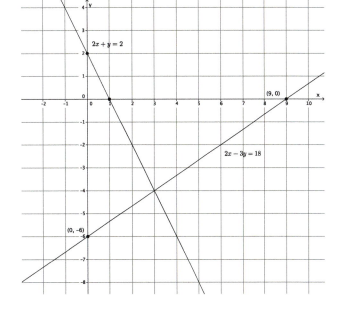

a. Name the ordered pair where the graphs of the two linear equations intersect.

$(3, -4)$

b. Verify that the ordered pair named in part (a) is a solution to $2x - 3y = 18$.

$$2(3) - 3(-4) = 18$$
$$6 + 12 = 18$$
$$18 = 18$$

The left and right sides of the equation are equal.

c. Verify that the ordered pair named in part (a) is a solution to $2x + y = 2$.

$$2(3) + (-4) = 2$$
$$6 - 4 = 2$$
$$2 = 2$$

The left and right sides of the equation are equal.

d. Could the point $(3, -1)$ be a solution to the system of linear equations? That is, would $(3, -1)$ make both equations true? Why or why not?

No. The graphs of the equations represent all of the possible solutions to the given equations. The point $(3, -1)$ is not on the graph of either line; therefore, it is not a solution to the system of linear equations.

5. Sketch the graphs of the linear system on a coordinate plane: $\begin{cases} y - x = 3 \\ y = -4x - 2 \end{cases}$.

For the equation $y - x = 3$:
$$y - 0 = 3$$
$$y = 3$$
The y-intercept point is $(0, 3)$.

$$0 - x = 3$$
$$-x = 3$$
$$x = -3$$
The x-intercept point is $(-3, 0)$.

For the equation $y = -4x - 2$:

The slope is $-\frac{4}{1}$, and the y-intercept point is $(0, -2)$.

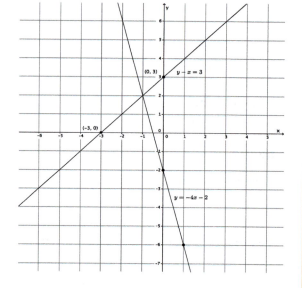

a. Name the ordered pair where the graphs of the two linear equations intersect.

$(-1, 2)$

b. Verify that the ordered pair named in part (a) is a solution to $y - x = 3$.
$$2 - (-1) = 3$$
$$3 = 3$$
The left and right sides of the equation are equal.

c. Verify that the ordered pair named in part (a) is a solution to $y = -4x - 2$.
$$2 = -4(-1) - 2$$
$$2 = 4 - 2$$
$$2 = 2$$
The left and right sides of the equation are equal.

d. Could the point $(-2, 6)$ be a solution to the system of linear equations? That is, would $(-2, 6)$ make both equations true? Why or why not?

No. The graphs of the equations represent all of the possible solutions to the given equations. The point $(-2, 6)$ is a solution to the equation $y = -4x - 2$ because it is on the graph of that equation. However, the point $(-2, 6)$ is not on the graph of the equation $y - x = 3$. Therefore, $(-2, 6)$ cannot be a solution to the system of equations.

Discussion (7 minutes)

The formal proof shown on the following page is optional. The Discussion can be modified by first asking students the questions in the first two bullet points and then having them make conjectures about why there is just one solution to the system of equations.

- How many points of intersection of the two lines were there?
  - There was only one point of intersection for each pair of lines.
- Was your answer to part (d) in Exercises 1–5 ever yes? Explain.
  - No. The answer was always no. Each time, the point given was either a solution to just one of the equations, or it was not a solution to either equation.
- So far, what you have observed strongly suggests that the solution of a system of linear equations is the point of intersection of two distinct lines. Let's show that this observation is true.

  THEOREM: Let

  $$\begin{cases} a_1x + b_1y = c_1 \\ a_2x + b_2y = c_2 \end{cases}$$

  be a system of linear equations, where $a_1, b_1, c_1, a_2, b_2,$ and c_2 are constants. At least one of a_1, b_1 is not equal to zero, and at least one of a_2, b_2 is not equal to zero. Let l_1 and l_2 be the lines defined by $a_1x + b_1y = c_1$ and $a_2x + b_2y = c_2$, respectively. If l_1 and l_2 are not parallel, then the solution of the system is exactly the point of intersection of l_1 and l_2.

- To prove the theorem, we have to show two things.
  (1) Any point that lies in the intersection of the two lines is a solution of the system.
  (2) Any solution of the system lies in the intersection of the two lines.
- We begin by showing that (1) is true. To show that (1) is true, use the definition of a solution. What does it mean to be a solution to an equation?
  - A solution is the ordered pair of numbers that makes an equation true. A solution is also a point on the graph of a linear equation.
- Suppose (x_0, y_0) is the point of intersection of l_1 and l_2. Since the point (x_0, y_0) is on l_1, it means that it is a solution to $a_1x + b_1y = c_1$. Similarly, since the point (x_0, y_0) is on l_2, it means that it is a solution to $a_2x + b_2y = c_2$. Therefore, since the point (x_0, y_0) is a solution to each linear equation, it is a solution to the system of linear equations, and the proof of (1) is complete.
- To prove (2), use the definition of a solution again. Suppose (x_0, y_0) is a solution to the system. Then, the point (x_0, y_0) is a solution to $a_1x + b_1y = c_1$ and, therefore, lies on l_1. Similarly, the point (x_0, y_0) is a solution to $a_2x + b_2y = c_2$ and, therefore, lies on l_2. Since there can be only one point shared by two distinct non-parallel lines, then (x_0, y_0) must be the point of intersection of l_1 and l_2. This completes the proof of (2).
- Therefore, given a system of two distinct non-parallel linear equations, there can be only one point of intersection, which means there is just one solution to the system.

Exercise 6 (3 minutes)

This exercise is optional. It requires students to write two different systems for a given solution. The exercise can be completed independently or in pairs.

> **Exercise 6**
>
> 6. Write two different systems of equations with $(1, -2)$ as the solution.
>
>   Answers will vary. Two sample solutions are provided:
>
> $\begin{cases} 3x + y = 1 \\ 2x - 3y = 8 \end{cases}$ and $\begin{cases} -x + 3y = -7 \\ 9x - 4y = 17 \end{cases}$.

Closing (5 minutes)

Summarize, or ask students to summarize, the main points from the lesson:

- We know how to sketch the graphs of a system of linear equations and find the point of intersection of the lines.
- We know that the point of intersection represents the solution to the system of linear equations.
- We know that two distinct non-parallel lines will intersect at only one point; therefore, the point of intersection is an ordered pair of numbers that makes both equations in the system true.

Lesson Summary

When the graphs of a system of linear equations are sketched, and if they are not parallel lines, then the point of intersection of the lines of the graph represents the solution to the system. Two distinct lines intersect at most at one point, if they intersect. The coordinates of that point (x, y) represent values that make both equations of the system true.

Example: The system $\begin{cases} x + y = 3 \\ x - y = 5 \end{cases}$ graphs as shown below.

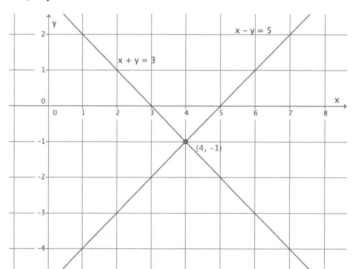

The lines intersect at $(4, -1)$. That means the equations in the system are true when $x = 4$ and $y = -1$.

$$x + y = 3$$
$$4 + (-1) = 3$$
$$3 = 3$$

$$x - y = 5$$
$$4 - (-1) = 5$$
$$5 = 5$$

Exit Ticket (5 minutes)

Name _____   Date _____

Lesson 25: Geometric Interpretation of the Solutions of a Linear System

Exit Ticket

Sketch the graphs of the linear system on a coordinate plane: $\begin{cases} 2x - y = -1 \\ y = 5x - 5 \end{cases}$.

a. Name the ordered pair where the graphs of the two linear equations intersect.

b. Verify that the ordered pair named in part (a) is a solution to $2x - y = -1$.

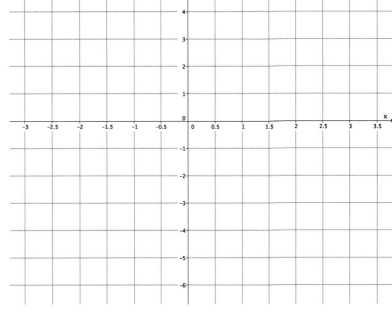

c. Verify that the ordered pair named in part (a) is a solution to $y = 5x - 5$.

Exit Ticket Sample Solutions

Sketch the graphs of the linear system on a coordinate plane: $\begin{cases} 2x - y = -1 \\ y = 5x - 5 \end{cases}$.

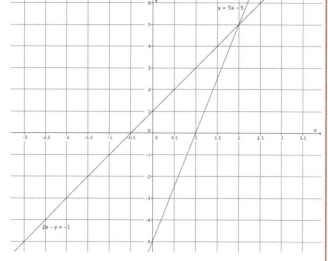

a. Name the ordered pair where the graphs of the two linear equations intersect.

$(2, 5)$

b. Verify that the ordered pair named in part (a) is a solution to $2x - y = -1$.

$2(2) - 5 = -1$
$4 - 5 = -1$
$-1 = -1$

The left and right sides of the equation are equal.

c. Verify that the ordered pair named in part (a) is a solution to $y = 5x - 5$.

$5 = 5(2) - 5$
$5 = 10 - 5$
$5 = 5$

The left and right sides of the equation are equal.

Problem Set Sample Solutions

1. Sketch the graphs of the linear system on a coordinate plane: $\begin{cases} y = \frac{1}{3}x + 1 \\ y = -3x + 11 \end{cases}$.

   For the equation $y = \frac{1}{3}x + 1$:

   The slope is $\frac{1}{3}$, and the y-intercept point is $(0, 1)$.

   For the equation $y = -3x + 11$:

   The slope is $-\frac{3}{1}$, and the y-intercept point is $(0, 11)$.

   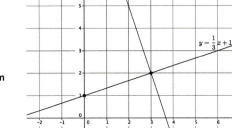

   a. Name the ordered pair where the graphs of the two linear equations intersect.

   $(3, 2)$

   b. Verify that the ordered pair named in part (a) is a solution to $y = \frac{1}{3}x + 1$.

   $$2 = \frac{1}{3}(3) + 1$$
   $$2 = 1 + 1$$
   $$2 = 2$$

   The left and right sides of the equation are equal.

   c. Verify that the ordered pair named in part (a) is a solution to $y = -3x + 11$.

   $$2 = -3(3) + 11$$
   $$2 = -9 + 11$$
   $$2 = 2$$

   The left and right sides of the equation are equal.

Lesson 25: Geometric Interpretation of the Solutions of a Linear System

2. Sketch the graphs of the linear system on a coordinate plane: $\begin{cases} y = \frac{1}{2}x + 4 \\ x + 4y = 4 \end{cases}$.

For the equation $y = \frac{1}{2}x + 4$:

The slope is $\frac{1}{2}$, and the y-intercept point is $(0, 4)$.

For the equation $x + 4y = 4$:
$$0 + 4y = 4$$
$$4y = 4$$
$$y = 1$$

The y-intercept point is $(0, 1)$.
$$x + 4(0) = 4$$
$$x = 4$$

The x-intercept point is $(4, 0)$.

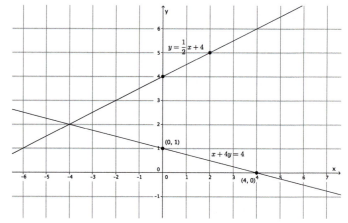

a. Name the ordered pair where the graphs of the two linear equations intersect.

$(-4, 2)$

b. Verify that the ordered pair named in part (a) is a solution to $y = \frac{1}{2}x + 4$.

$$2 = \frac{1}{2}(-4) + 4$$
$$2 = -2 + 4$$
$$2 = 2$$

The left and right sides of the equation are equal.

c. Verify that the ordered pair named in part (a) is a solution to $x + 4y = 4$.

$$-4 + 4(2) = 4$$
$$-4 + 8 = 4$$
$$4 = 4$$

The left and right sides of the equation are equal.

3. Sketch the graphs of the linear system on a coordinate plane: $\begin{cases} y = 2 \\ x + 2y = 10 \end{cases}$.

For the equation $x + 2y = 10$:
$$0 + 2y = 10$$
$$2y = 10$$
$$y = 5$$

The y-intercept point is $(0, 5)$.
$$x + 2(0) = 10$$
$$x = 10$$

The x-intercept point is $(10, 0)$.

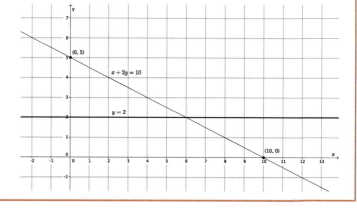

a. Name the ordered pair where the graphs of the two linear equations intersect.

   $(6, 2)$

b. Verify that the ordered pair named in part (a) is a solution to $y = 2$.

   $$2 = 2$$

   The left and right sides of the equation are equal.

c. Verify that the ordered pair named in part (a) is a solution to $x + 2y = 10$.

   $$6 + 2(2) = 10$$
   $$6 + 4 = 10$$
   $$10 = 10$$

   The left and right sides of the equation are equal.

4. Sketch the graphs of the linear system on a coordinate plane: $\begin{cases} -2x + 3y = 18 \\ 2x + 3y = 6 \end{cases}$.

For the equation $-2x + 3y = 18$:

$$-2(0) + 3y = 18$$
$$3y = 18$$
$$y = 6$$

The y-intercept point is $(0, 6)$.

$$-2x + 3(0) = 18$$
$$-2x = 18$$
$$x = -9$$

The x-intercept point is $(-9, 0)$.

For the equation $2x + 3y = 6$:

$$2(0) + 3y = 6$$
$$3y = 6$$
$$y = 2$$

The y-intercept point is $(0, 2)$.

$$2x + 3(0) = 6$$
$$2x = 6$$
$$x = 3$$

The x-intercept point is $(3, 0)$.

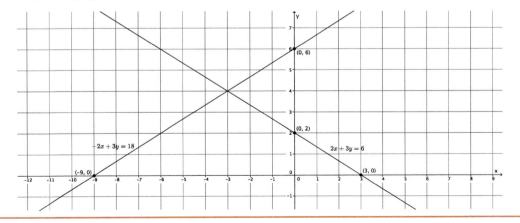

a. Name the ordered pair where the graphs of the two linear equations intersect.

$(-3, 4)$

b. Verify that the ordered pair named in part (a) is a solution to $-2x + 3y = 18$.

$$-2(-3) + 3(4) = 18$$
$$6 + 12 = 18$$
$$18 = 18$$

The left and right sides of the equation are equal.

c. Verify that the ordered pair named in part (a) is a solution to $2x + 3y = 6$.

$$2(-3) + 3(4) = 6$$
$$-6 + 12 = 6$$
$$6 = 6$$

The left and right sides of the equation are equal.

5. Sketch the graphs of the linear system on a coordinate plane: $\begin{cases} x + 2y = 2 \\ y = \frac{2}{3}x - 6 \end{cases}$.

For the equation $x + 2y = 2$:
$$0 + 2y = 2$$
$$2y = 2$$
$$y = 1$$

The y-intercept point is $(0, 1)$.
$$x + 2(0) = 2$$
$$x = 2$$

The x-intercept point is $(2, 0)$.

For the equation $y = \frac{2}{3}x - 6$:

The slope is $\frac{2}{3}$, and the y-intercept point is $(0, -6)$.

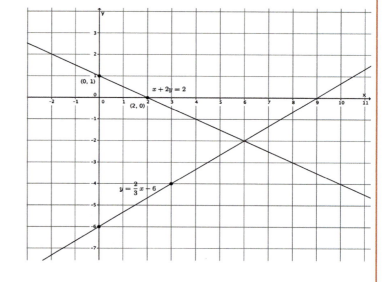

a. Name the ordered pair where the graphs of the two linear equations intersect.

$(6, -2)$

b. Verify that the ordered pair named in part (a) is a solution to $x + 2y = 2$.

$$6 + 2(-2) = 2$$
$$6 - 4 = 2$$
$$2 = 2$$

The left and right sides of the equation are equal.

c. Verify that the ordered pair named in part (a) is a solution to $y = \frac{2}{3}x - 6$.

$$-2 = \frac{2}{3}(6) - 6$$
$$-2 = 4 - 6$$
$$-2 = -2$$

The left and right sides of the equation are equal.

6. Without sketching the graph, name the ordered pair where the graphs of the two linear equations intersect.
$$\begin{cases} x = 2 \\ y = -3 \end{cases}$$

$(2, -3)$

Lesson 26: Characterization of Parallel Lines

Student Outcomes

- Students know that when a system of linear equations has no solution (i.e., no point of intersection of the lines), then the lines are parallel.

Lesson Notes

The Discussion is an optional proof of the theorem about parallel lines. Discuss the proof with students, or have students complete Exercises 4–10.

Classwork

Exercises 1–3 (10 minutes)

Students complete Exercises 1–3 independently. Once students are finished, debrief their work using the questions in the Discussion that follows the exercises.

Exercises

1. Sketch the graphs of the system.

$$\begin{cases} y = \frac{2}{3}x + 4 \\ y = \frac{4}{6}x - 3 \end{cases}$$

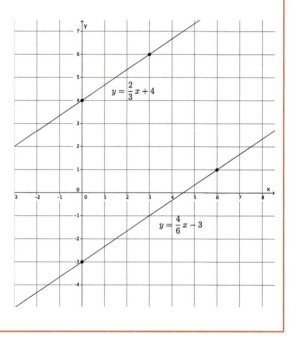

a. Identify the slope of each equation. What do you notice?

The slope of the first equation is $\frac{2}{3}$, and the slope of the second equation is $\frac{4}{6}$. The slopes are equal.

b. Identify the y-intercept point of each equation. Are the y-intercept points the same or different?

The y-intercept points are $(0, 4)$ and $(0, -3)$. The y-intercept points are different.

2. Sketch the graphs of the system.

$$\begin{cases} y = -\frac{5}{4}x + 7 \\ y = -\frac{5}{4}x + 2 \end{cases}$$

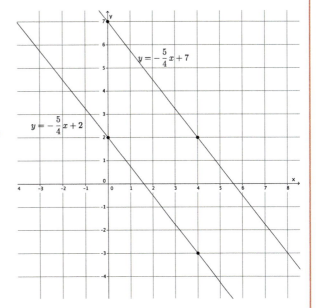

a. Identify the slope of each equation. What do you notice?

The slope of both equations is $-\frac{5}{4}$. The slopes are equal.

b. Identify the y-intercept point of each equation. Are the y-intercept points the same or different?

The y-intercept points are $(0, 7)$ and $(0, 2)$. The y-intercept points are different.

3. Sketch the graphs of the system.

$$\begin{cases} y = 2x - 5 \\ y = 2x - 1 \end{cases}$$

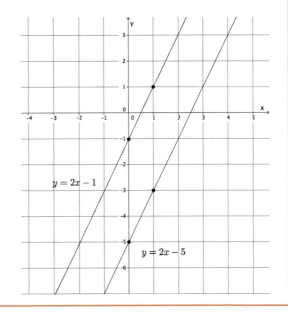

a. Identify the slope of each equation. What do you notice?

The slope of both equations is 2. The slopes are equal.

b. Identify the y-intercept point of each equation. Are the y-intercept points the same or different?

The y-intercept points are $(0, -5)$ and $(0, -1)$. The y-intercept points are different.

Discussion (10 minutes)

- What did you notice about each of the systems you graphed in Exercises 1–3?
  - *For each exercise, the graphs of the given linear equations look like parallel lines.*
- What did you notice about the slopes in each system?
  - *Each time, the linear equations of the system had the same slope.*

Lesson 26: Characterization of Parallel Lines

- What did you notice about the y-intercept points of the equations in each system?
  - In each case, the y-intercept points were different.
- If the equations had the same y-intercept point and the same slope, what would we know about the graphs of the lines?
  - There is only one line that can go through a given point with a given slope. If the equations had the same slope and y-intercept point, then their graphs are the same line.
- For that reason, when we discuss lines with the same slope, we must make sure to identify them as distinct lines.
- Write a summary of the conclusions you have reached by completing Exercises 1–3.

Provide time for students to write their conclusions. Share the theorem with them, and have students compare the conclusions that they reached to the statements in the theorem.

- What you observed in Exercises 1–3 can be stated as a theorem.

  THEOREM:

  (1) Two distinct, non-vertical lines in the plane are parallel if they have the same slope.

  (2) If two distinct, non-vertical lines have the same slope, then they are parallel.

- Suppose you have a pair of parallel lines on a coordinate plane. In how many places will those lines intersect?
  - By definition, parallel lines never intersect.
- Suppose you are given a system of linear equations whose graphs are parallel lines. How many solutions will the system have?
  - Based on work in the previous lesson, students learned that the solutions of a system lie in the intersection of the lines defined by the linear equations of the system. Since parallel lines do not intersect, then a system containing linear equations that graph as parallel lines will have no solution.
- What we want to find out is how to recognize when the lines defined by the equations are parallel. Then we would know immediately that we have a system with no solution as long as the lines are different.
- A system can easily be recognized as having no solutions when it is in the form of $\begin{cases} x = 2 \\ x = -7 \end{cases}$ or $\begin{cases} y = 6 \\ y = 15 \end{cases}$. Why is that so?
  - Because the system $\begin{cases} x = 2 \\ x = -7 \end{cases}$ graphs as two vertical lines. All vertical lines are parallel to the y-axis and, therefore, are parallel to one another. Similarly, the system $\begin{cases} y = 6 \\ y = 15 \end{cases}$ graphs as two horizontal lines. All horizontal lines are parallel to the x-axis and, therefore, are parallel to one another.
- We want to be able to recognize when we have a system of parallel lines that are not vertical or horizontal. What characteristics did each pair of equations have?
  - Each pair of equations had the same slope but different y-intercept points.

Discussion (15 minutes)

The following Discussion is an optional proof of the theorem about parallel lines. Having this discussion with students is optional. Instead, students may complete Exercises 4–10.

- We begin by proving (1). Recall that when we were shown that the slope between any two points would be equal to the same constant, m, we used what we knew about similar triangles. We will do something similar to prove (1).

- Suppose we have two non-vertical and non-horizontal parallel lines l_1 and l_2 in the coordinate plane. Assume that the y-intercept point of l_1 is greater than the y-intercept point of l_2. (We could assume the opposite is true; it does not make a difference with respect to the proof. We just want to say something clearly so our diagram will make sense.) Let's also assume that these lines are left-to-right inclining (i.e., they have a positive slope).

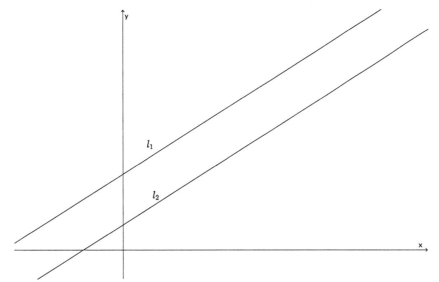

- Pick a point $P(p_1, p_2)$ on l_1, and draw a vertical line from P so that it intersects l_2 at point Q and the x-axis at point R. From points Q and R, draw horizontal lines so that they intersect lines l_1 and l_2 at points S and T, respectively.

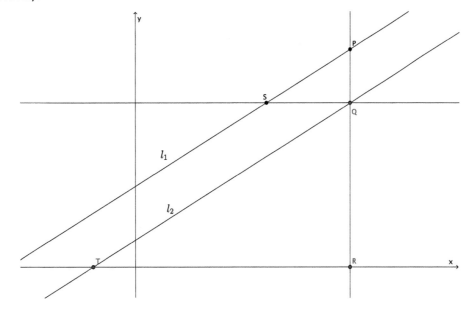

Lesson 26: Characterization of Parallel Lines

- By construction, ∠PQS and ∠QRT are right angles. How do we know for sure?
  - We drew a vertical line from point P. Therefore, the vertical line is parallel to the y-axis and perpendicular to the x-axis. Therefore, ∠QRT = 90°. Since we also drew horizontal lines, we know they are parallel. The vertical line through P is then a transversal that intersects parallel lines, which means corresponding angles are congruent. Since ∠QRT corresponds to ∠PQS, then ∠PQS = 90°.
- We want to show that △ PQS ~ △ QRT, which means we need another pair of equal angles in order to use the AA criterion. Do we have another pair of equal angles? Explain.
  - Yes. We know that lines l_1 and l_2 are parallel. By using the vertical line through P as the transversal, corresponding angles ∠TQR = ∠SPQ. Therefore, △ PQS ~ △ QRT.

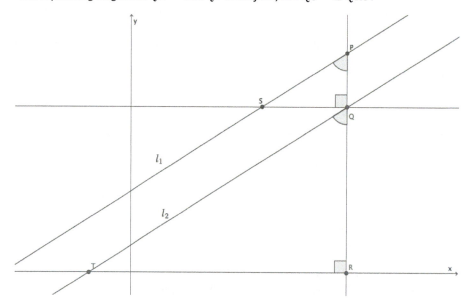

- To better see what we are doing, we will translate △ PQS along vector $\vec{QR}$ as shown.

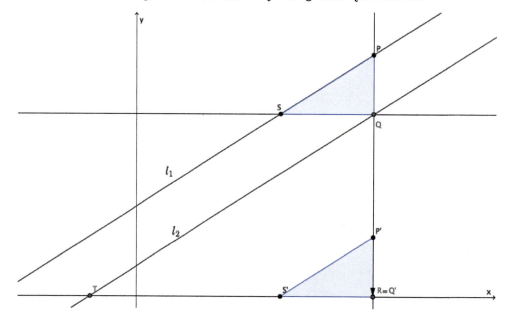

Lesson 26: Characterization of Parallel Lines

- By the definition of *dilation*, we know that:
$$\frac{|P'Q'|}{|QR|} = \frac{|Q'S'|}{|RT|}.$$

Equivalently, by the multiplication property of equality:
$$\frac{|P'Q'|}{|Q'S'|} = \frac{|QR|}{|RT|}.$$

Because translation preserves lengths of segments, we know that $|P'Q'| = |PQ|$ and $|Q'S'| = |QS|$, so we have
$$\frac{|PQ|}{|QS|} = \frac{|QR|}{|RT|}.$$

By definition of slope, $\frac{|PQ|}{|QS|}$ is the slope of l_1 and $\frac{|QR|}{|RT|}$ is the slope of l_2. Therefore, the slopes of l_1 and l_2 are equal, and (1) is proved.

- To prove (2), use the same construction as we did for (1). The difference this time is that we know we have side lengths that are equal in ratio because we are given that the slopes are the same, so we are trying to prove that the lines l_1 and l_2 are parallel. Since we do not know the lines are parallel, we also do not know that $\angle TQR = \angle SPQ$, but we do know that $\angle PQS$ and $\angle QRT$ are right angles.

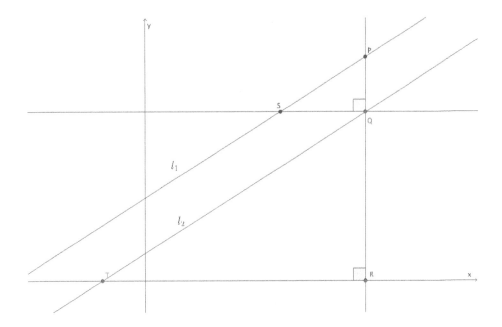

Lesson 26: Characterization of Parallel Lines

- Then, again, we translate △ PQS along vector $\overrightarrow{QR}$ as shown.

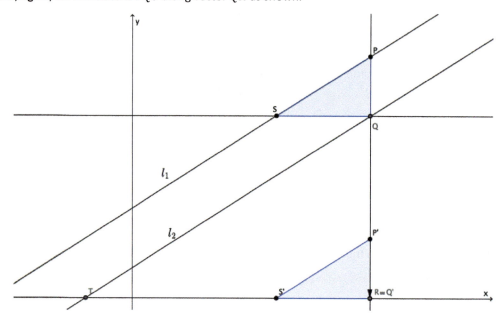

- Since the corresponding sides are equal in ratio to the scale factor $\frac{|\ |}{|QR|} = \frac{|\ |\ |}{|RT|}$ and share a common angle, ∠P'RS', by the fundamental theorem of similarity, we know that the lines containing $\overline{P'S'}$ and $\overline{QT}$ are parallel. Since the line containing $\overline{P'S'}$ is a translation of line PS, and translations preserve angle measures, we know that line PS is parallel to line QT. Since the line containing $\overline{PS}$ is line l_1, and the line containing $\overline{QT}$ is line l_2, we can conclude that $l_1 \parallel l_2$. This finishes the proof of the theorem.

Exercises 4–10 (15 minutes)

Students complete Exercises 4–10 independently. Once students are finished, debrief their work using the questions in the Discussion that follow the exercises.

---

4. Write a system of equations that has no solution.

   Answers will vary. Verify that the system that has been written has equations that have the same slope and unique y-intercept points. Sample student solution: $\begin{cases} y = \frac{3}{4}x + 1 \\ y = \frac{3}{4}x - 2 \end{cases}$

5. Write a system of equations that has $(2, 1)$ as a solution.

   Answers will vary. Verify that students have written a system where $(2, 1)$ is a solution to each equation. Sample student solution: $\begin{cases} 5x + y = 11 \\ y = \frac{1}{2}x \end{cases}$

Lesson 26: Characterization of Parallel Lines

6. How can you tell if a system of equations has a solution or not?

   If the slopes of the equations are different, the lines will intersect at some point, and there will be a solution to the system. If the slopes of the equations are the same, and the y-intercept points are different, then the equations will graph as parallel lines, which means the system will not have a solution.

7. Does the system of linear equations shown below have a solution? Explain.

   $$\begin{cases} 6x - 2y = 5 \\ 4x - 3y = 5 \end{cases}$$

   Yes, this system does have a solution. The slope of the first equation is 3, and the slope of the second equation is $\frac{4}{3}$. Since the slopes are different, these equations will graph as nonparallel lines, which means they will intersect at some point.

8. Does the system of linear equations shown below have a solution? Explain.

   $$\begin{cases} -2x + 8y = 14 \\ x = 4y + 1 \end{cases}$$

   No, this system does not have a solution. The slope of the first equation is $\frac{2}{8} = \frac{1}{4}$, and the slope of the second equation is $\frac{1}{4}$. Since the slopes are the same, but the lines are distinct, these equations will graph as parallel lines. Parallel lines never intersect, which means this system has no solution.

9. Does the system of linear equations shown below have a solution? Explain.

   $$\begin{cases} 12x + 3y = -2 \\ 4x + y = 7 \end{cases}$$

   No, this system does not have a solution. The slope of the first equation is $-\frac{12}{3} = -4$, and the slope of the second equation is -4. Since the slopes are the same, but the lines are distinct, these equations will graph as parallel lines. Parallel lines never intersect, which means this system has no solution.

10. Genny babysits for two different families. One family pays her $6 each hour and a bonus of $20 at the end of the night. The other family pays her $3 every half hour and a bonus of $25 at the end of the night. Write and solve the system of equations that represents this situation. At what number of hours do the two families pay the same for babysitting services from Genny?

    Let y represent the total amount Genny is paid for babysitting x hours. The first family pays $y = 6x + 20$. Since the other family pays by the half hour, $3 \cdot 2$ would represent the amount Genny is paid each hour. So, the other family pays $y = (3 \cdot 2)x + 25$, which is the same as $y = 6x + 25$.

    $$\begin{cases} y = 6x + 20 \\ y = 6x + 25 \end{cases}$$

    Since the equations in the system have the same slope and different y-intercept points, there will not be a point of intersection. That means that there will not be a number of hours for when Genny is paid the same amount by both families. The second family will always pay her $5 more than the first family.

Lesson 26: Characterization of Parallel Lines

Closing (5 minutes)

Summarize, or ask students to summarize, the main points from the lesson:

- We know that systems of linear equations whose graphs are two distinct vertical lines will have no solution because all vertical lines are parallel to the y-axis and, therefore, are parallel to one another. Similarly, systems of linear equations whose graphs are two distinct horizontal lines will have no solution because all horizontal lines are parallel to the x-axis and, therefore, are parallel to one another.
- We know that if a system contains linear equations whose graphs are distinct lines with the same slope, then the lines are parallel and, therefore, the system has no solution.

Lesson Summary

By definition, parallel lines do not intersect; therefore, a system of linear equations whose graphs are parallel lines will have no solution.

Parallel lines have the same slope but no common point. One can verify that two lines are parallel by comparing their slopes and their y-intercept points.

Exit Ticket (5 minutes)

Lesson 26: Characterization of Parallel Lines

Exit Ticket

Does each system of linear equations have a solution? Explain your answer.

1. $\begin{cases} y = \frac{5}{4}x - 3 \\ y + 2 = \frac{5}{4}x \end{cases}$

2. $\begin{cases} y = \frac{2}{3}x - 5 \\ 4x - 8y = 11 \end{cases}$

3. $\begin{cases} \frac{1}{3}x + y = 8 \\ x + 3y = 12 \end{cases}$

Exit Ticket Sample Solutions

Does each system of linear equations have a solution? Explain your answer.

1. $\begin{cases} y = \frac{5}{4}x - 3 \\ y + 2 = \frac{5}{4}x \end{cases}$

   No, this system does not have a solution. The slope of the first equation is $\frac{5}{4}$, and the slope of the second equation is $\frac{5}{4}$. Since the slopes are the same, and they are distinct lines, these equations will graph as parallel lines. Parallel lines never intersect; therefore, this system has no solution.

2. $\begin{cases} y = \frac{2}{3}x - 5 \\ 4x - 8y = 11 \end{cases}$

   Yes, this system does have a solution. The slope of the first equation is $\frac{2}{3}$, and the slope of the second equation is $\frac{1}{2}$. Since the slopes are different, these equations will graph as nonparallel lines, which means they will intersect at some point.

3. $\begin{cases} \frac{1}{3}x + y = 8 \\ x + 3y = 12 \end{cases}$

   No, this system does not have a solution. The slope of the first equation is $-\frac{1}{3}$, and the slope of the second equation is $-\frac{1}{3}$. Since the slopes are the same, and they are distinct lines, these equations will graph as parallel lines. Parallel lines never intersect; therefore, this system has no solution.

Problem Set Sample Solutions

Answer Problems 1–5 without graphing the equations.

1. Does the system of linear equations shown below have a solution? Explain.

   $$\begin{cases} 2x + 5y = 9 \\ -4x - 10y = 4 \end{cases}$$

   No, this system does not have a solution. The slope of the first equation is $-\frac{2}{5}$, and the slope of the second equation is $-\frac{4}{10}$, which is equivalent to $-\frac{2}{5}$. Since the slopes are the same, but the lines are distinct, these equations will graph as parallel lines. Parallel lines never intersect, which means this system has no solution.

2. Does the system of linear equations shown below have a solution? Explain.

   $$\begin{cases} \frac{3}{4}x - 3 = y \\ 4x - 3y = 5 \end{cases}$$

   Yes, this system does have a solution. The slope of the first equation is $\frac{3}{4}$, and the slope of the second equation is $\frac{4}{3}$. Since the slopes are different, these equations will graph as nonparallel lines, which means they will intersect at some point.

3. Does the system of linear equations shown below have a solution? Explain.

$$\begin{cases} x + 7y = 8 \\ 7x - y = -2 \end{cases}$$

Yes, this system does have a solution. The slope of the first equation is $-\frac{1}{7}$, and the slope of the second equation is 7. Since the slopes are different, these equations will graph as nonparallel lines, which means they will intersect at some point.

4. Does the system of linear equations shown below have a solution? Explain.

$$\begin{cases} y = 5x + 12 \\ 10x - 2y = 1 \end{cases}$$

No, this system does not have a solution. The slope of the first equation is 5, and the slope of the second equation is $\frac{10}{2}$, which is equivalent to 5. Since the slopes are the same, but the lines are distinct, these equations will graph as parallel lines. Parallel lines never intersect, which means this system has no solution.

5. Does the system of linear equations shown below have a solution? Explain.

$$\begin{cases} y = \frac{5}{3}x + 15 \\ 5x - 3y = 6 \end{cases}$$

No, this system does not have a solution. The slope of the first equation is $\frac{5}{3}$, and the slope of the second equation is $\frac{5}{3}$. Since the slopes are the same, but the lines are distinct, these equations will graph as parallel lines. Parallel lines never intersect, which means this system has no solution.

6. Given the graphs of a system of linear equations below, is there a solution to the system that we cannot see on this portion of the coordinate plane? That is, will the lines intersect somewhere on the plane not represented in the picture? Explain.

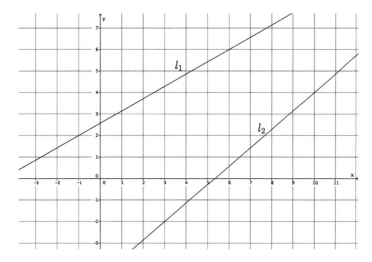

The slope of l_1 is $\frac{4}{7}$, and the slope of l_2 is $\frac{6}{7}$. Since the slopes are different, these lines are nonparallel lines, which means they will intersect at some point. Therefore, the system of linear equations whose graphs are the given lines will have a solution.

7. Given the graphs of a system of linear equations below, is there a solution to the system that we cannot see on this portion of the coordinate plane? That is, will the lines intersect somewhere on the plane not represented in the picture? Explain.

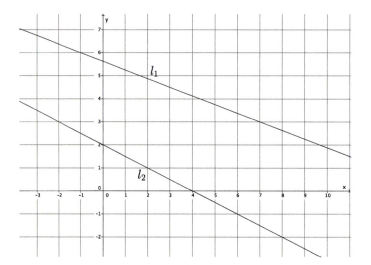

The slope of l_1 is $-\frac{3}{8}$, and the slope of l_2 is $-\frac{1}{2}$. Since the slopes are different, these lines are nonparallel lines, which means they will intersect at some point. Therefore, the system of linear equations whose graphs are the given lines will have a solution.

8. Given the graphs of a system of linear equations below, is there a solution to the system that we cannot see on this portion of the coordinate plane? That is, will the lines intersect somewhere on the plane not represented in the picture? Explain.

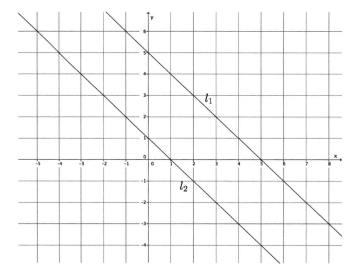

The slope of l_1 is -1, and the slope of l_2 is -1. Since the slopes are the same the lines are parallel lines, which means they will not intersect. Therefore, the system of linear equations whose graphs are the given lines will have no solution.

9. Given the graphs of a system of linear equations below, is there a solution to the system that we cannot see on this portion of the coordinate plane? That is, will the lines intersect somewhere on the plane not represented in the picture? Explain.

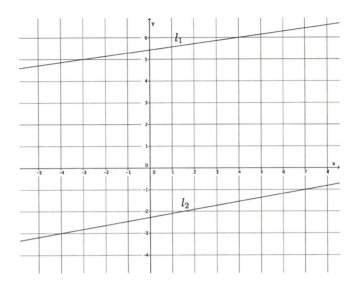

The slope of l_1 is $\frac{1}{7}$, and the slope of l_2 is $\frac{2}{11}$. Since the slopes are different, these lines are nonparallel lines, which means they will intersect at some point. Therefore, the system of linear equations whose graphs are the given lines will have a solution.

10. Given the graphs of a system of linear equations below, is there a solution to the system that we cannot see on this portion of the coordinate plane? That is, will the lines intersect somewhere on the plane not represented in the picture? Explain.

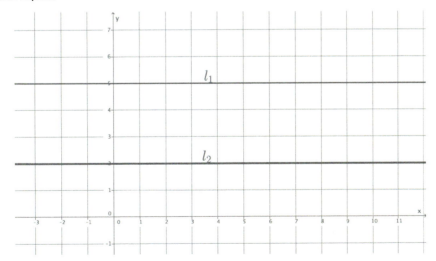

Lines l_1 and l_2 are horizontal lines. That means that they are both parallel to the x-axis and, thus, are parallel to one another. Therefore, the system of linear equations whose graphs are the given lines will have no solution.

Lesson 27: Nature of Solutions of a System of Linear Equations

Student Outcomes

- Students know that since two equations in the form $ax + by = c$ and $'x + b'y = c'$, when a, b, and c are nonzero numbers, graph as the same line when $\frac{a'}{a} = \frac{b'}{b} = \frac{c'}{c}$, then the system of linear equations has infinitely many solutions.
- Students know a strategy for solving a system of linear equations algebraically.

Classwork

Exercises 1–3 (5 minutes)

Students complete Exercises 1–3 independently in preparation for the Discussion that follows about infinitely many solutions.

Exercises

Determine the nature of the solution to each system of linear equations.

1. $\begin{cases} 3x + 4y = 5 \\ y = -\frac{3}{4}x + 1 \end{cases}$

   The slopes of these two distinct equations are the same, which means the graphs of these two equations are parallel lines. Therefore, this system will have no solution.

2. $\begin{cases} 7x + 2y = -4 \\ x - y = 5 \end{cases}$

   The slopes of these two equations are different. That means the graphs of these two equations are distinct nonparallel lines and will intersect at one point. Therefore, this system has one solution.

3. $\begin{cases} 9x + 6y = 3 \\ 3x + 2y = 1 \end{cases}$

   The lines defined by the graph of this system of equations are the same line because they have the same slope and the same y-intercept point.

A STORY OF RATIOS   Lesson 27   8•4

Discussion (7 minutes)

Ask students to summarize the nature of the solutions for each of the Exercises 1–3. Students should be able to state clearly what they observed in Exercise 1 and Exercise 2 as stated in (1) and (2) below.

- So far, our work with systems of linear equations has taught us that
  (1) If the lines defined by the equations in the system are parallel, then the system has no solution.
  (2) If the lines defined by the equations in the system are not parallel, then the system has exactly one solution, and it is the point of intersection of the two lines.

Have students discuss their "solution" to Exercise 3. Ask students what they noticed, if anything, about the equations in Exercise 3. If necessary, show that the first equation can be obtained from the second equation by multiplying each term by 3. Then the equations are exactly the same. Proceed with the second example below, along with the discussion points that follow.

- We now know there is a third possibility with respect to systems of equations. Consider the following system of linear equations below:

$$\begin{cases} 3x + 2y = 5 \\ 6x + 4y = 10 \end{cases}$$

What do you notice about the constants a, b, and c of the first equation, compared to the constants a', b', and c' of the second equation?

  ▫ When you compare $\dfrac{a'}{a} = \dfrac{b'}{b} = \dfrac{c'}{c}$, they are equal to the same constant, 2.

MP.2

- If you multiplied each term of the first equation by the constant 2, what is the result?

  ▫ Sample student work:

$$2(3x + 2y = 5)$$
$$6x + 4y = 10$$

  When you multiply each term of the first equation by the constant 2, the result is the second equation.

- What does this mean about the graphs of the equations in the system?

  ▫ It means that the lines defined by these equations are the same line. That is, they graph as the same line.

- When the lines defined by two linear equations are the same line, it means that every solution to one equation is also a solution to the other equation. Since we can find an infinite number of solutions to any linear equation in two variables, systems that comprise equations that define the same line have infinitely many solutions.

Provide students with time to verify the fact that the system $\begin{cases} 3x + 2y = 5 \\ 6x + 4y = 10 \end{cases}$ has infinitely many solutions. Instruct them to find at least two solutions to the first equation, and then show that the same ordered pairs satisfy the second equation as well.

  ▫ Sample student work:

$$\begin{cases} 3x + 2y = 5 \\ 6x + 4y = 10 \end{cases}$$

Lesson 27:   Nature of Solutions of a System of Linear Equations

Two solutions to $3x + 2y = 5$ are $(1, 1)$ and $(3, -2)$. Replacing the solutions in the second equation results in the true equations. When $x = 1$, and $y = 1$,

$$6(1) + 4(1) = 10$$
$$6 + 4 = 10$$
$$10 = 10,$$

and when $x = 3$, and $y = -2$,

$$6(3) + 4(-2) = 10$$
$$18 - 8 = 10$$
$$10 = 10.$$

- Since there is only one line passing through two distinct points, the lines defined by the two equations are the same line. That is, the system has infinitely many solutions.
- Therefore, the nature of the solutions of a system of linear equations is one of three possibilities: one solution, no solution, or infinitely many solutions. A system will have one solution when the graphs of the equations are distinct lines with different slopes. A system will have no solution when the equations have the same slope and different y-intercept points, in other words, when the graphs are parallel lines. A system will have infinitely many solutions when the lines defined by the equations are the same line; these equations will have the same slope and same y-intercept point.

Example 1 (7 minutes)

In this example, students realize that graphing a system of equations yields a solution, but the precise coordinates of the solution cannot be determined from the graph.

- The following figure contains the graphs of the system $\begin{cases} y = 3x + 5 \\ y = 8x + 3 \end{cases}$.

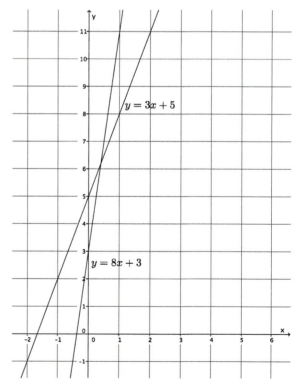

Lesson 27: Nature of Solutions of a System of Linear Equations

- Though the graphs of the equations were easy to sketch, it is not easy to identify the solution to the system because the intersection of the lines does not have integer coordinates. Estimate the solution based on the graph.
  - *The solution looks like it could be $\left(\frac{1}{3}, 6\right)$.*
- When we need a precise answer, we must use an alternative strategy for solving systems of linear equations.

Understanding the substitution method requires an understanding of the transitive property. One accessible way to introduce this to students at this level is through symbolic puzzles; one possible example follows:

If ! = \$\$ and \$\$ = &, is it true that ! = &? Why or why not? Allow students time to think about the symbol puzzle and share their thoughts with the class. Then, continue with the points below.

- When two linear expressions are equal to the same number, then the expressions are equal to each other, just like ! = & because both ! and & are equal to \$\$. Look at the system of equations given in this example:

$$\begin{cases} y = 3x + 5 \\ y = 8x + 3 \end{cases}$$

How is this like our puzzle about !, \$\$, and &?
  - *Both of the equations in the system are equal to y; therefore, $3x + 5$ must be equal to $8x + 3$.*
- The equation $3x + 5 = 8x + 3$ is a linear equation in one variable. Solve it for x.
  - *Sample student work:*

$$3x + 5 = 8x + 3$$
$$2 = 5x$$
$$\frac{2}{5} = x$$

- Keep in mind that we are trying to identify the point of intersection of the lines, in other words, the solution that is common to both equations. What we have just found is the x-coordinate of that solution, $\frac{2}{5}$, and we must now determine the y-coordinate. To do so, we can substitute the value of x into either of the two equations of the system to determine the value for y.

$$y = 3\left(\frac{2}{5}\right) + 5$$
$$y = \frac{6}{5} + 5$$
$$y = \frac{31}{5}$$

- Verify that we would get the same value for y using the second equation.
  - *Sample student work:*

$$y = 8\left(\frac{2}{5}\right) + 3$$
$$y = \frac{16}{5} + 3$$
$$y = \frac{31}{5}$$

Lesson 27: Nature of Solutions of a System of Linear Equations

- The solution to the system is $\left(\frac{2}{5}, \frac{31}{5}\right)$. Look at the graph. Does it look like this could be the solution?
  - Yes. The coordinates look correct, and they are close to our estimated solution.
- Our estimation was close to the actual answer but not as precise as when we solved the system algebraically.

Example 2 (4 minutes)

- Does the system $\begin{cases} y = 7x - 2 \\ 2y - 4x = 10 \end{cases}$ have a solution?
  - Yes. The slopes are different, which means they are not parallel and not the same line.
- Now that we know that the system has a solution, we will solve it without graphing.
- Notice that in this example we do not have two linear expressions equal to the same number. However, we do know what y is; it is equal to the expression $7x - 2$. Therefore, we can substitute the expression that y is equal to in the second equation. Since $y = 7x - 2$, then:

$$2y - 4x = 10$$
$$2(7x - 2) - 4x = 10$$
$$14x - 4 - 4x = 10$$
$$10x - 4 = 10$$
$$10x = 14$$
$$x = \frac{14}{10}$$
$$x = \frac{7}{5}.$$

- What does $x = \frac{7}{5}$ represent?
  - It represents the x-coordinate of the point of intersection of the graphs of the lines or the solution to the system.
- How can we determine the y-coordinate of the solution to the system?
  - Since we know the x-coordinate of the solution, we can substitute the value of x into either equation to determine the value of the y-coordinate.
- Determine the y-coordinate of the solution to the system.
  - Sample student work:

$$y = 7\left(\frac{7}{5}\right) - 2$$
$$y = \frac{49}{5} - 2$$
$$y = \frac{39}{5}$$

- The solution to this system is $\left(\frac{7}{5}, \frac{39}{5}\right)$. What does the solution represent?
  - The solution is the point of intersection of the graphs of the lines of the system. It is a solution that makes both equations of the system true.

Example 3 (8 minutes)

- Does the system $\begin{cases} 4y = 26x + 4 \\ y = 11x - 1 \end{cases}$ have a solution?
  - Yes. The slopes are different, which means they are not parallel and not the same line.
- Solve this system using substitution. Since we know what y is equal to, we can replace that value with the expression $11x - 1$ in the first equation.
  - Sample student work:

$$\begin{cases} 4y = 26x + 4 \\ y = 11x - 1 \end{cases}$$

$$4(11x - 1) = 26x + 4$$
$$44x - 4 = 26x + 4$$
$$44x - 26x - 4 = 26x - 26x + 4$$
$$18x - 4 = 4$$
$$18x - 4 + 4 = 4 + 4$$
$$18x = 8$$
$$x = \frac{8}{18}$$
$$x = \frac{4}{9}$$

$$y = 11\left(\frac{4}{9}\right) - 1$$
$$y = \frac{44}{9} - 1$$
$$y = \frac{35}{9}$$

The solution to the system is $\left(\frac{4}{9}, \frac{35}{9}\right)$.

- There is another option for solving this equation. We could multiply the second equation by 4. Then we would have two linear equations equal to $4y$ that could then be written as a single equation, as in Example 1. Or we could multiply the first equation by $\frac{1}{4}$. Then we would have two linear equations equal to y that could be written as a single equation, as in Example 1, again.
- For this system, let's multiply the second equation by 4. Then, we have

$$4(y = 11x - 1)$$
$$4y = 44x - 4.$$

Lesson 27: Nature of Solutions of a System of Linear Equations

- Now the system can be written as $\begin{cases} 4y = 26x + 4 \\ 4y = 44x - 4 \end{cases}$, and we can write the two linear expressions $26x + 4$ and $44x - 4$ as equal to one another. Solve the system.
    - *Sample student work:*

$$26x + 4 = 44x - 4$$
$$4 = 18x - 4$$
$$8 = 18x$$
$$\frac{8}{18} = x$$
$$\frac{4}{9} = x$$

$$y = 11\left(\frac{4}{9}\right) - 1$$
$$y = \frac{44}{9} - 1$$
$$y = \frac{35}{9}$$

The solution to the system is $\left(\frac{4}{9}, \frac{35}{9}\right)$.

- Compare the solution we got algebraically to the graph of the system of linear equations:

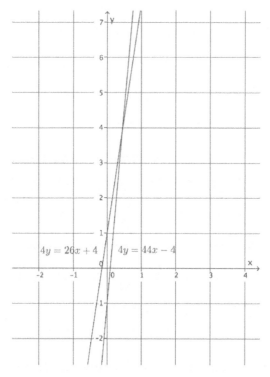

We can see that our answer is correct because it looks like the lines intersect at $\left(\frac{4}{9}, \frac{35}{9}\right)$.

Exercises 4–7 (7 minutes)

Students complete Exercises 4–7 independently.

> Determine the nature of the solution to each system of linear equations. If the system has a solution, find it algebraically, and then verify that your solution is correct by graphing.
>
> 4. $\begin{cases} 3x + 3y = -21 \\ x + y = -7 \end{cases}$
>
> *These equations define the same line. Therefore, this system will have infinitely many solutions.*
>
> 5. $\begin{cases} y = \frac{3}{2}x - 1 \\ 3y = x + 2 \end{cases}$
>
> *The slopes of these two equations are unique. That means they graph as distinct lines and will intersect at one point. Therefore, this system has one solution.*
>
> $$3\left(y = \frac{3}{2}x - 1\right)$$
> $$3y = \frac{9}{2}x - 3$$
> $$x + 2 = \frac{9}{2}x - 3$$
> $$2 = \frac{7}{2}x - 3$$
> $$5 = \frac{7}{2}x$$
> $$\frac{10}{7} = x$$
>
> $$y = \frac{3}{2}\left(\frac{10}{7}\right) - 1$$
> $$y = \frac{15}{7} - 1$$
> $$y = \frac{8}{7}$$
>
> *The solution is $\left(\frac{10}{7}, \frac{8}{7}\right)$.*

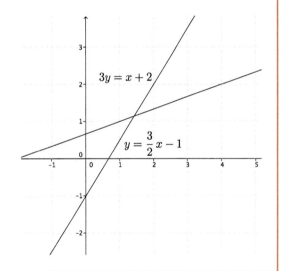

6. $\begin{cases} x = 12y - 4 \\ x = 9y + 7 \end{cases}$

   The slopes of these two equations are unique. That means they graph as distinct lines and will intersect at one point. Therefore, this system has one solution.

   $$12y - 4 = 9y + 7$$
   $$3y - 4 = 7$$
   $$3y = 11$$
   $$y = \frac{11}{3}$$

   $$x = 9\left(\frac{11}{3}\right) + 7$$
   $$x = 33 + 7$$
   $$x = 40$$

   The solution is $\left(40, \frac{11}{3}\right)$.

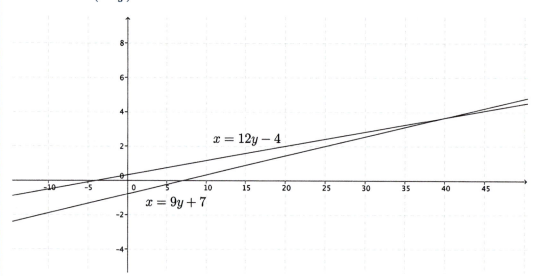

7. Write a system of equations with $(4, -5)$ as its solution.

   Answers will vary. Verify that students have written a system of equations where $(4, -5)$ is a solution to each equation in the system. Sample solution: $\begin{cases} y = x - 9 \\ x + y = -1 \end{cases}$

Closing (3 minutes)

Summarize, or ask students to summarize, the main points from the lesson:

- A system can have one solution, no solution, or infinitely many solutions. It will have one solution when the lines are distinct and their slopes are different; it will have no solution when the equations graph as distinct lines with the same slope; it will have infinitely many solutions when the equations define the same line.
- We learned a method for solving a system of linear equations algebraically. It requires us to write linear expressions equal to one another and substitution.

Lesson 27

Lesson Summary

A system of linear equations can have a unique solution, no solution, or infinitely many solutions.

Systems with a unique solution are comprised of two linear equations whose graphs have different slopes; that is, their graphs in a coordinate plane will be two distinct lines that intersect at only one point.

Systems with no solutions are comprised of two linear equations whose graphs have the same slope but different y-intercept points; that is, their graphs in a coordinate plane will be two parallel lines (with no intersection).

Systems with infinitely many solutions are comprised of two linear equations whose graphs have the same slope and the same y-intercept point; that is, their graphs in a coordinate plane are the same line (i.e., every solution to one equation will be a solution to the other equation).

A system of linear equations can be solved using a substitution method. That is, if two expressions are equal to the same value, then they can be written equal to one another.

Example:

$$\begin{cases} y = 5x - 8 \\ y = 6x + 3 \end{cases}$$

Since both equations in the system are equal to y, we can write the equation $5x - 8 = 6x + 3$ and use it to solve for x and then the system.

Example:

$$\begin{cases} 3x = 4y + 2 \\ x = y + 5 \end{cases}$$

Multiply each term of the equation $x = y + 5$ by 3 to produce the equivalent equation $3x = 3y + 15$. As in the previous example, since both equations equal $3x$, we can write $4y + 2 = 3y + 15$. This equation can be used to solve for y and then the system.

Exit Ticket (4 minutes)

Name _____ Date _____

Lesson 27: Nature of Solutions of a System of Linear Equations

Exit Ticket

Determine the nature of the solution to each system of linear equations. If the system has a solution, then find it without graphing.

1. $\begin{cases} y = \frac{1}{2}x + \frac{5}{2} \\ x - 2y = 7 \end{cases}$

2. $\begin{cases} y = \frac{2}{3}x + 4 \\ 2y + \frac{1}{2}x = 2 \end{cases}$

3. $\begin{cases} y = 3x - 2 \\ -3x + y = -2 \end{cases}$

Exit Ticket Sample Solutions

Determine the nature of the solution to each system of linear equations. If the system has a solution, then find it without graphing.

1. $\begin{cases} y = \frac{1}{2}x + \frac{5}{2} \\ x - 2y = 7 \end{cases}$

    The slopes of these two equations are the same, and the y-intercept points are different, which means they graph as parallel lines. Therefore, this system will have no solution.

2. $\begin{cases} y = \frac{2}{3}x + 4 \\ 2y + \frac{1}{2}x = 2 \end{cases}$

    The slopes of these two equations are unique. That means they graph as distinct lines and will intersect at one point. Therefore, this system has one solution.

    $$2\left(\frac{2}{3}x + 4\right) + \frac{1}{2}x = 2$$
    $$\frac{4}{3}x + 8 + \frac{1}{2}x = 2$$
    $$\frac{11}{6}x + 8 = 2$$
    $$\frac{11}{6}x = -6$$
    $$x = -\frac{36}{11}$$

    $$y = \frac{2}{3}\left(-\frac{36}{11}\right) + 4$$
    $$y = -\frac{24}{11} + 4$$
    $$y = \frac{20}{11}$$

    The solution is $\left(-\frac{36}{11}, \frac{20}{11}\right)$.

3. $\begin{cases} y = 3x - 2 \\ -3x + y = -2 \end{cases}$

    These equations define the same line. Therefore, this system will have infinitely many solutions.

Problem Set Sample Solutions

Students practice determining the nature of solutions of a system of linear equations and finding the solution for systems that have one.

Determine the nature of the solution to each system of linear equations. If the system has a solution, find it algebraically, and then verify that your solution is correct by graphing.

1. $\begin{cases} y = \frac{3}{7}x - 8 \\ 3x - 7y = 1 \end{cases}$

   The slopes of these two equations are the same, and the y-intercept points are different, which means they graph as parallel lines. Therefore, this system will have no solution.

2. $\begin{cases} 2x - 5 = y \\ -3x - 1 = 2y \end{cases}$

   $(2x - 5 = y)2$
   $4x - 10 = 2y$
   $\begin{cases} 4x - 10 = 2y \\ -3x - 1 = 2y \end{cases}$
   $4x - 10 = -3x - 1$
   $7x - 10 = -1$
   $7x = 9$
   $x = \frac{9}{7}$

   $y = 2\left(\frac{9}{7}\right) - 5$
   $y = \frac{18}{7} - 5$
   $y = -\frac{17}{7}$

   The solution is $\left(\frac{9}{7}, -\frac{17}{7}\right)$.

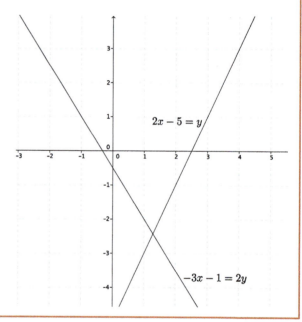

3. $\begin{cases} x = 6y + 7 \\ x = 10y + 2 \end{cases}$

$$6y + 7 = 10y + 2$$
$$7 = 4y + 2$$
$$5 = 4y$$
$$\frac{5}{4} = y$$

$$x = 6\left(\frac{5}{4}\right) + 7$$
$$x = \frac{15}{2} + 7$$
$$x = \frac{29}{2}$$

The solution is $\left(\frac{29}{2}, \frac{5}{4}\right)$.

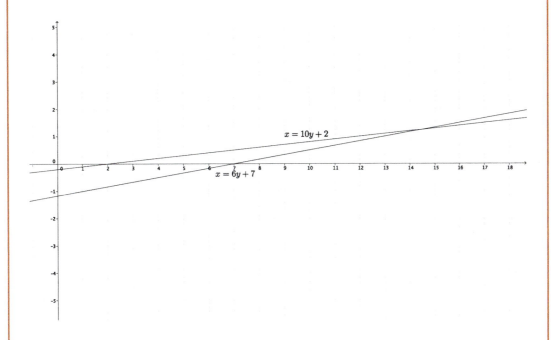

4. $\begin{cases} 5y = \frac{15}{4}x + 25 \\ y = \frac{3}{4}x + 5 \end{cases}$

These equations define the same line. Therefore, this system will have infinitely many solutions.

5. $\begin{cases} x + 9 = y \\ x = 4y - 6 \end{cases}$

$$4y - 6 + 9 = y$$
$$4y + 3 = y$$
$$3 = -3y$$
$$-1 = y$$

$$x + 9 = -1$$
$$x = -10$$

The solution is $(-10, -1)$.

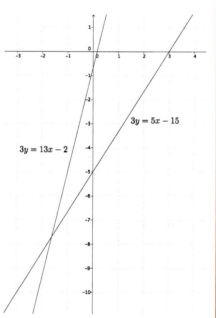

6. $\begin{cases} 3y = 5x - 15 \\ 3y = 13x - 2 \end{cases}$

$$5x - 15 = 13x - 2$$
$$-15 = 8x - 2$$
$$-13 = 8x$$
$$-\frac{13}{8} = x$$

$$3y = 5\left(-\frac{13}{8}\right) - 15$$
$$3y = -\frac{65}{8} - 15$$
$$3y = -\frac{185}{8}$$
$$y = -\frac{185}{24}$$

The solution is $\left(-\frac{13}{8}, -\frac{185}{24}\right)$.

7. $\begin{cases} 6x - 7y = \frac{1}{2} \\ 12x - 14y = 1 \end{cases}$

These equations define the same line. Therefore, this system will have infinitely many solutions.

8. $\begin{cases} 5x - 2y = 6 \\ -10x + 4y = -14 \end{cases}$

The slopes of these two equations are the same, and the y-intercept points are different, which means they graph as parallel lines. Therefore, this system will have no solution.

9. $\begin{cases} y = \frac{3}{2}x - 6 \\ 2y = 7 - 4x \end{cases}$

$$2\left(y = \frac{3}{2}x - 6\right)$$
$$2y = 3x - 12$$
$$\begin{cases} 2y = 3x - 12 \\ 2y = 7 - 4x \end{cases}$$
$$3x - 12 = 7 - 4x$$
$$7x - 12 = 7$$
$$7x = 19$$
$$x = \frac{19}{7}$$

$$y = \frac{3}{2}\left(\frac{19}{7}\right) - 6$$
$$y = \frac{57}{14} - 6$$
$$y = -\frac{27}{14}$$

The solution is $\left(\frac{19}{7}, -\frac{27}{14}\right)$.

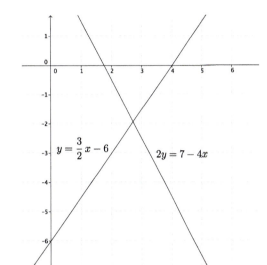

10. $\begin{cases} 7x - 10 = y \\ y = 5x + 12 \end{cases}$

$$7x - 10 = 5x + 12$$
$$2x - 10 = 12$$
$$2x = 22$$
$$x = 11$$

$$y = 5(11) + 12$$
$$y = 55 + 12$$
$$y = 67$$

The solution is $(11, 67)$.

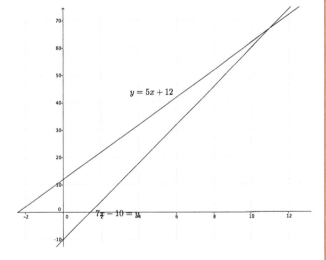

11. Write a system of linear equations with $(-3, 9)$ as its solution.

Answers will vary. Verify that students have written a system of equations where $(-3, 9)$ is a solution to each equation in the system. Sample solution: $\begin{cases} y = x + 12 \\ x + y = 6 \end{cases}$

Lesson 28: Another Computational Method of Solving a Linear System

Student Outcomes

- Students learn the elimination method for solving a system of linear equations.
- Students use properties of rational numbers to find a solution to a system, if it exists, through computation using substitution and elimination methods.

Lesson Notes

Throughout the lesson, students are asked to verify that their solution to a system is correct by graphing the system and comparing the point of intersection to their solution. For that reason, provide graph paper for student use for both the Exercises and the Problem Set. Graphs should be provided during the presentation of the Examples to discuss with students whether or not their estimated point of intersection verifies their solution.

Classwork

Discussion (5 minutes)

- In the last lesson, we saw that if a system of linear equations has a solution, it can be found without graphing. In each case, the first step was to eliminate one of the variables.
- Describe how you would solve this system algebraically: $\begin{cases} y = 3x + 5 \\ y = 8x + 3 \end{cases}$.
  - *Since both equations were equal to y, we could write $3x + 5 = 8x + 3$, thereby eliminating the y from the system.*
- Describe how you would solve this system algebraically: $\begin{cases} y = 7x - 2 \\ 2y - 4x = 10 \end{cases}$.
  - *We can substitute $7x - 2$ for y in the second equation, i.e., $2(7x - 2) - 4x = 10$, thereby eliminating the y again.*
- Describe how you would solve this system algebraically: $\begin{cases} x = 6y + 7 \\ x = 10y + 2 \end{cases}$.
  - *Since both equations are equal to x, we could write $6y + 7 = 10y + 2$, thereby eliminating the x.*
- In this lesson, we will learn a method for solving systems that requires us to eliminate one of the variables but in a different way from the last lesson.

Example 1 (8 minutes)

> **Example 1**
>
> Use what you noticed about adding equivalent expressions to solve the following system by elimination:
> $$\begin{cases} 6x - 5y = 21 \\ 2x + 5y = -5 \end{cases}$$

Show students the three examples of adding integer equations together. Ask students to verbalize what they notice in the examples and to generalize what they observe. The goal is for students to see that they can add equivalent expressions and still have an equivalence.

Example 1: If $2 + 5 = 7$ and $1 + 9 = 10$, does $2 + 5 + 1 + 9 = 7 + 10$?

Example 2: If $1 + 5 = 6$ and $7 - 2 = 5$, does $1 + 5 + 7 - 2 = 6 + 5$?

Example 3: If $-3 + 11 = 8$ and $2 + 1 = 3$, does $-3 + 11 + 2 + 1 = 8 + 3$?

- Use what you noticed about adding equivalent expressions to solve the following system by elimination:
$$\begin{cases} 6x - 5y = 21 \\ 2x + 5y = -5 \end{cases}$$

Provide students with time to attempt to solve the system by adding the equations together. Have students share their work with the class. If necessary, use the points belows to support students.

- Notice that terms $-5y$ and $5y$ are opposites; that is, they have a sum of zero when added. If we were to add the equations in the system, the y would be eliminated.

$$6x - 5y + 2x + 5y = 21 + (-5)$$
$$6x + 2x - 5y + 5y = 16$$
$$8x = 16$$
$$x = 2$$

- Just as before, now that we know what x is, we can substitute it into either equation to determine the value of y.

$$2(2) + 5y = -5$$
$$4 + 5y = -5$$
$$5y = -9$$
$$y = -\frac{9}{5}$$

The solution to the system is $\left(2, -\frac{9}{5}\right)$.

Lesson 28: Another Computational Method of Solving a Linear System

- We can verify our solution by sketching the graphs of the system.

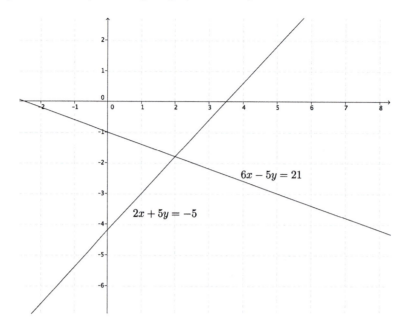

Example 2 (5 minutes)

> **Example 2**
>
> Solve the following system by elimination:
>
> $$\begin{cases} -2x + 7y = 5 \\ 4x - 2y = 14 \end{cases}$$

- We will solve the following system by elimination:

$$\begin{cases} -2x + 7y = 5 \\ 4x - 2y = 14 \end{cases}$$

- In this example, it is not as obvious which variable to eliminate. It will become obvious as soon as we multiply the first equation by 2.

$$2(-2x + 7y = 5)$$
$$-4x + 14y = 10$$

Now we have the system $\begin{cases} -4x + 14y = 10 \\ 4x - 2y = 14 \end{cases}$. It is clear that when we add $-4x + 4x$, the x will be eliminated. Add the equations of this system together, and determine the solution to the system.

Lesson 28: Another Computational Method of Solving a Linear System

▫ Sample student work:

$$-4x + 14y + 4x - 2y = 10 + 14$$
$$14y - 2y = 24$$
$$12y = 24$$
$$y = 2$$

$$4x - 2(2) = 14$$
$$4x - 4 = 14$$
$$4x = 18$$
$$x = \frac{18}{4}$$
$$x = \frac{9}{2}$$

The solution to the system is $\left(\frac{9}{2}, 2\right)$.

- We can verify our solution by sketching the graphs of the system.

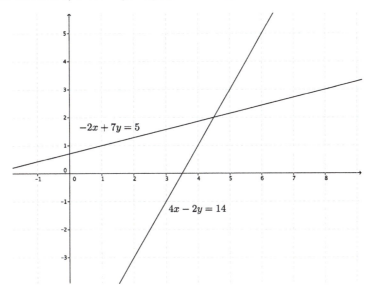

Example 3 (5 minutes)

Example 3

Solve the following system by elimination:
$$\begin{cases} 7x - 5y = -2 \\ 3x - 3y = 7 \end{cases}$$

- We will solve the following system by elimination:
$$\begin{cases} 7x - 5y = -2 \\ 3x - 3y = 7 \end{cases}$$

Provide time for students to solve this system on their own before discussing it as a class.

A STORY OF RATIOS
Lesson 28 8•4

- In this case, it is even less obvious which variable to eliminate. On these occasions, we need to rewrite both equations. We multiply the first equation by -3 and the second equation by 7.

$$-3(7x - 5y = -2)$$
$$-21x + 15y = 6$$

$$7(3x - 3y = 7)$$
$$21x - 21y = 49$$

Now we have the system $\begin{cases} -21x + 15y = 6 \\ 21x - 21y = 49 \end{cases}$, and it is obvious that the x can be eliminated.

- Look at the system again.

$$\begin{cases} 7x - 5y = -2 \\ 3x - 3y = 7 \end{cases}$$

What would we do if we wanted to eliminate the y from the system?

  - *We could multiply the first equation by 3 and the second equation by -5.*

Students may say to multiply the first equation by -3 and the second equation by 5. Whichever answer is given first, ask if the second is also a possibility. Students should answer yes. Then have students solve the system.

  - *Sample student work:*

$$\begin{cases} -21x + 15y = 6 \\ 21x - 21y = 49 \end{cases}$$

$$15y - 21y = 6 + 49$$
$$-6y = 55$$
$$y = -\frac{55}{6}$$

$$7x - 5\left(-\frac{55}{6}\right) = -2$$
$$7x + \frac{275}{6} = -2$$
$$7x = -\frac{287}{6}$$
$$x = -\frac{287}{42}$$

The solution to the system is $\left(-\frac{287}{42}, -\frac{55}{6}\right)$.

Lesson 28: Another Computational Method of Solving a Linear System

Lesson 28

Exercises (8 minutes)

Students complete Exercises 1–3 independently.

Exercises

Each of the following systems has a solution. Determine the solution to the system by eliminating one of the variables. Verify the solution using the graph of the system.

1. $\begin{cases} 6x - 7y = -10 \\ 3x + 7y = -8 \end{cases}$

$$6x - 7y + 3x + 7y = -10 + (-8)$$
$$9x = -18$$
$$x = -2$$

$$3(-2) + 7y = -8$$
$$-6 + 7y = -8$$
$$7y = -2$$
$$y = -\frac{2}{7}$$

The solution is $\left(-2, -\frac{2}{7}\right)$.

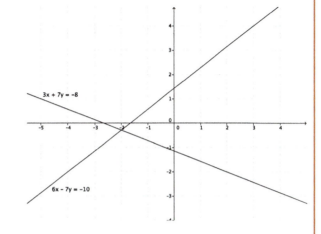

2. $\begin{cases} x - 4y = 7 \\ 5x + 9y = 6 \end{cases}$

$$-5(x - 4y = 7)$$
$$-5x + 20y = -35$$

$\begin{cases} -5x + 20y = -35 \\ 5x + 9y = 6 \end{cases}$

$$-5x + 20y + 5x + 9y = -35 + 6$$
$$29y = -29$$
$$y = -1$$

$$x - 4(-1) = 7$$
$$x + 4 = 7$$
$$x = 3$$

The solution is $(3, -1)$.

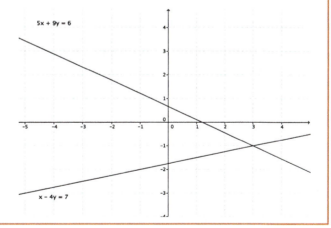

Lesson 28: Another Computational Method of Solving a Linear System

3. $\begin{cases} 2x - 3y = -5 \\ 3x + 5y = 1 \end{cases}$

$$-3(2x - 3y = -5)$$
$$-6x + 9y = 15$$
$$2(3x + 5y = 1)$$
$$6x + 10y = 2$$

$$\begin{cases} -6x + 9y = 15 \\ 6x + 10y = 2 \end{cases}$$

$$-6x + 9y + 6x + 10y = 15 + 2$$
$$19y = 17$$
$$y = \frac{17}{19}$$

$$2x - 3\left(\frac{17}{19}\right) = -5$$
$$2x - \frac{51}{19} = -5$$
$$2x = -5 + \frac{51}{19}$$
$$2x = -\frac{44}{19}$$
$$x = -\frac{44}{38}$$
$$x = -\frac{22}{19}$$

The solution is $\left(-\frac{22}{19}, \frac{17}{19}\right)$.

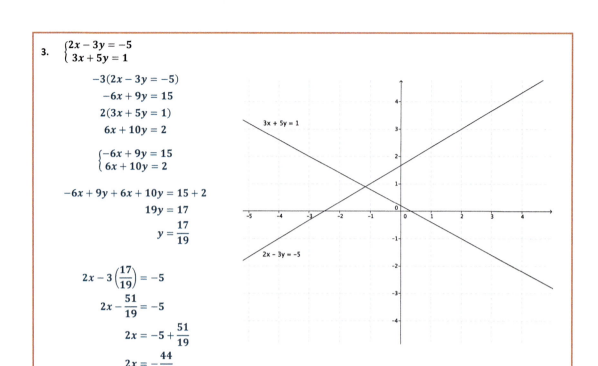

Discussion (6 minutes)

- Systems of linear equations can be solved by sketching the graphs of the lines defined by the equations of the system and looking for the intersection of the lines, substitution (as was shown in the last lesson), or elimination (as was shown in this lesson). Some systems can be solved more efficiently by elimination, while others by substitution. Which method do you think would be most efficient for the following system? Explain.

$$\begin{cases} y = 5x - 19 \\ 3x + 11 = y \end{cases}$$

  - Substitution would be the most efficient method. Since each equation is equal to y, it would be easiest to write the expressions $5x - 19$ and $3x + 11$ equal to one another; then, solve for y.

- What method would you use for the following system? Explain.

$$\begin{cases} 2x - 9y = 7 \\ x + 9y = 5 \end{cases}$$

  - Elimination would be the most efficient method because the terms $-9y + 9y$, when added, would eliminate the y from the equation.

Lesson 28: Another Computational Method of Solving a Linear System

- What method would you use for the following system? Explain.

$$\begin{cases} 4x - 3y = -8 \\ x + 7y = 4 \end{cases}$$

  □ *Elimination would likely be the most efficient method because we could multiply the second equation by -4 to eliminate the x from the equation.*

- What method would you use for the following system? Explain.

$$\begin{cases} x + y = -3 \\ 6x + 6y = 6 \end{cases}$$

  □ *Accept any reasonable answer students provide; then, remind them that the most efficient use of time is to check to see if the system has a solution at all. Since the slopes of the graphs of these lines are parallel, this system has no solution.*

Closing (4 minutes)

Summarize, or ask students to summarize, the main points from the lesson:

- We know how to solve a system by eliminating one of the variables. In some cases, we will have to multiply one or both of the given equations by a constant in order to eliminate a variable.
- We know that some systems are solved more efficiently by elimination than by other methods.

Lesson Summary

Systems of linear equations can be solved by eliminating one of the variables from the system. One way to eliminate a variable is by setting both equations equal to the same variable and then writing the expressions equal to one another.

Example: Solve the system $\begin{cases} y = 3x - 4 \\ y = 2x + 1 \end{cases}$.

Since the expressions $3x - 4$ and $2x + 1$ are both equal to y, they can be set equal to each other and the new equation can be solved for x:

$$3x - 4 = 2x + 1$$

Another way to eliminate a variable is by multiplying each term of an equation by the same constant to make an equivalent equation. Then, use the equivalent equation to eliminate one of the variables and solve the system.

Example: Solve the system $\begin{cases} 2x + y = 8 \\ x + y = 10 \end{cases}$.

Multiply the second equation by -2 to eliminate the x.

$$-2(x + y = 10)$$
$$-2x - 2y = -20$$

Now we have the system $\begin{cases} 2x + y = 8 \\ -2x - 2y = -20 \end{cases}$.

When the equations are added together, the x is eliminated.

$$2x + y - 2x - 2y = 8 + (-20)$$
$$y - 2y = 8 + (-20)$$

Once a solution has been found, verify the solution graphically or by substitution.

Lesson 28: Another Computational Method of Solving a Linear System

Exit Ticket (4 minutes)

The graphs have been provided in the Exit Ticket in order to allow students to check their solutions within the four minutes allotted.

Lesson 28: Another Computational Method of Solving a Linear System

Exit Ticket

Determine the solution, if it exists, for each system of linear equations. Verify your solution on the coordinate plane.

1. $\begin{cases} y = 3x - 5 \\ y = -3x + 7 \end{cases}$

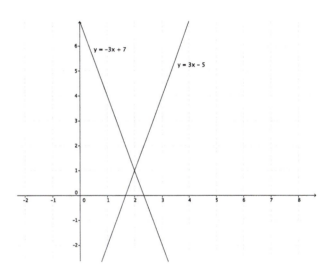

2. $\begin{cases} y = -4x + 6 \\ 2x - y = 11 \end{cases}$

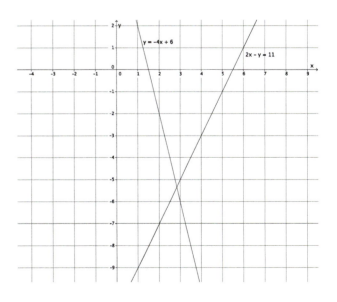

Exit Ticket Sample Solutions

Determine the solution, if it exists, for each system of linear equations. Verify your solution on the coordinate plane.

1. $\begin{cases} y = 3x - 5 \\ y = -3x + 7 \end{cases}$

$$3x - 5 = -3x + 7$$
$$6x = 12$$
$$x = 2$$

$$y = 3(2) - 5$$
$$y = 6 - 5$$
$$y = 1$$

The solution is $(2, 1)$.

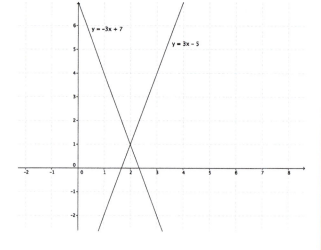

2. $\begin{cases} y = -4x + 6 \\ 2x - y = 11 \end{cases}$

$$2x - (-4x + 6) = 11$$
$$2x + 4x - 6 = 11$$
$$6x = 17$$
$$x = \frac{17}{6}$$

$$y = -4\left(\frac{17}{6}\right) + 6$$
$$y = -\frac{34}{3} + 6$$
$$y = -\frac{16}{3}$$

The solution is $\left(\frac{17}{6}, -\frac{16}{3}\right)$.

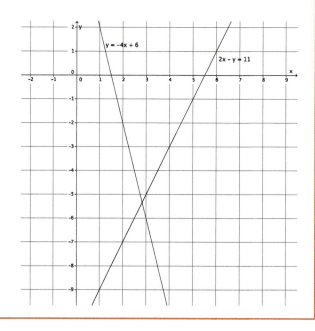

Problem Set Sample Solutions

Determine the solution, if it exists, for each system of linear equations. Verify your solution on the coordinate plane.

1. $\begin{cases} \frac{1}{2}x + 5 = y \\ 2x + y = 1 \end{cases}$

$$2x + \frac{1}{2}x + 5 = 1$$
$$\frac{5}{2}x + 5 = 1$$
$$\frac{5}{2}x = -4$$
$$x = -\frac{8}{5}$$

$$2\left(-\frac{8}{5}\right) + y = 1$$
$$-\frac{16}{5} + y = 1$$
$$y = \frac{21}{5}$$

The solution is $\left(-\frac{8}{5}, \frac{21}{5}\right)$.

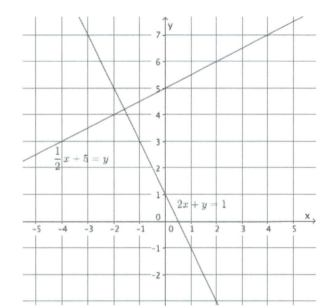

2. $\begin{cases} 9x + 2y = 9 \\ -3x + y = 2 \end{cases}$

$$3(-3x + y = 2)$$
$$-9x + 3y = 6$$
$$\begin{cases} 9x + 2y = 9 \\ -9x + 3y = 6 \end{cases}$$
$$9x + 2y - 9x + 3y = 15$$
$$5y = 15$$
$$y = 3$$

$$-3x + 3 = 2$$
$$-3x = -1$$
$$x = \frac{1}{3}$$

The solution is $\left(\frac{1}{3}, 3\right)$.

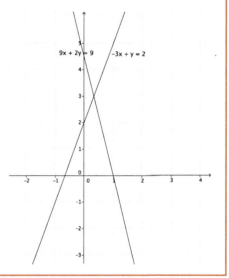

3. $\begin{cases} y = 2x - 2 \\ 2y = 4x - 4 \end{cases}$

These equations define the same line. Therefore, this system will have infinitely many solutions.

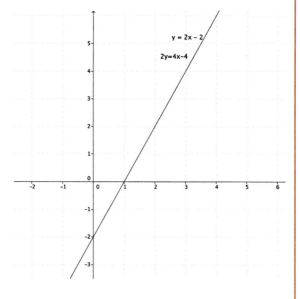

4. $\begin{cases} 8x + 5y = 19 \\ -8x + y = -1 \end{cases}$

$$8x + 5y - 8x + y = 19 - 1$$
$$5y + y = 18$$
$$6y = 18$$
$$y = 3$$

$$8x + 5(3) = 19$$
$$8x + 15 = 19$$
$$8x = 4$$
$$x = \frac{1}{2}$$

The solution is $\left(\frac{1}{2}, 3\right)$.

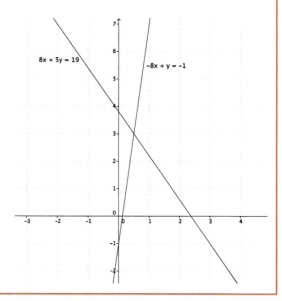

5. $\begin{cases} x + 3 = y \\ 3x + 4y = 7 \end{cases}$

$$3x + 4(x + 3) = 7$$
$$3x + 4x + 12 = 7$$
$$7x + 12 = 7$$
$$7x = -5$$
$$x = -\frac{5}{7}$$

$$-\frac{5}{7} + 3 = y$$
$$\frac{16}{7} = y$$

The solution is $\left(-\frac{5}{7}, \frac{16}{7}\right)$.

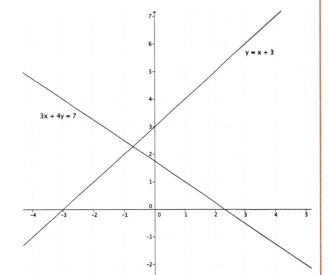

6. $\begin{cases} y = 3x + 2 \\ 4y = 12 + 12x \end{cases}$

The equations graph as distinct lines. The slopes of these two equations are the same, and the y-intercept points are different, which means they graph as parallel lines. Therefore, this system will have no solution.

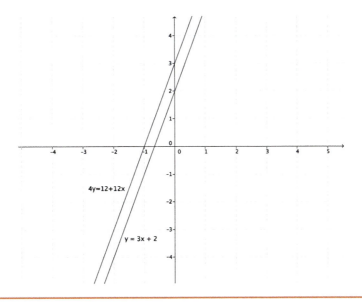

7. $\begin{cases} 4x - 3y = 16 \\ -2x + 4y = -2 \end{cases}$

$2(-2x + 4y = -2)$
$-4x + 8y = -4$

$\begin{cases} 4x - 3y = 16 \\ -4x + 8y = -4 \end{cases}$

$4x - 3y - 4x + 8y = 16 - 4$
$-3y + 8y = 12$
$5y = 12$
$y = \dfrac{12}{5}$

$4x - 3\left(\dfrac{12}{5}\right) = 16$
$4x - \dfrac{36}{5} = 16$
$4x = \dfrac{116}{5}$
$x = \dfrac{29}{5}$

The solution is $\left(\dfrac{29}{5}, \dfrac{12}{5}\right)$.

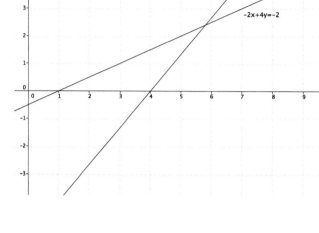

8. $\begin{cases} 2x + 2y = 4 \\ 12 - 3x = 3y \end{cases}$

The equations graph as distinct lines. The slopes of these two equations are the same, and the y-intercept points are different, which means they graph as parallel lines. Therefore, this system will have no solution.

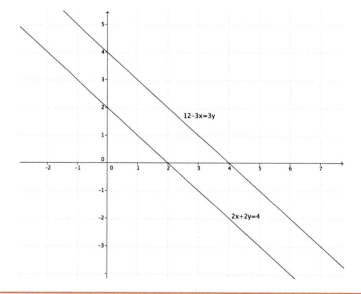

Lesson 28: Another Computational Method of Solving a Linear System

9. $\begin{cases} y = -2x + 6 \\ 3y = x - 3 \end{cases}$

$$3(y = -2x + 6)$$
$$3y = -6x + 18$$
$$\begin{cases} 3y = -6x + 18 \\ 3y = x - 3 \end{cases}$$
$$-6x + 18 = x - 3$$
$$18 = 7x - 3$$
$$21 = 7x$$
$$\frac{21}{7} = x$$
$$x = 3$$

$$y = -2(3) + 6$$
$$y = -6 + 6$$
$$y = 0$$

The solution is $(3, 0)$.

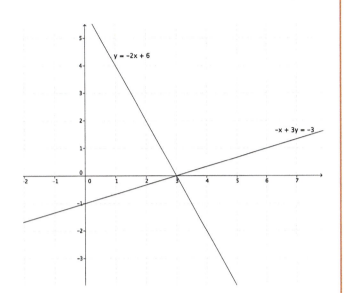

10. $\begin{cases} y = 5x - 1 \\ 10x = 2y + 2 \end{cases}$

These equations define the same line. Therefore, this system will have infinitely many solutions.

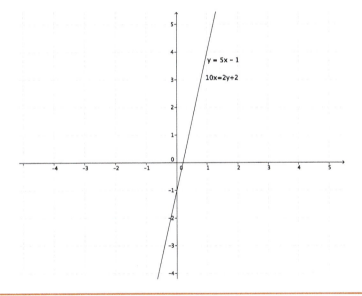

Lesson 28: Another Computational Method of Solving a Linear System

11. $\begin{cases} 3x - 5y = 17 \\ 6x + 5y = 10 \end{cases}$

$$3x - 5y + 6x + 5y = 17 + 10$$
$$9x = 27$$
$$x = 3$$

$$3(3) - 5y = 17$$
$$9 - 5y = 17$$
$$-5y = 8$$
$$y = -\frac{8}{5}$$

The solution is $\left(3, -\frac{8}{5}\right)$.

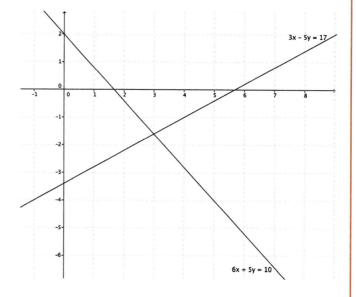

12. $\begin{cases} y = \frac{4}{3}x - 9 \\ y = x + 3 \end{cases}$

$$\frac{4}{3}x - 9 = x + 3$$
$$\frac{1}{3}x - 9 = 3$$
$$\frac{1}{3}x = 12$$
$$x = 36$$

$$y = 36 + 3$$
$$y = 39$$

The solution is $(36, 39)$.

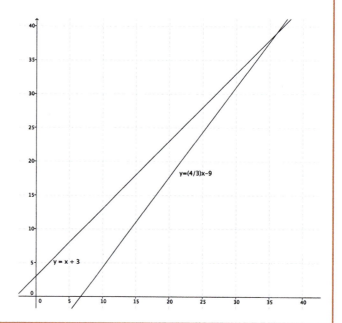

13. $\begin{cases} 4x - 7y = 11 \\ x + 2y = 10 \end{cases}$

$-4(x + 2y = 10)$
$-4x - 8y = -40$

$\begin{cases} 4x - 7y = 11 \\ -4x - 8y = -40 \end{cases}$

$4x - 7y - 4x - 8y = 11 - 40$
$-15y = -29$
$y = \dfrac{29}{15}$

$x + 2\left(\dfrac{29}{15}\right) = 10$
$x + \dfrac{58}{15} = 10$
$x = \dfrac{92}{15}$

The solution is $\left(\dfrac{92}{15}, \dfrac{29}{15}\right)$.

14. $\begin{cases} 21x + 14y = 7 \\ 12x + 8y = 16 \end{cases}$

The slopes of these two equations are the same, and the y-intercept points are different, which means they graph as parallel lines. Therefore, this system will have no solution.

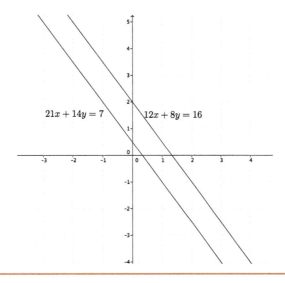

A STORY OF RATIOS   Lesson 29   8•4

 Lesson 29: Word Problems

Student Outcomes

- Students write word problems into systems of linear equations.
- Students solve systems of linear equations using elimination and substitution methods.

Lesson Notes

In this lesson, students use many of the skills learned in this module. For example, they begin by defining the variables described in a word problem. Next, students write a system of linear equations to represent the situation. Students then have to decide which method is most efficient for solving the system. Finally, they solve the system and check to make sure their answer is correct. For each of the examples in the lesson, pose the questions, and provide students time to answer them on their own. Then, select students to share their thoughts and solutions with the class.

Classwork

Example 1 (5 minutes)

> **Example 1**
>
> The sum of two numbers is 361, and the difference between the two numbers is 173. What are the two numbers?

- Together, we will read a word problem and work toward finding the solution.
- The sum of two numbers is 361, and the difference between the two numbers is 173. What are the two numbers?

Provide students time to work independently or in pairs to solve this problem. Have students share their solutions and explain how they arrived at their answers. Then, show how the problem can be solved using a system of linear equations.

- What do we need to do first?
    - *We need to define our variables.*
- If we define our variables, we can better represent the situation we have been given. What should the variables be for this problem?
    - *Let x represent one number, and let y represent the other number.*
- Now that we know the numbers are x and y, what do we need to do now?
    - *We need to write equations to represent the information in the word problem.*
- Using x and y, write equations to represent the information we are provided.
    - *The sum of two numbers is 361 can be written as $x + y = 361$. The difference between the two numbers is 173 can be written as $x - y = 173$.*

460   Lesson 29:   Word Problems

- We have two equations to represent this problem. What is it called when we have more than one linear equation for a problem, and how is it represented symbolically?
  - *We have a system of linear equations.*
  $$\begin{cases} x + y = 361 \\ x - y = 173 \end{cases}$$
- We know several methods for solving systems of linear equations. Which method do you think will be the most efficient, and why?
  - *We should add the equations together to eliminate the variable y because we can do that in one step.*
- Solve the system: $\begin{cases} x + y = 361 \\ x - y = 173 \end{cases}$.
  - *Sample student work:*
  $$\begin{cases} x + y = 361 \\ x - y = 173 \end{cases}$$
  $$x + x + y - y = 361 + 173$$
  $$2x = 534$$
  $$x = 267$$

  $$267 + y = 361$$
  $$y = 94$$

  The solution is $(267, 94)$.
- Based on our work, we believe the two numbers are 267 and 94. Check to make sure your answer is correct by substituting the numbers into both equations. If it makes a true statement, then we know we are correct. If it makes a false statement, then we need to go back and check our work.
  - *Sample student work:*
  $$267 + 94 = 361$$
  $$361 = 361$$

  $$267 - 94 = 173$$
  $$173 = 173$$
- Now we are sure that the numbers are 267 and 94. Does it matter which number is x and which number is y?
  - *Not necessarily, but we need their difference to be positive, so x must be the larger of the two numbers to make sense of our equation $x - y = 173$.*

Example 2 (7 minutes)

> **Example 2**
>
> There are 356 eighth-grade students at Euclid's Middle School. Thirty-four more than four times the number of girls is equal to half the number of boys. How many boys are in eighth grade at Euclid's Middle School? How many girls?

Lesson 29: Word Problems

- Again, we will work together to solve the following word problem.
- There are 356 eighth-grade students at Euclid's Middle School. Thirty-four more than four times the number of girls is equal to half the number of boys. How many boys are in eighth grade at Euclid's Middle School? How many girls? What do we need to do first?
    - *We need to define our variables.*
- If we define our variables, we can better represent the situation we have been given. What should the variables be for this problem?
    - *Let x represent the number of girls, and let y represent the number of boys.*

Whichever way students define the variables, ask them if it could be done the opposite way. For example, if students respond as stated above, ask them if we could let x represent the number of boys and y represent the number of girls. They should say that at this stage it does not matter if x represents girls or boys, but once the variable is defined, it does matter.

- Now that we know that x is the number of girls and y is the number of boys, what do we need to do now?
    - *We need to write equations to represent the information in the word problem.*
- Using x and y, write equations to represent the information we are provided.
    - *There are 356 eighth-grade students can be represented as $x + y = 356$. Thirty-four more than four times the number of girls is equal to half the number of boys can be represented as $4x + 34 = \frac{1}{2}y$.*
- We have two equations to represent this problem. What is it called when we have more than one linear equation for a problem, and how is it represented symbolically?
    - *We have a system of linear equations.*

$$\begin{cases} x + y = 356 \\ 4x + 34 = \frac{1}{2}y \end{cases}$$

- We know several methods for solving systems of linear equations. Which method do you think will be the most efficient and why?
    - *Answers will vary. There is no obvious "most efficient" method. Accept any reasonable responses as long as they are justified.*
- Solve the system: $\begin{cases} x + y = 356 \\ 4x + 34 = \frac{1}{2}y \end{cases}$.
    - Sample student work:

$$\begin{cases} x + y = 356 \\ 4x + 34 = \frac{1}{2}y \end{cases}$$

$$2\left(4x + 34 = \frac{1}{2}y\right)$$

$$8x + 68 = y$$

$$\begin{cases} x + y = 356 \\ 8x + 68 = y \end{cases}$$

A STORY OF RATIOS　　　　　　　　　　　　　　　　　　　　　　　　　　　Lesson 29　8•4

$$x + 8x + 68 = 356$$
$$9x + 68 = 356$$
$$9x = 288$$
$$x = 32$$

$$32 + y = 356$$
$$y = 324$$

The solution is $(32, 324)$.

- What does the solution mean in context?
  - *Since we let x represent the number of girls and y represent the number of boys, it means that there are 32 girls and 324 boys at Euclid's Middle School in eighth grade.*
- Based on our work, we believe there are 32 girls and 324 boys. How can we be sure we are correct?
  - *We need to substitute the values into both equations of the system to see if it makes a true statement.*

$$32 + 324 = 356$$
$$356 = 356$$

$$4(32) + 34 = \frac{1}{2}(324)$$
$$128 + 34 = 162$$
$$162 = 162$$

Example 3 (5 minutes)

> **Example 3**
>
> A family member has some five-dollar bills and one-dollar bills in her wallet. Altogether she has 18 bills and a total of $62. How many of each bill does she have?

- Again, we will work together to solve the following word problem.
- A family member has some five-dollar bills and one-dollar bills in her wallet. Altogether she has 18 bills and a total of $62. How many of each bill does she have? What do we do first?
  - *We need to define our variables.*
- If we define our variables, we can better represent the situation we have been given. What should the variables be for this problem?
  - *Let x represent the number of $5 bills, and let y represent the number of $1 bills.*

Again, whichever way students define the variables, ask them if it could be done the opposite way.

- Now that we know that x is the number of $5 bills and y is the number of $1 bills, what do we need to do now?
  - *We need to write equations to represent the information in the word problem.*

Lesson 29:　　Word Problems

- Using x and y, write equations to represent the information we are provided.
    - Altogether she has 18 bills and a total of $62 must be represented with two equations, the first being $x + y = 18$ to represent the total of 18 bills and the second being $5x + y = 62$ to represent the total amount of money she has.
- We have two equations to represent this problem. What is it called when we have more than one linear equation for a problem, and how is it represented symbolically?
    - We have a system of linear equations.

$$\begin{cases} x + y = 18 \\ 5x + y = 62 \end{cases}$$

- We know several methods for solving systems of linear equations. Which method do you think will be the most efficient and why?
    - Answers will vary. Students might say they could multiply one of the equations by -1, and then they would be able to eliminate the variable y when they add the equations together. Other students may say it would be easiest to solve for y in the first equation and then substitute the value of y into the second equation. After they have justified their methods, allow them to solve the system in any manner they choose.
- Solve the system: $\begin{cases} x + y = 18 \\ 5x + y = 62 \end{cases}$.
    - Sample student work:

$$\begin{cases} x + y = 18 \\ 5x + y = 62 \end{cases}$$

$$x + y = 18$$
$$y = -x + 18$$

$$\begin{cases} y = -x + 18 \\ 5x + y = 62 \end{cases}$$

$$5x + (-x) + 18 = 62$$
$$4x + 18 = 62$$
$$4x = 44$$
$$x = 11$$

$$11 + y = 18$$
$$y = 7$$

The solution is $(11, 7)$.

- What does the solution mean in context?
    - Since we let x represent the number of $5 bills and y represent the number of $1 bills, it means that the family member has 11 $5 bills, and 7 $1 bills.
- The next step is to check our work.
    - It is obvious that $11 + 7 = 18$, so we know the family member has 18 bills.

- It makes more sense to check our work against the actual value of those 18 bills in this case. Now we check the second equation.

$$5(11) + 1(7) = 62$$
$$55 + 7 = 62$$
$$62 = 62$$

Example 4 (9 minutes)

> **Example 4**
>
> A friend bought 2 boxes of pencils and 8 notebooks for school, and it cost him $11. He went back to the store the same day to buy school supplies for his younger brother. He spent $11.25 on 3 boxes of pencils and 5 notebooks. How much would 7 notebooks cost?

- Again, we will work together to solve the following word problem.
- A friend bought 2 boxes of pencils and 8 notebooks for school, and it cost him $11. He went back to the store the same day to buy school supplies for his younger brother. He spent $11.25 on 3 boxes of pencils and 5 notebooks. How much would 7 notebooks cost? What do we do first?
    - *We need to define our variables.*
- If we define our variables, we can better represent the situation we have been given. What should the variables be for this problem?
    - *Let x represent the cost of a box of pencils, and let y represent the cost of a notebook.*

Again, whichever way students define the variables, ask them if it could be done the opposite way.

- Now that we know that x is the cost of a box of pencils and y is the cost of a notebook, what do we need to do now?
    - *We need to write equations to represent the information in the word problem.*
- Using x and y, write equations to represent the information we are provided.
    - *A friend bought 2 boxes of pencils and 8 notebooks for school, and it cost him $11, which is represented by the equation $2x + 8y = 11$. He spent $11.25 on 3 boxes of pencils and 5 notebooks, which is represented by the equation $3x + 5y = 11.25$.*
- We have two equations to represent this problem. What is it called when we have more than one linear equation for a problem, and how is it represented symbolically?
    - *We have a system of linear equations.*

$$\begin{cases} 2x + 8y = 11 \\ 3x + 5y = 11.25 \end{cases}$$

- We know several methods for solving systems of linear equations. Which method do you think will be the most efficient and why?
    - *Answers will vary. Ask several students what they believe the most efficient method is, and have them share with the class a brief description of their plan. For example, a student may decide to multiply the first equation by 3 and the second equation by −2 to eliminate x from the system after adding the equations together. After several students have shared their plans, allow students to solve in any manner they choose.*

Lesson 29: Word Problems

- Solve the system: $\begin{cases} 2x + 8y = 11 \\ 3x + 5y = 11.25 \end{cases}$
  - *Sample student work:*

$$\begin{cases} 2x + 8y = 11 \\ 3x + 5y = 11.25 \end{cases}$$

$$3(2x + 8y = 11)$$
$$6x + 24y = 33$$
$$-2(3x + 5y = 11.25)$$
$$-6x - 10y = -22.50$$

$$\begin{cases} 6x + 24y = 33 \\ -6x - 10y = -22.50 \end{cases}$$

$$6x + 24y - 6x - 10y = 33 - 22.50$$
$$24y - 10y = 10.50$$
$$14y = 10.50$$
$$y = \frac{10.50}{14}$$
$$y = 0.75$$

$$2x + 8(0.75) = 11$$
$$2x + 6 = 11$$
$$2x = 5$$
$$x = 2.50$$

The solution is $(2.50, 0.75)$.
- What does the solution mean in context?
  - *It means that a box of pencils costs $2.50, and a notebook costs $0.75.*
- Before we answer the question that this word problem asked, check to make sure the solution is correct.
  - *Sample student work:*

$$2(2.50) + 8(0.75) = 11$$
$$5 + 6 = 11$$
$$11 = 11$$

$$3(2.50) + 5(0.75) = 11.25$$
$$7.50 + 3.75 = 11.25$$
$$11.25 = 11.25$$

- Now that we know we have the correct costs for the box of pencils and notebooks, we can answer the original question: How much would 7 notebooks cost?
  - *The cost of 7 notebooks is $7(0.75) = 5.25$. Therefore, 7 notebooks cost $7.25.*

- Keep in mind that some word problems require us to solve the system in order to answer a specific question, like this example about the cost of 7 notebooks. Other problems may just require the solution to the system to answer the word problem, like the first example about the two numbers and their sum and difference. It is always a good practice to reread the word problem to make sure you know what you are being asked to do.

Exercises (9 minutes)

Students complete Exercises 1–3 independently or in pairs.

Exercises

1. A farm raises cows and chickens. The farmer has a total of 42 animals. One day he counts the legs of all of his animals and realizes he has a total of 114. How many cows does the farmer have? How many chickens?

   Let x represent the number of cows and y represent the number of chickens. Then:

   $$\begin{cases} x + y = 42 \\ 4x + 2y = 114 \end{cases}$$

   $$-2(x + y = 42)$$
   $$-2x - 2y = -84$$

   $$\begin{cases} -2x - 2y = -84 \\ 4x + 2y = 114 \end{cases}$$

   $$-2x - 2y + 4x + 2y = -84 + 114$$
   $$-2x + 4x = 30$$
   $$2x = 30$$
   $$x = 15$$

   $$15 + y = 42$$
   $$y = 27$$

   The solution is $(15, 27)$.

   $$4(15) + 2(27) = 114$$
   $$60 + 54 = 114$$
   $$114 = 114$$

   The farmer has 15 cows and 27 chickens.

2. The length of a rectangle is 4 times the width. The perimeter of the rectangle is 45 inches. What is the area of the rectangle?

   Let x represent the length and y represent the width. Then:

   $$\begin{cases} x = 4y \\ 2x + 2y = 45 \end{cases}$$

   $$2(4y) + 2y = 45$$
   $$8y + 2y = 45$$
   $$10y = 45$$
   $$y = 4.5$$

   $$x = 4(4.5)$$
   $$x = 18$$

   The solution is $(18, 4.5)$.

   $$2(18) + 2(4.5) = 45$$
   $$36 + 9 = 45$$
   $$45 = 45$$

   Since $18 \times 4.5 = 81$, the area of the rectangle is 81 in^2.

Lesson 29: Word Problems

> 3. The sum of the measures of angles x and y is $127°$. If the measure of $\angle x$ is $34°$ more than half the measure of $\angle y$, what is the measure of each angle?
>
> Let x represent the measure of $\angle x$ and y represent the measure of $\angle y$. Then:
>
> $$\begin{cases} x + y = 127 \\ x = 34 + \frac{1}{2}y \end{cases}$$
>
> $$34 + \frac{1}{2}y + y = 127$$
>
> $$34 + \frac{3}{2}y = 127$$
>
> $$\frac{3}{2}y = 93$$
>
> $$y = 62$$
>
> $$x + 62 = 127$$
>
> $$x = 65$$
>
> The solution is $(65, 62)$.
>
> $$65 = 34 + \frac{1}{2}(62)$$
>
> $$65 = 34 + 31$$
>
> $$65 = 65$$
>
> The measure of $\angle x$ is $65°$, and the measure of $\angle y$ is $62°$.

Closing (5 minutes)

Summarize, or ask students to summarize, the main points from the lesson:

- We know how to write information from word problems into a system of linear equations.
- We can solve systems of linear equations using the elimination and substitution methods.
- When we solve a system, we must clearly define the variables we intend to use, consider which method for solving the system would be most efficient, check our answer, and think about what it means in context. Finally, we should ensure we have answered the question posed in the problem.

Exit Ticket (5 minutes)

Lesson 29: Word Problems

Exit Ticket

1. Small boxes contain DVDs, and large boxes contain one gaming machine. Three boxes of gaming machines and a box of DVDs weigh 48 pounds. Three boxes of gaming machines and five boxes of DVDs weigh 72 pounds. How much does each box weigh?

2. A language arts test is worth 100 points. There is a total of 26 questions. There are spelling word questions that are worth 2 points each and vocabulary word questions worth 5 points each. How many of each type of question are there?

Exit Ticket Sample Solutions

1. Small boxes contain DVDs, and large boxes contain one gaming machine. Three boxes of gaming machines and a box of DVDs weigh 48 pounds. Three boxes of gaming machines and five boxes of DVDs weigh 72 pounds. How much does each box weigh?

   Let x represent the weight of the gaming machine box, and let y represent the weight of the DVD box. Then:

   $$\begin{cases} 3x + y = 48 \\ 3x + 5y = 72 \end{cases}$$

   $$-1(3x + y = 48)$$
   $$-3x - y = -48$$

   $$\begin{cases} -3x - y = -48 \\ 3x + 5y = 72 \end{cases}$$

   $$3x - 3x + 5y - y = 72 - 48$$
   $$4y = 24$$
   $$y = 6$$

   $$3x + 6 = 48$$
   $$3x = 42$$
   $$x = 14$$

   The solution is $(14, 6)$.

   $$3(14) + 5(6) = 72$$
   $$72 = 72$$

   The box with one gaming machine weighs 14 pounds, and the box containing DVDs weighs 6 pounds.

2. A language arts test is worth 100 points. There is a total of 26 questions. There are spelling word questions that are worth 2 points each and vocabulary word questions worth 5 points each. How many of each type of question are there?

   Let x represent the number of spelling word questions, and let y represent the number of vocabulary word questions.

   $$\begin{cases} x + y = 26 \\ 2x + 5y = 100 \end{cases}$$

   $$-2(x + y = 26)$$
   $$-2x - 2y = -52$$

   $$\begin{cases} -2x - 2y = -52 \\ 2x + 5y = 100 \end{cases}$$

   $$2x - 2x + 5y - 2y = 100 - 52$$
   $$3y = 48$$
   $$y = 16$$

   $$x + 16 = 26$$
   $$x = 10$$

   The solution is $(10, 16)$.

   $$2(10) + 5(16) = 100$$
   $$100 = 100$$

   There are 10 spelling word questions and 16 vocabulary word questions.

Lesson 29: Word Problems

Problem Set Sample Solutions

1. Two numbers have a sum of $1,212$ and a difference of 518. What are the two numbers?

   Let x represent one number and y represent the other number.

   $$\begin{cases} x + y = 1212 \\ x - y = 518 \end{cases}$$

   $$x + y + x - y = 1212 + 518$$
   $$2x = 1730$$
   $$x = 865$$

   $$865 + y = 1212$$
   $$y = 347$$

   The solution is $(865, 347)$.

   $$865 - 347 = 518$$
   $$518 = 518$$

   The two numbers are 347 and 865.

2. The sum of the ages of two brothers is 46. The younger brother is 10 more than a third of the older brother's age. How old is the younger brother?

   Let x represent the age of the younger brother and y represent the age of the older brother.

   $$\begin{cases} x + y = 46 \\ x = 10 + \frac{1}{3}y \end{cases}$$

   $$10 + \frac{1}{3}y + y = 46$$
   $$10 + \frac{4}{3}y = 46$$
   $$\frac{4}{3}y = 36$$
   $$y = 27$$

   $$x + 27 = 46$$
   $$x = 19$$

   The solution is $(19, 27)$.

   $$19 = 10 + \frac{1}{3}(27)$$
   $$19 = 10 + 9$$
   $$19 = 19$$

   The younger brother is 19 years old.

3. One angle measures 54 more degrees than 3 times another angle. The angles are supplementary. What are their measures?

   Let x represent the measure of one angle and y represent the measure of the other angle.

   $$\begin{cases} x = 3y + 54 \\ x + y = 180 \end{cases}$$

   $$3y + 54 + y = 180$$
   $$4y + 54 = 180$$
   $$4y = 126$$
   $$y = 31.5$$

   $$x = 3(31.5) + 54$$
   $$x = 94.5 + 54$$
   $$x = 148.5$$

   The solution is $(148.5, 31.5)$.

   $$148.5 + 31.5 = 180$$
   $$180 = 180$$

   One angle measures $148.5°$, and the other measures $31.5°$.

4. Some friends went to the local movie theater and bought four large buckets of popcorn and six boxes of candy. The total for the snacks was $46.50. The last time you were at the theater, you bought a large bucket of popcorn and a box of candy, and the total was $9.75. How much would 2 large buckets of popcorn and 3 boxes of candy cost?

   Let x represent the cost of a large bucket of popcorn and y represent the cost of a box of candy.

   $$\begin{cases} 4x + 6y = 46.50 \\ x + y = 9.75 \end{cases}$$

   $$-4(x + y = 9.75)$$
   $$-4x - 4y = -39$$

   $$\begin{cases} 4x + 6y = 46.50 \\ -4x - 4y = -39 \end{cases}$$

   $$4x + 6y - 4x - 4y = 46.50 - 39$$
   $$6y - 4y = 7.50$$
   $$2y = 7.50$$
   $$y = 3.75$$

   $$x + 3.75 = 9.75$$
   $$x = 6$$

   The solution is $(6, 3.75)$.

   $$4(6) + 6(3.75) = 46.50$$
   $$24 + 22.50 = 46.50$$
   $$46.50 = 46.50$$

   Since one large bucket of popcorn costs $6 and one box of candy costs $3.75, then $2(6) + 3(3.75) = 12 + 11.25 = 23.25$, and two large buckets of popcorn and three boxes of candy will cost $23.25.

5. You have 59 total coins for a total of 12.05. You only have quarters and dimes. How many of each coin do you have?

   Let x represent the number of quarters and y represent the number of dimes.

   $$\begin{cases} x + y = 59 \\ 0.25x + 0.1y = 12.05 \end{cases}$$

   $$-4(0.25x + 0.1y = 12.05)$$
   $$-x - 0.4y = -48.20$$

   $$\begin{cases} x + y = 59 \\ -x - 0.4y = -48.20 \end{cases}$$

   $$x + y - x - 0.4y = 59 - 48.20$$
   $$y - 0.4y = 10.80$$
   $$0.6y = 10.80$$
   $$y = \frac{10.80}{0.6}$$
   $$y = 18$$

   $$x + 18 = 59$$
   $$x = 41$$

   The solution is $(41, 18)$.

   $$0.25(41) + 0.1(18) = 12.05$$
   $$10.25 + 1.80 = 12.05$$
   $$12.05 = 12.05$$

   I have 41 quarters and 18 dimes.

6. A piece of string is 112 inches long. Isabel wants to cut it into 2 pieces so that one piece is three times as long as the other. How long is each piece?

   Let x represent one piece and y represent the other.

   $$\begin{cases} x + y = 112 \\ 3y = x \end{cases}$$

   $$3y + y = 112$$
   $$4y = 112$$
   $$y = 28$$

   $$x + 28 = 112$$
   $$x = 84$$

   The solution is $(84, 28)$.

   $$3(28) = 84$$
   $$84 = 84$$

   One piece should be 84 inches long, and the other should be 28 inches long.

Lesson 30: Conversion Between Celsius and Fahrenheit

Student Outcomes

- Students learn a real-world application of linear equations with respect to the conversion of temperatures from Celsius to Fahrenheit and Fahrenheit to Celsius.

Classwork

Mathematical Modeling Exercise (20 minutes)

- There are two methods for measuring temperature: (1) Fahrenheit, which assigns the number 32 to the temperature of water freezing and the number 212 to the temperature of water boiling; and (2) Celsius, which assigns the numbers 0 and 100, respectively, to the same temperatures. These numbers will be denoted by 32°F, 212°F, 0°C, 100°C, respectively.

- Our goal is to address the following two questions:

> **Mathematical Modeling Exercise**
>
> (1) If t is a number, what is the degree in Fahrenheit that corresponds to $t°C$?
>
> (2) If t is a number, what is the degree in Fahrenheit that corresponds to $(-t)°C$?

- Instead of trying to answer these questions directly, let's try something simpler. With this in mind, can we find out what degree in Fahrenheit corresponds to 1°C? Explain.

- We can use the following diagram (double number line) to organize our thinking.

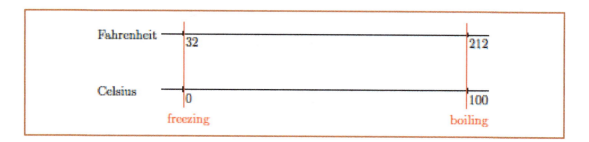

- At this point, the only information we have is that 0°C = 32°F, and 100°C = 212°F. We want to figure out what degree of Fahrenheit corresponds to 1°C. Where on the diagram would 1°C be located? Be specific.

Provide students time to talk to their partners about a plan, and then have them share. Ask them to make conjectures about what degree in Fahrenheit corresponds to 1°C, and have them explain their rationale for the numbers they chose. Consider recording the information, and have the class vote on which answer they think is closest to correct.

> □ We need to divide the Celsius number line from 0 to 100 into 100 equal parts. The first division to the right of zero will be the location of 1°C.

MP.3

Now that we know where to locate 1°C on the lower number line, we need to figure out what number it corresponds to on the upper number line representing Fahrenheit. Like we did with Celsius, we divide the number line from 32 to 212 into 100 equal parts. The number line from 32 to 212 is actually a length of 180 units (212 − 32 = 180). Now, how would we determine the precise number in Fahrenheit that corresponds to 1°C?

Provide students time to talk to their partners and compute the answer.

▫ We need to take the length 180 and divide it into 100 equal parts.

$$\frac{180}{100} = \frac{9}{5} = 1\frac{4}{5} = 1.8$$

- If we look at a magnified version of the number line with this division, we have the following diagram:

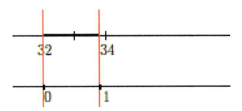

- Based on your computation, what number falls at the intersection of the Fahrenheit number line and the red line that corresponds to 1°C? Explain.
    ▫ Since we know that each division on the Fahrenheit number line has a length of 1.8, then when we start from 32 and add 1.8, we get 33.8. Therefore, 1°C is equal to 33.8°F.

Revisit the conjecture made at the beginning of the activity, and note which student came closest to guessing 33.8°F. Ask the student to explain how he arrived at such a close answer.

- Eventually, we want to revisit the original two questions. But first, let's look at a few more concrete questions. What is 37°C in Fahrenheit? Explain.

Provide students time to talk to their partners about how to answer the question. Ask students to share their ideas and explain their thinking.

▫ Since the unit length on the Celsius scale is equal to the unit length on the Fahrenheit scale, then 37°C means we need to multiply (37 × 1.8) to determine the corresponding location on the Fahrenheit scale. But, because 0 on the Celsius scale is 32 on the Fahrenheit scale, we will need to add 32 to our answer. In other words, 37°C = (32 + 37 × 1.8)°F = (32 + 66.6)°F = 98.6°F.

Exercises (8 minutes)

Have students work in pairs or small groups to determine the corresponding Fahrenheit temperature for each given Celsius temperature. The goal is for students to be consistent in their use of repeated reasoning to lead them to the general equation for the conversion between Celsius and Fahrenheit.

Exercises

Determine the corresponding Fahrenheit temperature for the given Celsius temperatures in Exercises 1–5.

1. How many degrees Fahrenheit is 25°C?

    $25°C = (32 + 25 × 1.8)°F = (32 + 45)°F = 77°F$

2. How many degrees Fahrenheit is 42°C?

   $42°C = (32 + 42 \times 1.8)°F = (32 + 75.6)°F = 107.6°F$

3. How many degrees Fahrenheit is 94°C?

   $94°C = (32 + 94 \times 1.8)°F = (32 + 169.2)°F = 201.2°F$

4. How many degrees Fahrenheit is 63°C?

   $63°C = (32 + 63 \times 1.8)°F = (32 + 113.4)°F = 145.4°F$

5. How many degrees Fahrenheit is $t°C$?

   $t°C = (32 + 1.8t)°F$

Discussion (10 minutes)

Have students share their answers from Exercise 5. Select several students to explain how they derived the equation to convert between Celsius and Fahrenheit. Close that part of the Discussion by letting them know that they answered Question (1) that was posed at the beginning of the lesson:

(1) If t is a number, what is the degree in Fahrenheit that corresponds to $t°C$?

The following discussion answers Question (2):

(2) If t is a number, what is the degree in Fahrenheit that corresponds to $(-t)°C$?

- Now that Question (1) has been answered, let's begin thinking about Question (2). Where on the number line would we find a negative Celsius temperature?
  - *A negative Celsius temperature will be to the left of zero on the number line.*

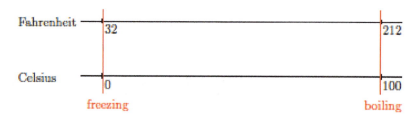

- Again, we will start simply. How can we determine the Fahrenheit temperature that corresponds to $-1°C$?

Provide students time to think, confirm with a partner, and then share with the class.

  - *We know that each unit on the Celsius scale is equal to $1.8°F$. Then $(-1)°C$ will equal $(32 - 1.8)°F = 30.2°F$.*

- How many degrees Fahrenheit corresponds to $(-15)°C$?

Provide students time to think, confirm with a partner, and then share with the class.

  - $(-15)°C = (32 - 15 \times 1.8)°F = (32 - 27)°F = 5°F$

- How many degrees Fahrenheit corresponds to $(-36)°C$?

Provide students time to think, confirm with a partner, and then share with the class.

  □ $(-36)°C = (32 - 36 \times 1.8)°F = (32 - 64.8)°F = -32.8°F$

- How many degrees Fahrenheit corresponds to $(-t)°C$?

Provide students time to think, confirm with a partner, and then share with the class.

  □ $(-t)°C = (32 - 1.8t)°F$

- Each of the previous four temperatures was negative. Then $(32 - 1.8t)°F$ can be rewritten as $(32 + 1.8(-t))°F$ where the second equation looks a lot like the one we wrote for $t°C$; that is, $t°C = (32 + 1.8t)°F$. Are they the same equation? In other words, given any number t, positive or negative, would the result be the correct answer? We already know that the equation works for positive Celsius temperatures, so now let's focus on negative Celsius temperatures. Use $t°C = (32 + 1.8t)°F$ where $t = -15$. We expect the same answer as before: $(-15)°C = (32 - 15 \times 1.8)°F = (32 - 27)°F = 5°F$. Show that it is true.

  □ $(-15)°C = (32 + 1.8(-15))°F = (32 + (-27)) = 5°F$

- Therefore, the equation $t°C = (32 + 1.8t)°F$ will work for any t.

- On a coordinate plane, if we let x be the given temperature, which in each case above has been given in Celsius and y be the temperature in Celsius, then we have the equation $y = x$. But when we let x be the given temperature in Celsius and y be the temperature in Fahrenheit, we have the equation $y = 32 + 1.8x$.

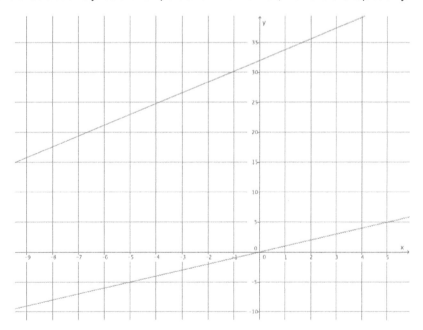

- Will these lines intersect? Explain?
  □ *Yes. They have different slopes, so at some point they will intersect.*

- What will that point of intersection represent?
  - *It represents the solution to the system* $\begin{cases} y = x \\ y = 1.8x + 32 \end{cases}$. *That point will represent when the given temperature is the same number in Celsius and in Fahrenheit.*
- Solve the system of equations algebraically to determine at what number $t°C = t°F$.
  - *Sample student work:*

$$\begin{cases} y = x \\ y = 1.8x + 32 \end{cases}$$

$$x = 1.8x + 32$$
$$-0.8x = 32$$
$$x = -40$$

At -40 degrees, the temperatures will be equal in both units. In other words, at -40 degrees Celsius, the temperature in Fahrenheit will also be -40 degrees.

Closing (3 minutes)

Summarize, or ask students to summarize, the main points from the lesson:

- We know how to use a linear equation in a real-world situation like converting between Celsius and Fahrenheit.
- We can use the computations we make for specific numbers to help us determine a general linear equation for a situation.

Exit Ticket (4 minutes)

Name _____ Date _____

Lesson 30: Conversion Between Celsius and Fahrenheit

Exit Ticket

Use the equation developed in class to answer the following questions:

1. How many degrees Fahrenheit is 11°C?

2. How many degrees Fahrenheit is −3°C?

3. Graph the equation developed in class, and use it to confirm your results from Problems 1 and 2.

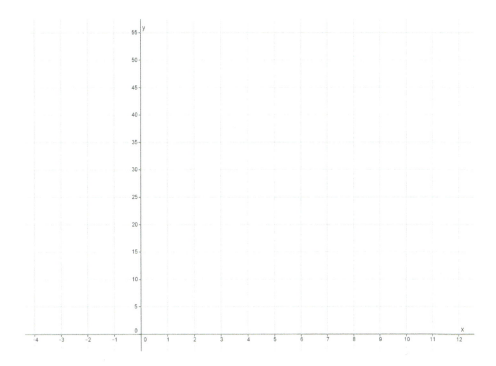

Exit Ticket Sample Solutions

Use the equation developed in class to answer the following questions:

1. How many degrees Fahrenheit is $11°C$?

$$11°C = (32 + 11 \times 1.8)°F$$
$$11°C = (32 + 19.8)°F$$
$$11°C = 51.8°F$$

2. How many degrees Fahrenheit is $-3°C$?

$$-3°C = (32 + (-3) \times 1.8)°F$$
$$-3°C = (32 - 5.4)°F$$
$$-3°C = 26.6°F$$

3. Graph the equation developed in class, and use it to confirm your results from Problems 1 and 2.

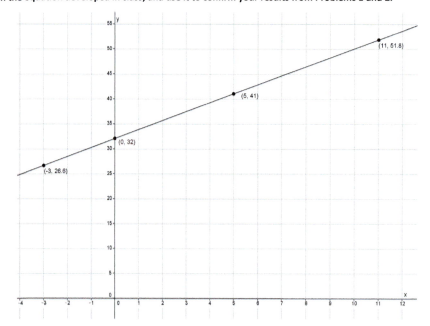

When I graph the equation developed in class, $t°C = (32 + 1.8t)°F$, the results from Problems 1 and 2 are on the line, confirming they are solutions to the equation.

A STORY OF RATIOS   Lesson 30   8•4

Problem Set Sample Solutions

1. Does the equation $t°C = (32 + 1.8t)°F$ work for any rational number t? Check that it does with $t = 8\frac{2}{3}$ and $t = -8\frac{2}{3}$.

$$\left(8\frac{2}{3}\right)°C = \left(32 + 8\frac{2}{3} \times 1.8\right)°F = (32 + 15.6)°F = 47.6°F$$

$$\left(-8\frac{2}{3}\right)°C = \left(32 + \left(-8\frac{2}{3}\right) \times 1.8\right)°F = (32 - 15.6)°F = 16.4°F$$

2. Knowing that $t°C = \left(32 + \frac{9}{5}t\right)°F$ for any rational t, show that for any rational number d, $d°F = \left(\frac{5}{9}(d - 32)\right)°C$.

   Since $d°F$ can be found by $\left(32 + \frac{9}{5}t\right)$, then $d = \left(32 + \frac{9}{5}t\right)$, and $d°F = t°C$. Substituting $d = \left(32 + \frac{9}{5}t\right)$ into $d°F$, we get

$$d°F = \left(32 + \frac{9}{5}t\right)°F$$
$$d = 32 + \frac{9}{5}t$$
$$d - 32 = \frac{9}{5}t$$
$$\frac{5}{9}(d - 32) = t.$$

   Now that we know $t = \frac{5}{9}(d - 32)$, then $d°F = \left(\frac{5}{9}(d - 32)\right)°C$.

3. Drake was trying to write an equation to help him predict the cost of his monthly phone bill. He is charged $35 just for having a phone, and his only additional expense comes from the number of texts that he sends. He is charged $0.05 for each text. Help Drake out by completing parts (a)–(f).

   a. How much was his phone bill in July when he sent 750 texts?

   $35 + 750(0.05) = 35 + 37.5 = 72.5$

   His bill in July was $72.50.

   b. How much was his phone bill in August when he sent 823 texts?

   $35 + 823(0.05) = 35 + 41.15 = 76.15$

   His bill in August was $76.15.

   c. How much was his phone bill in September when he sent 579 texts?

   $35 + 579(0.05) = 35 + 28.95 = 63.95$

   His bill in September was $63.95.

   d. Let y represent the total cost of Drake's phone bill. Write an equation that represents the total cost of his phone bill in October if he sends t texts.

   $y = 35 + t(0.05)$

Lesson 30:   Conversion Between Celsius and Fahrenheit   481

This work is derived from Eureka Math ™ and licensed by Great Minds. ©2015 Great Minds. eureka-math.org
G8-M4-TE-B3-1.3.0-07.2015

e. Another phone plan charges $20 for having a phone and $0.10 per text. Let y represent the total cost of the phone bill for sending t texts. Write an equation to represent his total bill.

$y = 20 + t(0.10)$

f. Write your equations in parts (d) and (e) as a system of linear equations, and solve. Interpret the meaning of the solution in terms of the phone bill.

$$\begin{cases} y = 35 + t(0.05) \\ y = 20 + t(0.10) \end{cases}$$

$$35 + (0.05)t = 20 + (0.10)t$$
$$15 + (0.05)t = (0.10)t$$
$$15 = 0.05t$$
$$300 = t$$

$$y = 20 + 300(0.10)$$
$$y = 50$$

The solution is $(300, 50)$, meaning that when Drake sends 300 texts, the cost of his bill will be $\$50$ using his current phone plan or the new one.

A STORY OF RATIOS

Mathematics Curriculum

GRADE 8 • MODULE 4

Optional Topic E
Pythagorean Theorem

8.G.B.7, 8.EE.C.8

Focus Standards:	8.G.B.7	Apply the Pythagorean Theorem to determine unknown side lengths in right triangles in real-world and mathematical problems in two and three dimensions.
	8.EE.C.8	Analyze and solve pairs of simultaneous linear equations.
		a. Understand that solutions to a system of two linear equations in two variables correspond to points of intersection of their graphs, because points of intersection satisfy both equations simultaneously.
		b. Solve systems of two linear equations in two variables algebraically, and estimate solutions by graphing the equations. Solve simple cases by inspection. *For example, $3x + 2y = 5$ and $3x + 2y = 6$ have no solution because $3x + 2y$ cannot simultaneously be 5 and 6.*
		c. Solve real-world and mathematical problems leading to two linear equations in two variables. *For example, given coordinates for two pairs of points, determine whether the line through the first pair of points intersects the line through the second pair.*
Instructional Days:	1	
	Lesson 31:	System of Equations Leading to Pythagorean Triples (S)[1]

Lesson 31 shows students how to apply what they learned about systems of linear equations to find a Pythagorean triple (**8.G.B.7**, **8.EE.C.8b**). The Babylonian method of generating Pythagorean triples, described in Lesson 31, uses a system of linear equations.

---

[1]Lesson Structure Key: **P**-Problem Set Lesson, **M**-Modeling Cycle Lesson, **E**-Exploration Lesson, **S**-Socratic Lesson

Lesson 31: System of Equations Leading to Pythagorean Triples

Student Outcomes

- Students know that a Pythagorean triple can be obtained by multiplying any known triple by a common whole number. Students use this method to generate Pythagorean triples.
- Students use a system of equations to find three numbers, a, b, and c, so that $a^2 + b^2 = c^2$.

Lesson Notes

This lesson is optional as it includes content related to the Pythagorean theorem. The purpose of this lesson is to demonstrate an application of systems of linear equations to other content in the curriculum. Though Pythagorean triples are not part of the standard for the grade, it is an interesting topic and should be shared with students if time permits.

Classwork

Discussion (10 minutes)

- A New York publicist, George Arthur Plimpton, bought a clay tablet from an archaeological dealer for $10 in 1922. This tablet was donated to Columbia University in 1936 and became known by its catalog number, Plimpton 322. What made this tablet so special was not just that it was 4,000 years old but that it showed a method for finding Pythagorean triples. It was excavated near old Babylonia, which is now Iraq.

Plimpton 322, Rare Book & Manuscript Library, Columbia University in the City of New York. Photo C. Proust

- Any three numbers, a, b, c, that satisfy $a^2 + b^2 = c^2$ are considered a triple, but when the three numbers are positive integers, then they are known as *Pythagorean triples*. It is worth mentioning that one of the Pythagorean triples found on the tablet was 12,709, 13,500, 18,541.
- An easy-to-remember Pythagorean triple is 3, 4, 5. (Quickly verify for students that 3, 4, 5 is a triple.) To generate another Pythagorean triple, we need only to multiply each of the numbers 3, 4, 5 by the same whole number. For example, the numbers 3, 4, 5 when each is multiplied by 2, the result is the triple 6, 8, 10. (Again, quickly verify that 6, 8, 10 is a triple.) Let's think about why this is true in a geometric context.

  Shown below are the two right triangles.

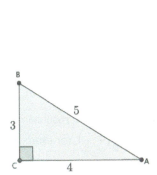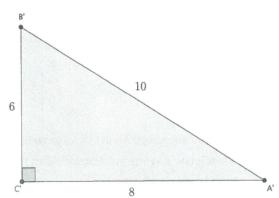

- Discuss with your partners how the method for finding Pythagorean triples can be explained mathematically.
  - △ 'B'C' can be obtained by dilating △ BC by a scale factor of 2. Each triangle has a right angle with corresponding sides that are equal in ratio to the same constant, 2. That is how we know that these triangles are similar. The method for finding Pythagorean triples can be directly tied to our understanding of dilation and similarity. Each triple is just a set of numbers that represent a dilation of △ BC by a whole-number scale factor.
- Of course, we can also find triples by using a scale factor $0 < r < 1$, but since it produces a set of numbers that are not whole numbers, they are not considered to be Pythagorean triples. For example, if $r = \frac{1}{10}$, then a triple using side lengths 3, 4, 5 is 0.3, 0.4, 0.5.

Exercises 1–3 (5 minutes)

Students complete Exercises 1–3 independently. Allow students to use a calculator to verify that they are identifying triples.

> **Exercises**
>
> 1. Identify two Pythagorean triples using the known triple 3, 4, 5 (other than 6, 8, 10).
>
>   *Answers will vary. Accept any triple that is a whole number multiple of 3, 4, 5.*
>
> 2. Identify two Pythagorean triples using the known triple 5, 12, 13.
>
>   *Answers will vary. Accept any triple that is a whole number multiple of 5, 12, 13.*
>
> 3. Identify two triples using either 3, 4, 5 or 5, 12, 13.
>
>   *Answers will vary.*

Lesson 31: System of Equations Leading to Pythagorean Triples

Discussion (10 minutes)

- Pythagorean triples can also be explained algebraically. For example, assume a, b, c represent a Pythagorean triple. Let m be a positive integer. Then, by the Pythagorean theorem, $a^2 + b^2 = c^2$:

$$(ma)^2 + (mb)^2 = m^2a^2 + m^2b^2 \quad \text{By the second law of exponents}$$
$$= m^2(a^2 + b^2) \quad \text{By the distributive property}$$
$$= m^2c^2 \quad \text{By substitution } (a^2 + b^2 = c^2)$$
$$= (mc)^2$$

Our learning of systems of linear equations leads us to another method for finding Pythagorean triples, and it is actually the method that was discovered on the tablet Plimpton 322.

- Consider the system of linear equations:

$$\begin{cases} x + y = \dfrac{t}{s} \\ x - y = \dfrac{s}{t} \end{cases}$$

where s and t are positive integers and $t > s$. Incredibly, the solution to this system results in a Pythagorean triple. When the solution is written as fractions with the same denominator, $\left(\dfrac{c}{b}, \dfrac{a}{b}\right)$, for example, the numbers a, b, c are a Pythagorean triple.

- To make this simpler, let's replace s and t with 1 and 2, respectively. Then we have

$$\begin{cases} x + y = \dfrac{2}{1} \\ x - y = \dfrac{1}{2} \end{cases}.$$

- Which method should we use to solve this system? Explain.
    - We should add the equations together to eliminate the variable y.
- By the elimination method, we have

$$x + y + x - y = 2 + \dfrac{1}{2}$$
$$2x = \dfrac{5}{2}$$
$$x = \dfrac{5}{4}.$$

Now we can substitute x into one of the equations to find y.

$$\dfrac{5}{4} + y = 2$$
$$y = 2 - \dfrac{5}{4}$$
$$y = \dfrac{3}{4}$$

Then the solution to the system is $\left(\dfrac{5}{4}, \dfrac{3}{4}\right)$. When a solution is written as fractions with the same denominator, $\left(\dfrac{c}{b}, \dfrac{a}{b}\right)$, for example, it represents the Pythagorean triple a, b, c. Therefore, our solution yields the triple 3, 4, 5.

The remaining time can be used to complete Exercises 4–7 where students practice finding triples using the system of linear equations just described or with the Discussion below, which shows the solution to the general system (without using concrete numbers for s and t).

Exercises 4–7 (10 minutes)

These exercises are to be completed in place of the Discussion. Have students complete Exercises 4–7 independently.

Use the system $\begin{cases} x+y = \dfrac{t}{s} \\ x-y = \dfrac{s}{t} \end{cases}$ to find Pythagorean triples for the given values of s and t. Recall that the solution in the form of $\left(\dfrac{c}{b}, \dfrac{a}{b}\right)$ is the triple a, b, c.

4. $s = 4, t = 5$

$$\begin{cases} x+y = \dfrac{5}{4} \\ x-y = \dfrac{4}{5} \end{cases}$$

$$x+y+x-y = \dfrac{5}{4} + \dfrac{4}{5}$$
$$2x = \dfrac{5}{4} + \dfrac{4}{5}$$
$$2x = \dfrac{41}{20}$$
$$x = \dfrac{41}{40}$$

$$\dfrac{41}{40} + y = \dfrac{5}{4}$$
$$y = \dfrac{5}{4} - \dfrac{41}{40}$$
$$y = \dfrac{9}{40}$$

Then the solution is $\left(\dfrac{41}{40}, \dfrac{9}{40}\right)$, and the triple is 9, 40, 41.

5. $s = 7, t = 10$

$$\begin{cases} x+y = \dfrac{10}{7} \\ x-y = \dfrac{7}{10} \end{cases}$$

$$x+y+x-y = \dfrac{10}{7} + \dfrac{7}{10}$$
$$2x = \dfrac{149}{70}$$
$$x = \dfrac{149}{140}$$

$$\dfrac{149}{140} + y = \dfrac{10}{7}$$
$$y = \dfrac{10}{7} - \dfrac{149}{140}$$
$$y = \dfrac{51}{140}$$

Then the solution is $\left(\dfrac{149}{140}, \dfrac{51}{140}\right)$, and the triple is 51, 140, 149.

Lesson 31: System of Equations Leading to Pythagorean Triples

6. $s = 1, t = 4$

$$\begin{cases} x + y = \dfrac{4}{1} \\ x - y = \dfrac{1}{4} \end{cases}$$

$x + y + x - y = 4 + \dfrac{1}{4}$

$2x = \dfrac{17}{4}$

$x = \dfrac{17}{8}$

$\dfrac{17}{8} + y = \dfrac{4}{1}$

$y = 4 - \dfrac{17}{8}$

$y = \dfrac{15}{8}$

Then the solution is $\left(\dfrac{17}{8}, \dfrac{15}{8}\right)$, and the triple is 15, 8, 17.

7. Use a calculator to verify that you found a Pythagorean triple in each of the Exercises 4–6. Show your work below.

For the triple 9, 40, 41:

$$9^2 + 40^2 = 41^2$$
$$81 + 1600 = 1681$$
$$1681 = 1681$$

For the triple 51, 140, 149:

$$51^2 + 140^2 = 149^2$$
$$2601 + 19600 = 22201$$
$$22201 = 22201$$

For the triple 15, 8, 17:

$$15^2 + 8^2 = 17^2$$
$$225 + 64 = 289$$
$$289 = 289$$

Discussion (10 minutes)

This Discussion is optional and replaces Exercises 4–7.

- Now we solve the system generally.

$$\begin{cases} x + y = \dfrac{t}{s} \\ x - y = \dfrac{s}{t} \end{cases}$$

- Which method should we use to solve this system? Explain.
    - *We should add the equations together to eliminate the variable y.*
- By the elimination method, we have

$$x + y + x - y = \dfrac{t}{s} + \dfrac{s}{t}$$

$$2x = \dfrac{t}{s} + \dfrac{s}{t}.$$

To add the fractions, we will need the denominators to be the same. So, we use what we know about equivalent fractions and multiply the first fraction by $\frac{t}{t}$ and the second fraction by $\frac{s}{s}$:

$$2x = \frac{t}{s}\left(\frac{t}{t}\right) + \frac{s}{t}\left(\frac{s}{s}\right)$$

$$2x = \frac{t^2}{st} + \frac{s^2}{st}$$

$$2x = \frac{t^2 + s^2}{st}$$

Now we multiply both sides of the equation by $\frac{1}{2}$:

$$\frac{1}{2}(2x) = \frac{1}{2}\left(\frac{t^2 + s^2}{st}\right)$$

$$x = \frac{t^2 + s^2}{2st}$$

Now that we have a value for x, we can solve for y as usual, but it is simpler to go back to the system:

$$\begin{cases} x + y = \frac{t}{s} \\ x - y = \frac{s}{t} \end{cases}$$

It is equivalent to the system

$$\begin{cases} x = \frac{t}{s} - y \\ x = \frac{s}{t} + y \end{cases}$$

$$\frac{t}{s} - y = \frac{s}{t} + y$$

$$\frac{t}{s} = \frac{s}{t} + 2y$$

$$\frac{t}{s} - \frac{s}{t} = 2y,$$

which is very similar to what we have done before when we solved for x. Therefore,

$$y = \frac{t^2 - s^2}{2st}.$$

The solution to the system is $\left(\frac{t^2+s^2}{2st}, \frac{t^2-s^2}{2st}\right)$. When a solution is written as fractions with the same denominator, $\left(\frac{c}{b}, \frac{a}{b}\right)$, for example, it represents the Pythagorean triple a, b, c. Therefore, our solution yields the triple $t^2 - s^2, 2st, t^2 + s^2$.

Closing (5 minutes)

Summarize, or ask students to summarize, the main points from the lesson:

- We know how to find an infinite number of Pythagorean triples: Multiply a known triple by a whole number.
- We know that if the numbers a, b, c are not whole numbers, they can still be considered a triple, just not a Pythagorean triple.
- We know how to use a system of linear equations, just like the Babylonians did 4,000 years ago, to find Pythagorean triples.

Lesson Summary

A Pythagorean triple is a set of three positive integers that satisfies the equation $a^2 + b^2 = c^2$.

An infinite number of Pythagorean triples can be found by multiplying the numbers of a known triple by a whole number. For example, 3, 4, 5 is a Pythagorean triple. Multiply each number by 7, and then you have 21, 28, 35, which is also a Pythagorean triple.

The system of linear equations, $\begin{cases} x + y = \dfrac{t}{s} \\ x - y = \dfrac{s}{t} \end{cases}$, can be used to find Pythagorean triples, just like the Babylonians did 4,000 years ago.

Exit Ticket (5 minutes)

A STORY OF RATIOS   Lesson 31   8•4

Name _____   Date _____

Lesson 31: System of Equations Leading to Pythagorean Triples

Exit Ticket

Use a calculator to complete Problems 1–3.

1. Is 7, 20, 21 a Pythagorean triple? Is $1, \frac{15}{8}, \frac{17}{8}$ a Pythagorean triple? Explain.

2. Identify two Pythagorean triples using the known triple 9, 40, 41.

3. Use the system $\begin{cases} x + y = \frac{t}{s} \\ x - y = \frac{s}{t} \end{cases}$ to find Pythagorean triples for the given values of $s = 2$ and $t = 3$. Recall that the solution in the form of $\left(\frac{c}{b}, \frac{a}{b}\right)$ is the triple a, b, c. Verify your results.

A STORY OF RATIOS | Lesson 31 | 8•4

Exit Ticket Sample Solutions

Use a calculator to complete Problems 1–3.

1. Is 7, 20, 21 a Pythagorean triple? Is $1, \frac{15}{8}, \frac{17}{8}$ a Pythagorean triple? Explain.

   The set of numbers 7, 20, 21 is not a Pythagorean triple because $7^2 + 20^2 \neq 21^2$.

   The set of numbers $1, \frac{15}{8}, \frac{17}{8}$ is not a Pythagorean triple because the numbers $\frac{15}{8}$ and $\frac{17}{8}$ are not whole numbers.

   But they are a triple because $1^2 + \left(\frac{15}{8}\right)^2 = \left(\frac{17}{8}\right)^2$.

2. Identify two Pythagorean triples using the known triple 9, 40, 41.

   Answers will vary. Accept any triple that is a whole number multiple of 9, 40, 41.

3. Use the system $\begin{cases} x + y = \frac{t}{s} \\ x - y = \frac{s}{t} \end{cases}$ to find Pythagorean triples for the given values of $s = 2$ and $t = 3$. Recall that the solution in the form of $\left(\frac{c}{b}, \frac{a}{b}\right)$ is the triple a, b, c. Verify your results.

   $$\begin{cases} x + y = \frac{3}{2} \\ x - y = \frac{2}{3} \end{cases}$$

   $$x + y + x - y = \frac{3}{2} + \frac{2}{3}$$
   $$2x = \frac{13}{6}$$
   $$x = \frac{13}{12}$$

   $$\frac{13}{12} + y = \frac{3}{2}$$
   $$y = \frac{3}{2} - \frac{13}{12}$$
   $$y = \frac{5}{12}$$

   Then the solution is $\left(\frac{13}{12}, \frac{5}{12}\right)$, and the triple is 5, 12, 13.

   $$5^2 + 12^2 = 13^2$$
   $$25 + 144 = 169$$
   $$169 = 169$$

Problem Set Sample Solutions

Students practice finding triples using both methods discussed in this lesson.

1. Explain in terms of similar triangles why it is that when you multiply the known Pythagorean triple 3, 4, 5 by 12, it generates a Pythagorean triple.

   The triangle with lengths 3, 4, 5 is similar to the triangle with lengths 36, 48, 60. They are both right triangles whose corresponding side lengths are equal to the same constant.

   $$\frac{36}{3} = \frac{48}{4} = \frac{60}{5} = 12$$

   Therefore, the triangles are similar, and we can say that there is a dilation from some center with scale factor $r = 12$ that makes the triangles congruent.

492   Lesson 31:   System of Equations Leading to Pythagorean Triples

2. Identify three Pythagorean triples using the known triple 8, 15, 17.

   Answers will vary. Accept any triple that is a whole number multiple of 8, 15, 17.

3. Identify three triples (numbers that satisfy $a^2 + b^2 = c^2$, but a, b, c are not whole numbers) using the triple 8, 15, 17.

   Answers will vary. Accept any triple that is not a set of whole numbers.

Use the system $\begin{cases} x + y = \frac{t}{s} \\ x - y = \frac{s}{t} \end{cases}$ to find Pythagorean triples for the given values of s and t. Recall that the solution, in the form of $\left(\frac{c}{b}, \frac{a}{b}\right)$, is the triple, a, b, c.

4. $s = 2, t = 9$

$$\begin{cases} x + y = \frac{9}{2} \\ x - y = \frac{2}{9} \end{cases}$$

$$x + y + x - y = \frac{9}{2} + \frac{2}{9}$$
$$2x = \frac{85}{18}$$
$$x = \frac{85}{36}$$

$$\frac{85}{36} + y = \frac{9}{2}$$
$$y = \frac{9}{2} - \frac{85}{36}$$
$$y = \frac{77}{36}$$

Then the solution is $\left(\frac{85}{36}, \frac{77}{36}\right)$, and the triple is 77, 36, 85.

5. $s = 6, t = 7$

$$\begin{cases} x + y = \frac{7}{6} \\ x - y = \frac{6}{7} \end{cases}$$

$$x + y + x - y = \frac{7}{6} + \frac{6}{7}$$
$$2x = \frac{85}{42}$$
$$x = \frac{85}{84}$$

$$\frac{85}{84} + y = \frac{7}{6}$$
$$y = \frac{7}{6} - \frac{85}{84}$$
$$y = \frac{13}{84}$$

Then the solution is $\left(\frac{85}{84}, \frac{13}{84}\right)$, and the triple is 13, 84, 85.

6. $s = 3, t = 4$

$$\begin{cases} x + y = \dfrac{4}{3} \\ x - y = \dfrac{3}{4} \end{cases}$$

$$x + y + x - y = \dfrac{4}{3} + \dfrac{3}{4}$$
$$2x = \dfrac{25}{12}$$
$$x = \dfrac{25}{24}$$

$$\dfrac{25}{24} + y = \dfrac{4}{3}$$
$$y = \dfrac{4}{3} - \dfrac{25}{24}$$
$$y = \dfrac{7}{24}$$

Then the solution is $\left(\dfrac{25}{24}, \dfrac{7}{24}\right)$, and the triple is 7, 24, 25.

7. Use a calculator to verify that you found a Pythagorean triple in each of the Problems 4–6. Show your work.

    For the triple 77, 36, 85:

    $$77^2 + 36^2 = 85^2$$
    $$5929 + 1296 = 7225$$
    $$7225 = 7225$$

    For the triple 13, 84, 85:

    $$13^2 + 84^2 = 85^2$$
    $$169 + 7056 = 7225$$
    $$7225 = 7225$$

    For the triple 7, 24, 25:

    $$7^2 + 24^2 = 25^2$$
    $$49 + 576 = 625$$
    $$625 = 625$$

Name _____   Date _____

1. Use the graph below to answer parts (a)–(c).

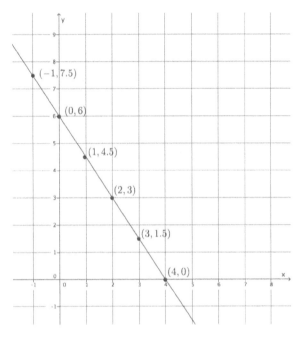

   a. Use any pair of points to calculate the slope of the line.

   b. Use a different pair of points to calculate the slope of the line.

   c. Explain why the slopes you calculated in parts (a) and (b) are equal.

2. Jeremy rides his bike at a rate of 12 miles per hour. Below is a table that represents the number of hours and miles Kevin rides. Assume both bikers ride at a constant rate.

Time in Hours (x)	Distance in Miles (y)
1.5	17.25
2	23
3.5	40.25
4	46

a. Which biker rides at a greater speed? Explain your reasoning.

b. Write an equation for a third biker, Lauren, who rides twice as fast as Kevin. Use y to represent the number of miles Lauren travels in x hours. Explain your reasoning.

c. Create a graph of the equation in part (b).

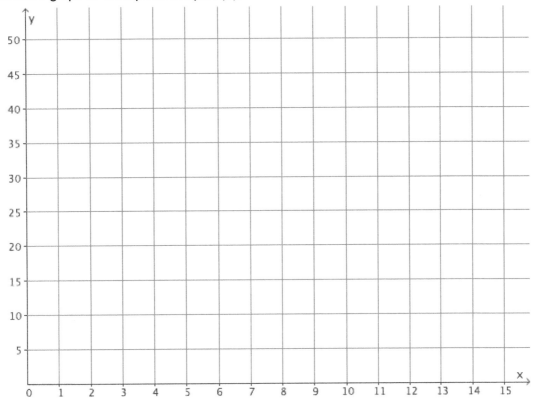

d. Calculate the slope of the line in part (c), and interpret its meaning in this situation.

3. The cost of five protractors is $14.95 at Store A. The graph below compares the cost of protractors at Store A with the cost at Store B.

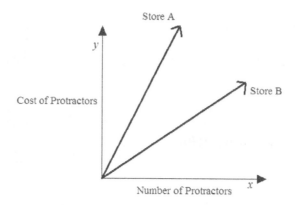

Estimate the cost of one protractor at Store B. Use evidence from the graph to justify your answer.

4. Given the equation $3x + 9y = -8$, write a second linear equation to create a system that:

   a. Has exactly one solution. Explain your reasoning.

   b. Has no solution. Explain your reasoning.

   c. Has infinitely many solutions. Explain your reasoning.

   d. Interpret the meaning of the solution, if it exists, in the context of the graph of the following system of equations.

   $$\begin{cases} -5x + 2y = 10 \\ 10x - 4y = -20 \end{cases}$$

5. Students sold 275 tickets for a fundraiser at school. Some tickets are for children and cost $3, while the rest are adult tickets that cost $5. If the total value of all tickets sold was $1,025, how many of each type of ticket was sold?

6.

   a. Determine the equation of the line connecting the points $(0, -1)$ and $(2, 3)$.

   b. Will the line described by the equation in part (a) intersect the line passing through the points $(-2, 4)$ and $(-3, 3)$? Explain why or why not.

7. Line l_1 and line l_2 are shown on the graph below. Use the graph to answer parts (a)–(f).

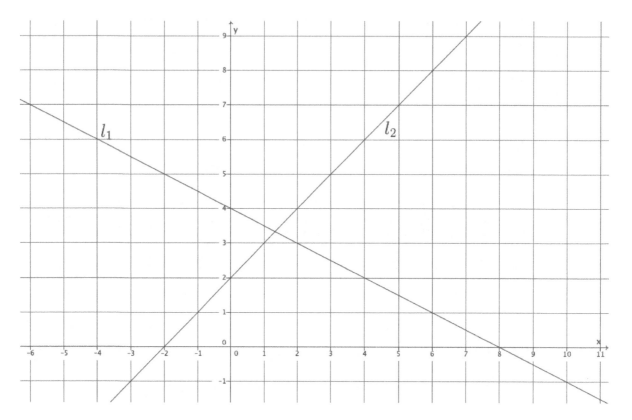

a. What is the y-intercept of l_1?

b. What is the y-intercept of l_2?

c. Write a system of linear equations representing lines l_1 and l_2.

d. Use the graph to estimate the solution to the system.

e. Solve the system of linear equations algebraically.

f. Show that your solution from part (e) satisfies both equations.

A STORY OF RATIOS   End-of-Module Assessment Task   8•4

A Progression Toward Mastery					
Assessment Task Item		STEP 1 Missing or incorrect answer and little evidence of reasoning or application of mathematics to solve the problem.	STEP 2 Missing or incorrect answer but evidence of some reasoning or application of mathematics to solve the problem.	STEP 3 A correct answer with some evidence of reasoning or application of mathematics to solve the problem, OR an incorrect answer with substantial evidence of solid reasoning or application of mathematics to solve the problem.	STEP 4 A correct answer supported by substantial evidence of solid reasoning or application of mathematics to solve the problem.
1	a–b 8.EE.B.5	Student makes no attempt to find the slope in part (a) and/or part (b).	Student computes the slope in parts (a) and (b) but makes computational errors leading to slopes that are not equal. Student may have used the same two points for both parts (a) and (b).	Student computes slope both times but may have forgotten to include the negative sign or makes another simple computational error. Student finds the slopes in both parts (a) and (b) to be equal.	Student correctly computes the slope both times and finds $m = -\frac{3}{2}$ (or an equivalent fraction). Student finds the slopes in both parts (a) and (b) to be equal.
	c 8.EE.B.6	Student makes no attempt to answer the question.	Student states that the slopes in parts (a) and (b) are not equal.	Student makes a weak argument by stating that the slopes are equal because the fractions are equal or that the fractions representing the slope are equivalent.	Student makes a convincing argument and references similar triangles to explain why the slopes between any two points on a line are equal.
2	a 8.EE.B.5	Student makes no attempt to answer the question or writes "Kevin" or "Jeremy" with no evidence of an application of mathematics to solve the problem.	Student writes an incorrect answer but shows some evidence of reasoning in the explanation.	Student writes the correct answer that Jeremy rides at a greater speed. Student explanation lacks precision or is incorrect. For example, student may have written that Jeremy travels a farther distance in two hours instead of referencing the rates of each biker.	Student writes the correct answer that Jeremy rides at a greater speed. Student provides a strong mathematical explanation as to which biker rides at a greater speed by referencing a graph (slopes of each line where one slope is steeper) or a numerical comparison of their rates.

504   Module 4:   Linear Equations

	b–d 8.EE.B.5	Student makes little or no attempt to complete parts (b)–(d). Student may have plotted points on the graph with no relevance to the problem.	Student writes an incorrect equation in part (b) and/or graphs the equation incorrectly and/or calculates the slope incorrectly. Student does not connect the slope of the line to Lauren's rate.	Student correctly identifies the equation, graphs and calculates the slope, and identifies it as Lauren's rate, but the answer shows no evidence of reasoning in part (b). OR Student makes a mistake in writing the equation for part (b), which leads to an incorrect graph and slope in parts (c)–(d).	Student correctly writes the equation, $y = 23x$, in part (b) or writes an equivalent equation. Student explains "twice as fast" in terms of distance traveled for a given time interval compared to the data for Kevin given in the table. Student correctly graphs the situation in part (c) and correctly identifies the slope of the line, 23, as the rate that Lauren rides for part (d).
3	8.EE.B.5	Student makes no attempt to answer the question. OR Student writes a dollar amount with no explanation.	Student may or may not have correctly calculated the unit rate of protractors for Store A. Student writes an estimate for Store B but does not justify the estimate using evidence from the graph.	Student uses the information provided about Store A to determine the unit rate of protractors but may have made a computational error leading to a poor estimate. Student may or may not have used evidence from the graph to justify the estimate or makes a weak connection between the estimate and the graph.	Student uses the information provided about Store A to determine the unit rate of protractors and references the unit rate at Store A in the justification of the estimate. Student writes an estimate that makes sense (e.g., less than Store A, about half as much) and uses evidence from the graph in the justification (e.g., comparison of slopes, size of angles).
4	a–c 8.EE.C.7a 8.EE.C.8	Student makes no attempt to answer any parts of (a)–(c). OR Student only rewrites the given equation as an answer.	Student answers at least one part of (a)–(c) correctly. Student may have left two parts blank. Answer may or may not show evidence of reasoning.	Student answers at least two parts of (a)–(c) correctly. Student may have left one part blank. Student explains reasoning in at least two parts of (a)–(c), noting the characteristics required to achieve the desired number of solutions.	Student provides a correct equation and explanation for each of the parts of (a)–(c). Specifically, for part (a), an equation that represents a distinct line from the given equation has a slope different from $-\frac{1}{3}$; for part (b), an equation that represents a line parallel to the given equation has the same slope; and for part (c), an equation that represents the same line as the given equation whose graphs coincide.

Module 4: Linear Equations

505

	d 8.EE.C.8	Student makes little or no attempt to answer the question. Student does not provide a mathematical explanation or apply any mathematical reasoning to support the answer.	Student gives an incorrect answer. Student may have said that the point of intersection of the lines is the solution to the system or that there is no solution because the lines are parallel.	Student may have tried to find the solution algebraically. Student states that there are infinitely many solutions to the system but may not have referenced what the graph would look like (i.e., each equation produced the same line on the graph). Student supplies weak mathematical reasoning to support the answer.	Student correctly states that the graphs of the equations produce the same line. Student explains that one equation can be obtained by the other by multiplying the first equation by -2 or the second equation by $-\frac{1}{2}$. OR Student explains that both lines had the same slope of $\frac{5}{2}$ and the same y-intercept of $(0, 5)$. Student supplies strong mathematical reasoning to support the answer.
5	8.EE.C.8	Student makes little or no attempt to write and solve a system of linear equations.	Student may have written an incorrect system of equations to represent the situation. Student may or may not have defined the variables. Student may have used another strategy to determine the numbers of tickets of each type that were sold. There is some evidence of mathematical reasoning.	Student correctly writes a system of linear equations to represent the situation but makes a computational error leading to an incorrect solution. OR Student correctly writes and solves a system but does not define the variables.	Student correctly writes and solves a system of linear equations to solve the problem. Student defines the variables used in the system. Student states clearly that 175 children's tickets and 100 adults' tickets were sold.
6	a 8.EE.C.8	Student makes little or no attempt to write the equation.	Student incorrectly computes the slope of the line as something other than 2. Student does not write the correct equation of the line.	Student uses the points to correctly determine the slope of the line as 2 but may have written an incorrect equation.	Student uses the points to determine the slope of the line and then writes the equation of the line passing through those two points as $y = 2x - 1$ or equivalent.

	b 8.EE.C.8	Student makes little or no attempt to answer the question. OR Student responds with *yes* or *no* only.	Student incorrectly computes the slope of the line as something other than 1 and may or may not have drawn an incorrect conclusion about whether or not the lines would intersect.	Student uses the points to correctly determine the slope of the line as 1 but makes an incorrect conclusion about whether or not the lines would intersect. OR Student makes a computational error for the slope and draws the wrong conclusion about the lines. OR Student says the lines would intersect but does not provide an explanation.	Students uses the points to determine the slope as 1 and correctly concludes that the lines intersect because the slopes are different.
7	**a–b** 8.EE.C.8	Student leaves both parts (a) and (b) blank. OR Student identifies coordinates that do not fall on either the x- or y-axis.	Student identifies one of the two y-intercepts but may have inversed the coordinates. Student leaves either (a) or (b) blank.	Student identifies the y-intercepts of l_1 and l_2 but switches the coordinates, i.e., $(4,0)$ and $(2,0)$, or identifies the y-intercept of l_1 as $(0,2)$ and l_2 as $(0,4)$.	Student correctly identifies the y-intercepts of l_1 and l_2 as $(0,4)$ and $(0,2)$, respectively.
	c–d 8.EE.C.8	Student leaves the item blank. OR Student only writes one equation that may or may not have represented one of the lines on the graph. Student may or may not have written an estimate or writes an estimate where the x-value is not between 1 and 2 and the y-value is not between 3 and 4.	Student writes two equations to represent the system, but the equations do not represent the lines on the graph. Student may or may not have written an estimate or writes an estimate where the x-value is not between 1 and 2 and the y-value is not between 3 and 4.	Student writes a system of equations, but one of the equations is written incorrectly. Student writes an estimate where the x-value is between 1 and 2 and the y-value is between 3 and 4.	Student correctly writes the system as $\begin{cases} x - y = -2 \\ x + 2y = 8 \end{cases}$ or a system equivalent to this given one. Student writes an estimate where the x-value is between 1 and 2 and the y-value is between 3 and 4.
	e–f 8.EE.C.8	Student is unable to solve the system algebraically.	Student solves the system algebraically but makes serious computational errors leading to an incorrect solution. Student is unable to complete part (f) or notices an error and does not correct it.	Student solves the system but may have made a computational error leading to an incorrect x- or y-coordinate. Student verifies the solution in part (f). Student makes a computational error and believes the solution is correct.	Student correctly solves the system and identifies the solution as $\left(\frac{4}{3}, \frac{10}{3}\right)$. Student verifies the solution in part (f).

Module 4:   Linear Equations

Name _____   Date _____

1. Use the graph below to answer parts (a)–(c).

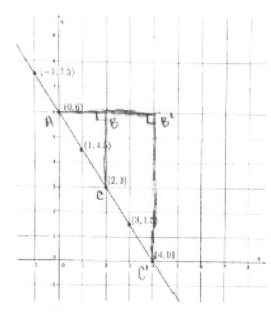

   a. Use any pair of points to calculate the slope of the line.

   $$m = \frac{6-3}{0-2} = \frac{3}{-2} = -\frac{3}{2}$$

   b. Use a different pair of points to calculate the slope of the line.

   $$m = \frac{6-0}{0-4} = \frac{6}{-4} = -\frac{3}{2}$$

   c. Explain why the slopes you calculated in parts (a) and (b) are equal.

   THE SLOPES ARE EQUAL BECAUSE THE SLOPE TRIANGLES ARE SIMILAR, $\triangle ABC \sim \triangle AB'C'$. EACH TRIANGLE HAS A 90° ANGLE AT $\angle ABC$ & $\angle AB'C'$, RESPECTIVELY. THEY ARE 90° BECAUSE THEY ARE AT THE INTERSECTION OF THE GRID LINES. BOTH TRIANGLES SHARE $\angle BAC$. BY THE AA CRITERION $\triangle ABC \sim \triangle AB'C'$, WHICH MEANS THEIR CORRESPONDING SIDES ARE EQUAL IN RATIO:

   $\frac{|B'C'|}{|BC|} = \frac{|AB'|}{|AB|}$ WHICH IS THE SAME AS $\frac{|B'C'|}{|AB'|} = \frac{|BC|}{|AB|}$ WHERE

   $-\frac{|B'C'|}{|AB'|}$ IS THE SLOPE IN (b) AND $-\frac{|BC|}{|AB|}$ IS THE SLOPE IN (a).

2. Jeremy rides his bike at a rate of 12 miles per hour. Below is a table that represents the number of hours and miles Kevin rides. Assume both bikers ride at a constant rate.

Time in Hours (x)	Distance in Miles (y)
1.5	17.25
2	23
3.5	40.25
4	46

a. Which biker rides at a greater speed? Explain your reasoning.

LET y BE THE DISTANCE TRAVELED AND x BE THE NUMBER OF HOURS. THEN FOR JEREMY, $\frac{y}{x} = \frac{12}{1} \Rightarrow y = 12x$.

FOR KEVIN, $\frac{46-23}{4-2} = \frac{23}{2} = 11.5$, THEN $y = 11.5x$

WHEN YOU COMPARE THEIR RATES, $12 > 11.5$, THEREFORE JEREMY RIDES AT A GREATER SPEED.

GRAPHICALLY:

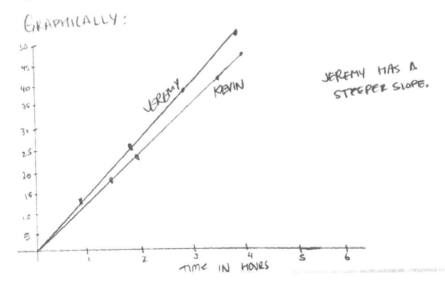

JEREMY HAS A STEEPER SLOPE.

b. Write an equation for a third biker, Lauren, who rides twice as fast as Kevin. Use y to represent the number of miles Lauren travels in x hours. Explain your reasoning.

"TWICE AS FAST" MEANS LAUREN GOES TWICE THE DISTANCE IN THE SAME TIME. THEN IN 2 HOURS SHE RIDES 46 MILES AND IN 4 HOURS, 92 MILES. IF y IS THE TOTAL DISTANCE IN x HOURS, $y = \frac{46}{2}x$

$y = 23x$.

c. Create a graph of the equation in part (b).

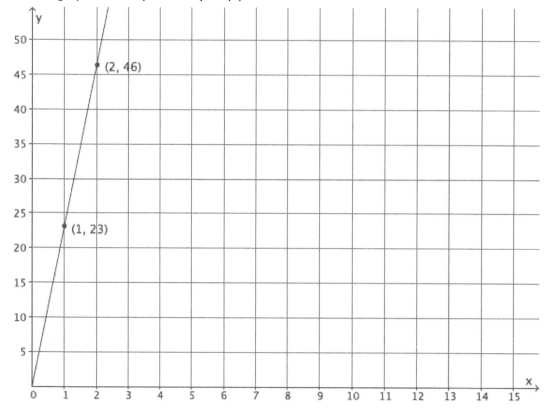

d. Calculate the slope of the line in part (c), and interpret its meaning in this situation.

$m = \frac{46-23}{2-1} = \frac{23}{1}$

THE SLOPE IS THE RATE THAT LAUREN RIDES, 23 MILES PER HOUR.

3. The cost of five protractors is $14.95 at Store A. The graph below compares the cost of protractors at Store A with the cost at Store B.

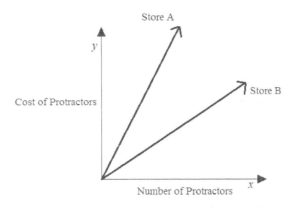

Estimate the cost of one protractor at Store B. Use evidence from the graph to justify your answer.

$$5 \overline{)14.95} = 2.99$$

THE COST OF PROTRACTORS AT STORE B IS PROBABLY ABOUT $1.50. STORE A CHARGES $2.99 PER PROTRACTOR AND IT LOOKS LIKE THE SLOPE FOR STORE B IS ABOUT HALF OF THE SLOPE FOR STORE A.

* ANSWERS WILL VARY

4. Given the equation $3x + 9y = -8$, write a second linear equation to create a system that:

   a. Has exactly one solution. Explain your reasoning.

   $4x + 9y = -10$ THIS EQUATION HAS A SLOPE DIFFERENT FROM $3x + 9y = -8$ SO THE GRAPHS OF THE EQUATIONS WILL INTERSECT.

   b. Has no solution. Explain your reasoning.

   $x + 3y = 10$ THIS EQUATION HAS THE SAME SLOPE AS $3x + 9y = -8$, AND NO COMMON POINTS (SOLUTIONS) THEREFORE THE GRAPHS OF THE EQUATIONS ARE PARALLEL LINES.

   c. Has infinitely many solutions. Explain your reasoning.

   $6x + 18y = -16$ THIS EQUATION DEFINES THE SAME LINE AS $3x + 9y = -8$ AND THE GRAPHS OF THE EQUATIONS WILL COINCIDE.

   d. Interpret the meaning of the solution, if it exists, in the context of the graph of the following system of equations.

   $$\begin{cases} -5x + 2y = 10 \\ 10x - 4y = -20 \end{cases} \quad \begin{matrix} m = \frac{5}{2} & (0,5) \\ m = \frac{5}{2} & (0,5) \end{matrix}$$

   THIS SYSTEM WILL HAVE INFINITELY MANY SOLUTIONS BECAUSE THE GRAPHS OF THESE LINEAR EQUATIONS ARE THE SAME LINE. EACH EQUATION HAS A SLOPE OF $m = \frac{5}{2}$ AND A Y-INTERCEPT AT $(0,5)$. THERE EXISTS ONLY ONE LINE THROUGH A POINT AND A GIVEN SLOPE. THEREFORE THIS SYSTEM GRAPHS AS THE SAME LINE AND HAS INFINITELY MANY SOLUTIONS.

5. Students sold 275 tickets for a fundraiser at school. Some tickets are for children and cost $3, while the rest are adult tickets that cost $5. If the total value of all tickets sold was $1,025, how many of each type of ticket was sold?

Let x be the # of kids tickets
Let y be the # of adults tickets

$$\begin{cases} x + y = 275 \\ 3x + 5y = 1025 \end{cases}$$

$$3x + 5y = 1025$$
$$-3x - 3y = -825$$

$$2y = 200$$
$$y = 100$$

$$x + 100 = 275$$
$$x = 175$$

$$(175, 100)$$

175 children's tickets and 100 adult's tickets were sold.

6.

   a. Determine the equation of the line connecting the points $(0, -1)$ and $(2, 3)$.

   $m = \dfrac{3-(-1)}{2-0} = \dfrac{4}{2} = 2$

   $y = 2x - 1$

   b. Will the line described by the equation in part (a) intersect the line passing through the points $(-2, 4)$ and $(-3, 3)$? Explain why or why not.

   $m = \dfrac{4-3}{-2-(-3)} = \dfrac{1}{1}$

   YES, THE LINES WILL INTERSECT BECAUSE THEY HAVE DIFFERENT SLOPES. THEY WILL EVENTUALLY INTERSECT.

7. Line l_1 and line l_2 are shown on the graph below. Use the graph to answer parts (a)–(f).

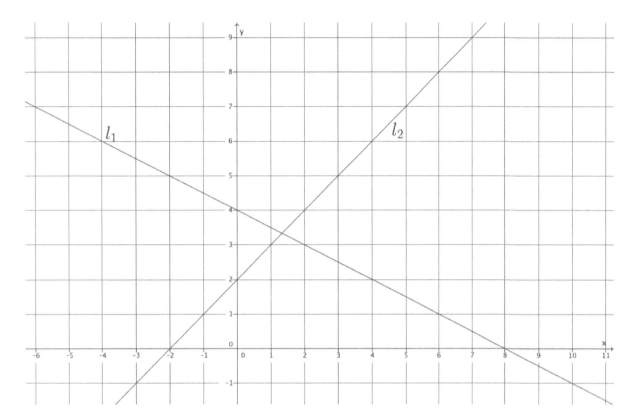

a. What is the y-intercept of l_1?

   (0,4)

b. What is the y-intercept of l_2?

   (0,2)

c. Write a system of linear equations representing lines l_1 and l_2.

   $l_1: y = -\frac{1}{2}x + 4$
   $l_2: y = x + 2$

   $\begin{cases} y = -\frac{1}{2}x + 4 \\ y = x + 2 \end{cases}$

d. Use the graph to estimate the solution to the system.

   (1.2, 3.3)

e. Solve the system of linear equations algebraically.

$$\begin{cases} y = -\frac{1}{2}x + 4 \\ y = x + 2 \end{cases}$$

$$-\frac{1}{2}x + 4 = x + 2$$
$$4 = \frac{3}{2}x + 2$$
$$2 = \frac{3}{2}x$$
$$\frac{4}{3} = x$$

$$y = \frac{4}{3} + 2 = \frac{10}{3}$$

$$\left(\frac{4}{3}, \frac{10}{3}\right)$$

f. Show that your solution from part (e) satisfies both equations.

$$\frac{10}{3} = -\frac{1}{2}\left(\frac{4}{3}\right) + 4$$
$$\frac{10}{3} = -\frac{2}{3} + 4$$
$$\frac{10}{3} = \frac{10}{3}$$

$$\frac{10}{3} = \frac{4}{3} + 2$$
$$\frac{10}{3} = \frac{10}{3}$$